Ageing, Financial Markets and Monetary Policy

Springer
Berlin
Heidelberg
New York
Barcelona
Hong Kong
London
Milan
Paris
Tokyo

Alan J. Auerbach · Heinz Herrmann
(Editors)

Ageing, Financial Markets and Monetary Policy

With 32 Figures and 38 Tables

Springer

Professor Alan J. Auerbach
University of California
Department of Economics
549 Evans Hall
Berkeley, CA 94720-3880
USA
auerbach@econ.berkeley.edu

Dr. Heinz Herrmann
Deutsche Bundesbank
Wilhelm-Epstein-Straße 14
60431 Frankfurt am Main
Germany
heinz.herrmann@bundesbank.de

ISBN 3-540-42727-9 Springer-Verlag Berlin Heidelberg New York

Library of Congress Cataloging-in-Publication Data applied for
Die Deutsche Bibliothek – CIP-Einheitsaufnahme
Ageing, financial markets and monetary policy / ed.: Alan Auerbach; Heinz Herrmann. – Berlin; Heidelberg; New York; Barcelona; Hong Kong; London; Milan; Paris; Tokyo: Springer, 2002
ISBN 3-540-42727-9

Springer-Verlag Berlin Heidelberg New York
a member of BertelsmannSpringer Science+Business Media GmbH

http://www.springer.de

Hardcover-Design: Erich Kirchner, Heidelberg

SPIN 10854752 42/2202-5 4 3 2 1 0 – Printed on acid-free paper

Contents

Ageing, financial markets and monetary policy – a summary

Alan Auerbach and Heinz Herrmann

On May 4 and 5, 2001 the Bundesbank and the Burch Center of the University of California, Berkeley held a joint conference on "Ageing, financial markets, and monetary policy". Economic problems generated by an ageing society have already been discussed extensively in the past. Such issues concern, for example, the effects on labour markets, capital formation and growth. Another issue focuses on whether existing pension systems (and public transfer systems in general) can be maintained in the long run in the event of a shift in the population's age profile. In this regard, a consensus has evolved that an ageing population will make a reform of old-age pension systems indispensable in most countries. This applies, in particular, to those countries in which generous public pay-as-you-go pension schemes still exist. One approach to reform aims at strengthening the funded elements of old-age pension provision. This kind of transformation of pension systems is expected to lead to a number of changes in financial systems. The emerging changes in both underlying macroeconomic conditions and financial systems will have an impact on central bank policy in the future. This was the reason for holding such a conference. This volume reproduces the conference papers and discussions as well as the introductory statements to two panel discussions.

One key question that is the starting point for a number of further considerations in this field is how the expected shift in the age profile will affect the saving decisions of private households. To provide for old age is an important reason for people to save. How easy it will be to effectively handle the problem of ageing will depend on people's willingness to build up such savings. Particularly in Europe, however, little is known about the relationship between saving and old-age pension provision. In the opening chapter, "*Ageing and Saving in Europe*", *Agar Brugiavini* first provides an overview of how saving patterns have developed in important European countries and then attempts to link those developments to demographic trends and a number of other relevant features. These include, for instance, the current conditions of statutory pension schemes and the present state of development of financial markets in the different countries. In the second part, she presents an empirical approach that is intended to shed light on how statutory changes in claims against the state pension scheme influence private saving behaviour. This question arises naturally when examining the relationship between saving and ageing. The empirical analysis is based on a life-cycle model and regresses individual saving rates on the ratio of future pension wealth to current net income. It draws on the experience gained when the social security system in Italy was reformed in the early 1990s, which led to drastic reductions in the number of claims against the statutory pension scheme. The various sectors of the population (elderly people, young people, public employees as opposed to private-sector employees, etc.) were affected to differing degrees. From Brugiavini's estimates, it becomes clear that a reduction in claims against the statutory pension system has led to a markedly increased willingness to save, particularly in the sectors of the population which were most affected.

Reinhold Schnabel starts by discussing possible consequences of a pension cut from a German perspective. He discusses possible effects on saving and labour supply. In the second part of his comments he questions whether all of Brugiavini's interpretations of the "Italian experiment" are convincing. He doubts in particular, whether the reform was unexpected.

The problem of ageing populations affects most industrial countries. However, this problem is more acute in some economies than in others. Germany belongs to that group of countries, with considerable implications for social security systems in the future. Corresponding scenarios based on closed economy models have often been played through in the past. In the case of free movement of capital, however, the foreseeable ageing process - due to differences in intensity and the staggered phasing of developments across countries - can be expected to lead to a pattern of capital exports and imports. According to this pattern, households in the initially or more strongly affected countries invest some of their savings intended to provide an old-age pension abroad, with a view to transferring those funds back in the later "consumption phase". In the second chapter, "*Ageing and international capital flows*", *Axel Börsch-Supan, Alexander Ludwig* and *Joachim Winter* use a multi-country overlapping generations model to display the significance of this kind of international movement of capital. The simulations focus on the implications for Germany, although the model also allows the same calculations to be made for other countries, some of which are also referred to. According to these model calculations, substantial international capital flows can be expected over the coming decades as a result of the shifts in the age profile. Even taking a rather conservative model variant, an annual capital export of more than 7 % of GDP can be expected. Those exports are even higher if the reform of the current social security system entails moving towards a more extensively funded system. The assumptions with regard to capital mobility are another important element. Even though capital mobility is assumed to be high in Europe only - with a large proportion of the exported capital being invested here - the movement of capital is considerable. Hence Börsch-Supan and his co-authors have high hopes that the opportunity for international diversification of capital used to provide old-age pensions will alleviate the problem of an ageing society, particularly in the case of a more extensively funded system.

In his remarks, *Gary Hufbauer* points out that the assumption of ultra-rational households in the model leads to saving for retirement being overestimated if reforms fail to materialise. In his opinion, the authors are therefore exaggerating the advantages of capital mobility should the status quo be maintained. The differences between the reform scenario and the status quo scenario for capital movements are, however, understated. In his observations, *Ulrich Grosch* claims that the Börsch-Supan et al. model neglects a number of aspects, with the result that the paper overstates the importance of capital flows. These aspects include, for example, the deterministic structure of the model, which leaves no room for risk aversion.

Chapters 3 to 5 deal with possible implications for the financial markets. In particular, risks resulting from the problem of population ageing are highlighted from different perspectives, and proposals are made as to how these risks should be managed. For the individual who has to make provision for his old age and who can no longer rely (completely) on the statutory pension scheme, it is important to find ways of safeguarding himself against uncertainty during his lifetime. Annuities are an appropriate instrument in this respect; they can guarantee either a nominal or a price-indexed life-long pension. Such instruments have played an important role in some countries for quite some time, and in others, they are becoming increasingly important. Against this backdrop, *Olivia Mitchell*, in the chapter entitled *"Developments in decumulation: The role of annuity products in financing retirement"*, begins by describing how life annuities work. She analyses a number of pitfalls which have to be taken into account when determining a fair price for such products. These include problems relating to investors' risk aversion (and which are particularly relevant, for example, in terms of the extent to which investors are able to obtain a safe pension from other sources) and difficulties in calculating life expectancies. In this connection, Mitchell highlights a special adverse selection problem that arises if investors with especially long life expectancies purchase annuities. Overall, she concludes that annuities should play an important role in pensioners' portfolios. The chapter then goes on to analyse the extent to which government regulations and supervisory provisions can hamper or foster this market.

In his commentary, *David Blake* mainly describes the British experience with annuities. In the United Kingdom, this market has reached the highest level of development by international standards. Blake compares different saving schemes for old-age pension provision in the United Kingdom and confirms the benefits of annuities, particularly for risk-averse pensioners. *Friedrich Breyer* points out in his observations that problems relating to asymmetric information also exist outside the voluntary annuity systems described. He refers to moral hazard problems in (public) compulsory annuity systems. A combination of public and private pension systems might help to overcome such problems.

While, in the third chapter, a risk is considered against which the individual can take out insurance, in the fourth chapter *Henning Bohn* deals with the macroeconomic question of how aggregate risk arising in connection with old-age pension provision might best be distributed among the generations. In his paper

"Retirement savings in an ageing society. A case for innovative government debt management", Bohn applies different variants of an overlapping generations model. In the first period, households work, consume, pay pension contributions and save for their old age. The second generation consumes and receives public transfers. The government arranges a public pension scheme and can levy a tax on capital and incur debt by issuing bonds. There are four types of risk in the economy: uncertain productivity, uncertainty about the development of the younger generation and about the longevity of the older generation and, finally, uncertainty about the profitability of the existing capital stock. On the basis of this model, Bohn examines different intergenerational transfer schemes to determine how those risks are distributed among the generations and which schemes result in optimal risk sharing, i.e. as even a distribution of risks as possible. This chapter contains a number of interesting conclusions. For example, it shows how important it is for governments to have an overall pension scheme concept, since, in the final analysis, the distribution of risks results from taking all the instruments together - including the distributable taxes. In Bohn's model, for example, a tax on capital may fulfil an important function, leading to an increase in potential welfare. On the other hand, Bohn points out that different innovative elements in debt management may help to achieve better results in terms of the question posed. Those elements include wage-indexed and longevity-indexed bonds.

Alan Auerbach stresses that he considers Bohn's general approach to be very useful. He doubts, however, whether all his conclusions could be upheld if the model were modified so as to make it more realistic. In particular, he is sceptical about the possibility of achieving more favourable risk distributions by levying high taxes on capital and labour (compulsory contributions to the pension scheme). In his opinion, Bohn underestimates the distorting effects of such taxes in real life. *Philipp Rother* is critical of the fact that Bohn places too much emphasis on the relevance of the asset risk to the generation of pensioners. Temporary fluctuations in the market value of the capital stock are of little relevance to pensioners, who are concerned with financing their retirement income through the real return on their capital stock. He also considers that Bohn overestimates the risk of unforeseen developments within the younger generation (population risk), because, in reality, society is able to adapt to such developments in good time.

In the fifth chapter, *"Ageing and Financial Stability"*, *Philip Davis* deals with the question of why the process of population ageing may lead to increased risks to the stability of the financial system. Davis distinguishes between two cases. In the first case, he assumes that there will be no timely reforms aimed at putting old-age pension provision on a sounder footing. This would lead to the risk of increased uncertainty and unsatisfactory ad hoc solutions attempted by the government, which might ultimately also undermine the functioning of the financial system. An example of the former would arise from suboptimal forms of saving in an environment where households no longer trust the social security system but on their own encounter an inadequately developed financial structure. In the end this may endanger the stability of the system. An example of the latter is attempts by the government to finance the statutory pension scheme by incurring ever more debt. The second part of the paper focuses on problems that may arise if funding elements are increasingly introduced into old-age pension provision, with institutional investors gaining in importance. Although Davis considers the basic problems to be less significant in this case, he does not rule out systemic risks. Their magnitude also depends on the government's and/or the central banks' regulatory policies.

Franklin Allen categorises the different types of financial crises discussed in the literature and identifies those with links to the ageing problem. The types of crises that are most likely to be triggered by the ageing problem are those due to the business cycle where fixed promises are made but there are insufficient resources to meet them, those due to inconsistent government macro policies and those due to bubble collapse. The first type is more relevant for a bank-base system and the third one for a system with significant market funding of pensions and institutional investors. Considerations of political economy lead *Martin Hellwig* to a rather sceptical view about the possibilities of solving the future problems of ageing with a pay-as-you-go system or by stimulating saving and investment. Furthermore he stresses that it is probably neither optimal nor feasible to immunize old-age provisions from shocks. Finally he explains why and how regulatory interventions in ageing-provision should play a role.

The sixth chapter deals directly with possible implications of population ageing for monetary policy. In *"Should monetary policy be different in a greyer world"*, *David Miles* focuses on a number of aspects that may become relevant for the monetary transmission process. He bases his analysis on three versions of an OLG model. First, this type of model is used to examine how population ageing changes the optimal capital stock and, thus, the real equilibrium interest rate. Miles regards the equilibrium interest rate as an important reference variable for a central bank which, for example, is adhering to a Taylor rule. In this model, the equilibrium interest rate in an elderly population (with a small labour supply) is far lower than in a young population. This is true, in particular, if the current pension system promotes saving. In a second version of the model, Miles turns to the question of the different impact of a (permanent) exogenous change in interest rates on the consumption decisions of young households with a long life expectancy and on those of elderly, wealthy households. In this case, too, the results depend, *inter alia*, on the generosity of the pension system. In a third version of the model, different types of uncertainty are taken into account (e.g. about labour income or mortality rates). At the same time, there are two assets – a risky one and one without risk. Depending on the type of asset, households are affected to a differing degree by different uncertainties and therefore prefer different types of portfolios. As in the first two cases discussed, Miles again emphasises that the existing pension system is of crucial significance. It determines both how much precautionary saving is made and to what extent risky forms of investment are preferred. This means that monetary policy effects in an older population cannot be discussed without giving consideration to future changes in pension systems.

Hiroshi Fujiki discusses the findings of Miles against the background of the experience gained in Japan. He points to the fact that a falling equilibrium interest rate may aggravate the problem of the zero bound on interest rates. *Benoit Mojon* points out that the empirical relations between interest rate changes and saving behaviour and between increased ageing and the interest rate sensitivity of saving have not yet been clarified. A number of model calculations, presented as examples, show that it does not generally hold true that the interest rate sensitivity rises in line with an increasingly ageing society, as Miles suggests. Finally, he is doubtful whether the shift in the equilibrium interest rate postulated by Miles is of major relevance to monetary policy, since it materialises only very gradually.

During the conference, two panel discussions supplemented the presentations of the papers. The introductory statements to these discussions are reproduced in this volume.

The first discussion was on *"Ageing as a challenge for economic policy"*. *David Wise* opened the debate by presenting his view of the major challenges arising from the ageing process in industrial countries. Financing of social security programs will require substantial reforms in the future. In this connection he emphasised that it is important to encourage private saving and work at older age. Another, related challenge will be health care. *Martin Werding* described the need to reform the social security systems in Europe and the possible ways of doing so. He warned against hoping that the current problems in the labour markets, which have led to high unemployment rates, will be resolved automatically when the labour supply shrinks in the future. Rather he pleaded for reforms in this field, too. *Claus Hofmann* described past reforms in Germany in the area of pension insurance and plans for the future. A role will be played here by modified pension adjustments and supplementary funded systems. In addition, he stressed the importance of the higher participation of certain groups in the labour market. *Platon Tinios* explained the specific problems in some southern European countries and in Greece, in particular. He emphasised the role of the European Monetary Union in fostering reforms in the pension system.

The topic of the second panel discussion was *"How pension reforms are changing financial systems"*. *Hans Blommestein* described how pension systems with greater funding will lead to a modernisation of the capital market infrastructure in OECD countries, including greater globalisation of the financial markets. He highlighted the advantages and possible risks of these developments. *Olivier Davanne* gave an overview of the current state of the French pension reform, which also gives more prominence to the funded component. In this connection, he expressed some concern about the way pension money is managed today. *Jürgen Pfister* explained the plans to encourage saving for retirement in

Germany. He saw a certain amount of over-regulation in this reform. Subsequently, he gave his view on how these reforms may change saving, financial markets and the financial system in Germany in general. He stressed that the structure of the system, and the role of banks in particular, will not be fundamentally changed. *Joseph Bisignano* emphasised that it is not easy to forecast exactly how pension reforms will affect the financial landscape of a country. He assumed that in the end it will lead to a sizeable demand for longer term finance and to major changes in the structure in European finance. The experience of the past teaches us that this will bring new challenges for European central bankers.

Ageing and Saving in Europe

Agar Brugiavini[*]

1 Introduction

Saving is the result of intertemporal decisions taken by individuals. Despite the recent research effort, economists find it hard to provide a satisfactory explanation for the cross-country variation in the level of private saving and for the observed trends. Even more demanding is the task of attributing precise amounts of savings to specific types of future conditional consumption and to specific risks. Most economists would agree, however, that providing for resources in old age is one major motive for household saving. In countries where the state provides for old age via a public pay-as-you-go (PAYG) pension system[1] the need to save for consumption during retirement should be mitigated. Alternatively, old-age insurance can be obtained through the capital market via institutional investors in group insurance (second pillar) or through individual contracts (third pillar), in these cases provision for retirement is normally part of private accumulation. Whether, from an accounting point of view, saving for retirement is part of private saving or not, it is clear that one of the most - if not *the* most - important interaction between savings and ageing is centred around pensions. In Europe, ongoing or impending pension-reform attempts have made even more apparent that ageing has an impact on saving, and have made it even more compelling to analyse the link between pensions and savings.

The mixture between PAYG and funded old-age provision varies quite remarkably across countries. In most Continental European countries, notably in the three largest countries France, Germany and Italy, the PAYG mechanism is the most important instrument. A move towards private funded schemes, whether participation in the scheme is made compulsory or voluntary, requires some

---

* I am in debt to my co-authors for letting me draw freely on our joint work. I wish to thank Alan Auerbach, Axel Börsch-Supan, Heinz Herrmann, Olivia Mitchell and Reinhold Schnabel for constructive comments. Participants at the Bundesbank meeting on "Ageing, financial markets and monetary policy" in Eltville contributed to the final version of this paper through a lively discussion. Massimo Garbuio provided excellent research assistance. Financial support from the EU-DGXII-TMR programme "Savings and Pensions" is gratefully acknowledged.

1 Normally referred to as a "first pillar". In this paper I use, rather loosely, the terms "first pillar", "public pensions" or "social security" to indicate public pension provision.

knowledge of the degree of substitutability between "saving for retirement" and "other saving". Two kinds of substitution may be relevant: (1) substitution between bequeathable wealth and wealth that is annuitized for retirement, and (2) substitution within "retirement wealth".[2] Both substitution mechanisms are important: while all forms of private saving (whether annuitized or not) constitute private wealth accumulation, and are usually recorded that way in national accounts, the difference between bequeathable (fungible) wealth and retirement wealth rests on the basic motives for saving. Hence, in theoretical models or in simulation exercises, one is usually concerned with describing changes in total capital accumulation (i.e. total private wealth, even if annuitized), but in the empirical work the extent of substitutability is usually assessed by comparing bequeathable wealth on the one hand, and retirement wealth (or just public retirement wealth), on the other.

This paper briefly reviews the reasons for relating saving to ageing. It presents some facts on ageing and saving in Europe, and it mainly focuses attention on the empirical work on substitutability between pension wealth and private wealth for European households.

2 Ageing and saving: facts and theory

2.1 Saving in Europe

European countries, and particularly countries in continental Europe, have traditionally been described as "high-saving-rate" economies, when compared to the US. Although the saving rates have declined in many European countries, the "high-saving-rate" paradigm still seems confirmed by the data for the late 1990s (Table 1).[3] Both the high level of the personal saving rates and the recent

---

2 Bernheim (1987a and 1987b) first addressed this issue.

3 OECD Economic Outlook (2000b) provides a long-time series of saving rates on a consistent basis for a number of countries used in Table 1. However one should be careful in carrying out cross-country comparisons, and should rather focus the attention on the time variation for each individual country, as the definitions adopted are not fully consistent. In fact, some countries have provided data on gross saving rates (gross of capital depreciation) and some have provided the net figure. However, the gross and net figures do not seem to correspond to the labels indicated by the OECD. The difference could be marked: for example, according to calculations based on National Accounts Data, the German gross saving rate in the year 1999 is 15.5%, while the net saving rate is 9.9%. Despite these caveats, there is no doubt that the European saving rate (particularly continental Europe) and the North-American are two extreme cases.

declining trends pose a puzzle. Why do households save in countries where the welfare state is relatively generous and it strongly conforms to the *Bismarckian* tradition?[4] This is a hard question, which I do not try to develop fully here.

The debate in the US is posed in terms of "is the saving rate too low?" The microeconomic reasons are related to the basic task of saving: provision for future consumption. If people are myopic or underestimate future risks, saving will be too low. Paternalistic arguments then generate the provision of "forced saving" through public pension systems. These arguments rest on the assumption that individuals are myopic (and eventually free riders); a milder version of this argument tends to favour some tax-relief for retirement-saving, worrying about a lack of capital accumulation and growth. There is a growing literature that tries to assess the existence and the extent of myopia, based on survey data and through "experiments". However, this is still an area of research in its infancy: see Kotlikoff, Spivak and Summers (1982) and Hamermesh (1982) for early contributions, and the recent work on "mental accounts" and "hyperbolic discounting" for structural models (Laibson, Repetto and Tobacman, 1998 provide a useful example, a survey of this literature can be found in Bernheim, 1999). From a European perspective, we may be tempted to ask whether saving is too high: this may be a ill-posed question, but it has significance over and above the problem of capital accumulation and growth. The question of the adequacy of saving to provide for retirement, or the extent of over-annuitization, remains in fact an important argument to justify the concern of economists, precisely because European countries are ageing considerably and are ageing fast.

Looking at the dynamics: what has determined the declining trends? The household-saving rate fell in Italy, particularly during the late 1970s and 1980s: some authors (Rossi and Visco, 1995) argue that this decline is almost completely explained by the increasing generosity of the social security system (see Börsch-Supan and Brugiavini, 2001). However, it is hard to find a one-dimensional explanation for these trends in each country. In fact, the development of formal credit and mortgage markets, the generalized access to financial markets where a larger range of products is available, surely affected personal saving throughout Europe, possibly changing also the national saving rate. In addition, the decline in the share of composite households observed in many Southern-European Countries may have contributed to the decline in household saving: change in household head-ship can also partly explain the patterns observed in the data, so that what appears as a decline in the saving rate is partly an artefact of the data (see section 2.3 below). Obviously, the tax system and its evolution over time also play a major role (see Poterba, 1994 and Browing and Lusardi, 1996). Finally,

---

4 So that the standard of living for most workers is preserved under most negative contingencies. In these countries, social security provides old-age protection, insurance against disability and to survivors. Quite often, health care and unemployment insurance are also provided to workers through a National Programme.

some authors suggest that economic liberalization overall, i.e. the reduction in tariffs, import quotas on goods and capital controls, may have contributed to reducing private saving rates in all OECD countries (Saint-Paul, 1992 and Mansori, 2001). Attanasio (1997) and Banks and Tanner (1999) argue that no convincing explanation for the fluctuations in personal saving rates observed in the UK has been singled out. I would add that, probably, only partial explanations can be given for the rest of Europe, too, but ageing and old-age provision have a prominent role.

In order to support the idea that ageing is the driving force behind the observed changes in saving rates, it is important to refer to a theory of saving which can provide the link between facts and intuition.

Table 1. Household-Saving Rates in selected countries and for selected years[(a)]

	1982	1985	1990	1991	1992	1993	1994	1995	1996	1997	1998	1999	2000*	2001*
United States [(b)]	10.9	9.2	7.8	8.3	8.7	7.1	6.1	5.6	4.8	4.5	3.7	2.4	1.1	1.5
Japan	16.7	15.6	12.1	13.2	13.1	13.4	13.3	13.7	13.4	12.6	13.4	13.1	12.2	12.0
Germany [(b)]	10.9	9.5	16.1	12.2	12.0	11.8	11.0	10.3	9.9	9.5	9.1	8.6	8.4	8.8
France [(b)]	16.2	12.3	12.6	13.5	14.7	15.2	14.8	15.9	14.8	16.0	15.6	15.7	15.2	15.0
Italy [(b)]	22.5	21.0	18.4	18.7	18.4	17.2	17.2	16.6	16.0	14.6	13.4	12.7	12.7	12.7
United Kingdom [(b)]	10.8	9.6	7.7	9.7	11.8	11.2	9.6	10.5	9.7	9.6	6.3	6.2	6.1	6.2
Canada [(b)]	19.0	14.2	11.5	11.7	11.4	10.3	7.7	7.5	5.2	2.8	2.4	1.4	1.0	1.4
Belgium [(c)]	12.5	10.5	13.9	15.6	16.6	17.9	15.8	15.1	13.4	12.5	12.2	12.9	12.7	12.8
Denmark	..	..	11.4	11.1	9.9	8.6	4.5	7.1	5.6	4.8	5.7	5.1	5.0	5.1
Finland	3.6	3.2	2.9	7.8	10.1	7.6	2.7	6.0	2.0	4.4	3.9	4.6	3.8	3.6
Ireland	16.4	13.7	9.8	11.0	9.8	11.8	7.8	9.9	9.1	11.7	11.7	11.3	11.7	11.6
Netherlands [(d)]	9.5	5.5	11.9	7.2	8.4	6.8	7.1	6.5	5.7	5.7	4.2	2.8	2.4	3.7
Norway	4.4	-1.8	2.2	4.2	5.9	6.9	5.9	5.7	4.7	4.8	6.6	6.7	6.6	6.4
Portugal	24.1	24.3	16.4	17.0	14.8	12.6	10.2	10.3	10.2	10.4	10.6	9.5	8.7	8.2
Spain [(b)]	12.2	10.6	11.8	12.7	11.2	13.9	11.4	13.4	12.6	11.4	11.1	10.2	10.0	10.0
Sweden	3.2	3.9	1.0	4.7	9.2	9.7	9.3	6.9	5.1	2.0	2.4	1.6	1.0	0.2

Source: OECD (2000b), Table 26. Saving rate as percentage of disposable income. a) National definition except United States. b) Gross saving; c) Data are on ESA 79 basis; d) Excluding mandatory saving through occupational pension schemes; *) Projections.

2.2 The theoretical framework

Saving is the result of consumers' decision to provide for future consumption. Most economists would refer to the life-cycle theory of consumption (Modigliani and Brumberg, 1954) to describe consumers' behaviour. Although this is not the only model which deals with saving decisions, and although it has been challenged in many respects, it still remains a very useful theoretical approach to analyse the intertemporal allocation of resources (see Deaton, 1995; Attanasio, 1998 and Bernheim, 1999 for surveys of the literature). It is particularly so if one takes the model just as a framework for the analysis: it is flexible enough to allow for numerous generalizations, it is coherent with the literature on labour supply and portfolio choices, and it produces a number of interesting implications.

In order to use it as a general structure, rather than just a model, one has to go beyond the basic result that consumers are "smoothers", in view of predictable and exogenous shocks. In more general terms: consumers equalize discounted marginal utilities across periods and across states of nature, hence, decisions about labour supply (retirement), portfolio allocation (including home purchases) and family composition are taken jointly with saving decisions. The life-cycle model could also be extended to allow for some form of bequest motives. But even when attention is confined to intertemporal allocation of consumption as the only decision that consumers face, the approach is quite general, in that the existence of liquidity constraints and precautionary motives can be modelled explicitly within the life-cycle framework (Deaton, 1991; Carroll, 1992), making it a very useful set-up.

The well-known implication of the stripped-down version of the life-cycle theory (in the absence of any public provision for old age) is that the age-saving profile is hump-shaped in view of an uneven distribution of resources over the life cycle. In fact, labour income typically grows during the working years, and peters out after retirement. This simple version of the theory delivers one important messages: the main motive for saving is provision of resources in old age. The non-trivial consideration emerging from this result is that pension and social security provisions, and more generally old-age-insurance markets, are one important aspect to take account of when dealing with saving.

From a methodological viewpoint, the main tension in the recent literature rests on a dichotomy between structural models consistent with the general life-cycle approach, on the one hand, which are fully equipped to answer the important questions on saving, but are often hardly tractable and, on the other hand, studies looking at the effects of fiscal incentives on saving, which are appealing in their descriptive simplicity but at times inadequate. Two promising strands of the literature stem from simulations, and from "approximations" to the structural model (Attanasio, 1997). Simulations make use of numerical solutions to solve for the structural models, and have added more realistic features and uncovered a number of interesting twists to the basic life-cycle paradigm (see Attanasio, Banks, Meghir and Weber, 1999; Deaton, 1991; Engen, Gale and Uccello, 1999; Hubbard, Skinner and Zeldes, 1995, to quote a few in this area[5]); in these cases, one can directly control for the effects of the different assumptions regarding policy decisions and basic parameters on saving rates. "Approximations" is a loose term to indicate the possibility of deriving a solution to the individuals' maximization problem (i.e. to write down the marginal utility of wealth), which makes the model tractable and which lends itself to empirical analysis without losing sight of the general problem and of the assumptions used in the implementation. There exist numerous examples of this approach, both in the consumption/saving literature and in the retirement-decision literature, and I am not quoting them here. The second part of this paper will focus on the latter approach, and it will look at a very specific aspect within this approach.

2.3 Ageing, demographics and saving in Europe

Many changes have taken place in Europe over the past decades, and I will try to hint at some possible effects of these changes on the personal-saving rate. In particular, I will focus on changes in personal saving which can be explained by making reference to the relationship between age and saving. From a microeconomic point of view, the textbook version of the life-cycle theory gives an immediate implication that saving, and the saving rate, varies by age. In particular, it implies dissaving in old-age, hence one might expect a decline in the aggregate saving rate in response to population ageing: but this conclusion is not granted. First, ageing may affect individuals' behaviour, i.e. age-specific saving rates, not the least through changes in the pension system or tax system; note that this change may be more marked for younger cohorts than for older cohorts if individuals are indeed forward-looking, as the life-cycle theory maintains. Furthermore, a shift in the demographic composition of a country alters the aggregate saving rate even if the economic behaviour of each individual remains unaffected by population ageing: this is a pure composition effect. In addition, there are macroeconomic feedback effects, for instance through changes in the rate of return on capital (general equilibrium effects). Once again, one could take these feedback effects one step further and link ageing to policy decisions (e.g. pension reforms), which in turn affect saving decisions.

In order to establish the link between ageing and aggregate saving rates one should address all the issues I have listed above. Instead, the focus of this paper is on individual's behaviour and, in particular, on the effects that ageing has on saving via provision for retirement. Even making sense of this piece of the puzzle

---

5 Of the many contributions which follow this approach, I have quoted just a few examples, which stress different aspects: from precautionary saving and market imperfections to the role of demographics.

is not a simple task, as it requires to identify and to quantify behavioural responses, which are notoriously hard to pinpoint. Furthermore, even confining the analysis to a very narrow microeconomic aspect of the problem, i.e. substitutability between private wealth and pension (annuitized) wealth, one has to take account of the compositional effects and the feedback effects mentioned above. Finally, in order to carry out the analysis at the micro level, one has to recall that the basic implications of the life-cycle theory build on extreme assumptions on the features of the underlying economy. Since this "ideal" setting is hardly found in reality, it is useful to document the variation in institutional arrangements and in financial and insurance-market structures existing in Europe.[6]

After having discussed trends in saving rates, a compulsory step in this brief overview of the European landscape is to document the extent of ageing in Europe. The EU-average of the dependency ratio in 1995 is approximately 45%; Italy is above this figure, while Ireland is below, but dispersion around the EU-mean is not marked. Demographic projections are surrounded by a heated debate, and I am not discussing the issue in this paper, but most demographers and economists would agree that the picture for the future is worrisome, particularly because the demographic trend is a permanent feature of the data, and it is not idiosyncratic. Italy is projected to have more people aged 60 and over than working-age individuals by the year 2050; the other EU countries are closer to the EU-average but all EU-countries will be hit by ageing at the same time. The demographic shock is permanent because it is not just the ageing of the baby-boom generation causing these trends, more importantly, it is permanent increases in longevity and low fertility rates (which may revert in the long run, but not enough to change the basic pattern). Public pension systems financed through a Pay-As-You-Go (PAYG) mechanism are particularly vulnerable to this trend as the growth-rate of the population is the engine behind this mechanism. Hence many countries are facing the prospect of a possible financial crisis of public pension systems, and some have already reacted by implementing pension reforms. However, these reforms are usually "parametric", in that they do not change the un-funded nature of the pensions system (Disney, 2000).

Parallel to demographic trends, labour-market trends are also important both directly and indirectly, in analysing saving. The direct effect comes from the general life-cycle approach, where labour-supply decisions and consumption

6 Obviously, the tax system is also crucial. However, describing taxation of assets in Europe and its effects on saving decisions is a vast research area in its own right, and it is not the focus of attention in this paper. I will only briefly touch on tax incentives in section 3 below. Paradoxically, there exists very little research on this issue in Europe (Poterba (1994) contains contributions from several European countries on this topic, and Börsch-Supan and Brugiavini (2001) review some of the literature).

Figure 1. Dependency ratios in EU

Source:: Eurostat, 1996, Ration op population aged 60 and over on population of working age.

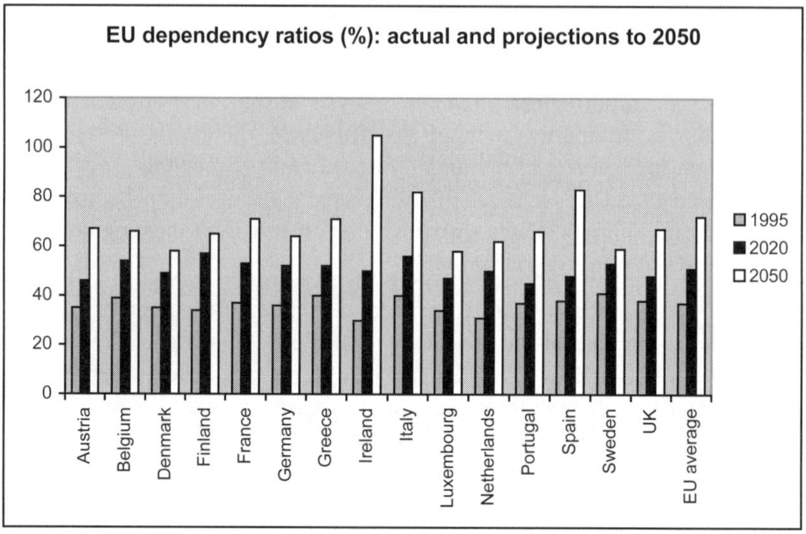

decisions are taken jointly over the life cycle. The indirect effects may come through many routes, one is particularly important and goes, again, via public pension systems. EU-countries have witnessed a decline in the labour force participation of older men, particularly for the age groups 50-55 and 56-60. The evidence for female labour force participation is somewhat different due to the labour-market experience of younger cohorts who "catch up" with respect to male workers. Note that this pattern in female labour-force participation is good news for public pension systems, but is in turn to be correlated with decreasing fertility rates. The decline in labour force participation of older men may be the optimal response by individuals to retirement incentives built into public pension systems (Gruber and Wise, 1999): these incentives are widespread, and they are very powerful, also in explaining early retirement in many developed countries.[7] Early exits into retirement magnify the demographic shock and contribute to the financial distress of public pension systems.

A description of the variety of institutional arrangements existing in different countries to provide income for retirement is given in Table 2, which shows overall replacement rates and the composition by sources of retirement income in the late 1990s. If one focuses attention on the composition between the first pillar

---

7 Both the ageing trend and the rapidly-increasing early retirement trend are present in most EU-countries. The UK is no exception but it is at the lower end of the spectrum (the dependency ratio will not grow as much as elsewhere, and early retirement does not reach such extreme values as in France or Belgium, for example).

(public pension schemes, normally PAYG-financed) and the second pillar (occupational pensions, normally funded), a clear difference emerges between countries like Italy and Germany, where the second pillar is negligible, and countries like the UK and the Netherlands, where the second pillar is very important. Since countries where the PAYG-financed public system is prevalent are most likely to suffer the economic consequences of an ageing population, it is also likely that, in these countries, there will be a move towards the UK model (the Dutch model), achieved by some form of pre-funding of public pension systems.[8] However, the top row of Table 2 shows similarity in the overall replacement rates. Adding up the income from all pillars, we find that, just after retirement, households receive approximately 80% of the income they enjoyed shortly before retirement. This suggests that either the different States somehow impose substitutability between public pension provision, and private pension provision, by making coverage (in the second or third pillar) compulsory, or that consumers perceive different forms of pension provisions as substitutes when given choice.

Before turning to the issue of substitutability, it is useful to provide the overall picture on the features of financial markets and insurance markets on which European consumers take decisions.

---

8 Indeed, in many countries throughout Europe this reform process has started or is under discussion.

Table 2. Replacement rates and composition of retirement income in selected countries

	D	F	I	E	NL	CH	GB	US
Replacement [a]	82%	79%	80%	./.	78%	81%	69%	./.
First Pillar [b]	85%	51%	74%	92%	50%	42%	65% [c]	45%
Second Pillar [d]	5%	34%[e]	1%	4%	40%	32%	25%	13%
Third Pillar [f]	10%	15%	25%	4%	10%	26%	10%	42% [g]

Notes: The first row (a) gives the replacement rate of total retirement income. This is based on all income sources of an average two-person retiree household just after retirement as percent of the total income of an average two-person household just before retirement. The other rows (summing to 100%) show the composition of retirement income according to sources. (b) Public retirement income (public pensions, social assistance, civil servants' pensions, etc.) as percent of the total income of a two-person retiree household. (c) In Great Britain, first-pillar income also includes SERPS. (d) Private occupational pension income as percent of the total income of a two-person retiree household. (e) In France, almost all second-pillar income is pay-as-you-go financed. (d) All other retirement income (asset income, net transfers received, earnings, etc.) as per cent of the total income of a two-person retiree household. (g) 25 percentage points of this figure are earnings.

Source: Börsch-Supan and Miegel (2001), calculations also based on: Disney, Mira and Scherer (1998), Gruber and Wise (1999).

3 Formal and informal financial and insurance markets in Europe

Continental Europe exhibits a financial landscape that is quite different from the Anglo-Saxon countries. In particular, stock-market capitalization in the large Continental European countries is still much smaller than in the Anglo-Saxon countries. However, the growth of institutional investors documented in Blommestein and Funke (1998), and the liberalization of markets - coupled with the advent of EMU (European Monetary Union) - has often been referred to as a new engine for growth in European capital markets.

The channels through which developments of financial markets, and in particular, the advent of institutional investors, take place are determined by demand conditions and the features of the supply of financial services (Blommestein and Funke, 1998). Pension reforms may imply a shift toward private provision for retirement and, in turn, this requires well functioning financial markets. In this perspective, the introduction of new financial products, mainly through institutional investors, is directly related to population ageing. However, there exists a complex relationship among the reform of public pension provisions, the onset of private pension funds and the growth of capital markets. One scenario is that the advent of pension funds should prompt "thick market externalities", i.e. positive externalities, which, also because of economies of scale, reduce transaction costs and increase the willingness to invest in risky assets.[9] Technological progress in information management, which reduces the costs of providing complex financial products, should result in increased competition and growth of the market - in turn speeding up technological progress. This process could be further accelerated by European financial integration as a result of specific EC directives (particularly in credit and mortgage markets). The other dimension of rapid financial and monetary integration could go via the reduction of differentials in the taxation of financial products among member countries.

Parallel to these reforms, changes of mortgage arrangements and consumer-credit availability, may contribute to creating a different choice for savings of the next generations. The interactions between changes in household-saving behaviour and developments of markets obviously work in both directions.

3.1 Pension funds in Continental Europe

In Europe, overall, total assets of pension funds have grown at an annual growth rate (between 1990 and 1996) of 9.6%, as opposed to a growth of 12.1% in North America. Furthermore, while private pensions and other forms of income from capital account for up to 70% of household income for older households of the higher income group (60% in Canada and in the UK), these are negligible in Italy and account for 20% of total income in Germany. Net social transfers (inclusive of public pensions) account for more than 90% of resources in the low-income group in Italy.[10] Overall, pension funds are still underdeveloped in Europe, as opposed to North America, and even "rich" people still shun this type of investments.[11]

What is clear is that most people in Continental Europe are not used to stocks or stock-related funds. There has, however, been a strong increase recently in the share of stocks in household portfolios, see Eymann and Börsch-Supan (2001) for Germany; Alessie, Hochguertel and van Soest (2001) for the Netherlands. This tendency has large potential benefits, but also carries risks that might requires policy attention. First, the lack of information has been documented extensively; and the UK mis-selling experience, has shown the risks involved in a partial

---

9 Pagano (1993) provides a theoretical model to explain this mechanism. The onset of pension funds and pension-related financial instruments may also lead to qualitative changes in supply, by stimulating the creation of new products or even prompting a deregulation process (Davies, 1998)

10 Source: OECD (2000a), table 6.1.

11 Note: there are measurement issues, and a lack of unified accounting rules, even within the EU. Harmonization of accounting rules is debated (see EC 1999).

transition that is not accompanied by government regulations on disclosure, information, and the possibility of revoking contracts within a given period of time.

Finally, the importance of encouraging competition in the pension industry (or even of encouraging direct participation in equity markets as in the U.S.) or the need to apply the prudent-man principle in a global environment, rather than adopting home-biased pension-investment policies, are the two issues often discussed in relation to the return-risk trade-off characterizing the choices by pension-fund managers.

Hence pension funds will probably grow, particularly in those countries which need catching up, but the transition process does not seem likely to be smooth.

3.2 Credit availability and informal markets

Households may need to borrow during their lifetime. The standard life-cycle motive is that young households borrow in order to finance current consumption, in the expectation of increasing age-earnings profiles. More generally, households may save in order to maintain a given target of the wealth-to-income ratio or as a buffer. Increased credit availability may reduce savings as the result of removing a liquidity constraint.

House purchases and durable-goods purchases are a common reason for starting up a mortgage or a consumer loan or for signing consumer-purchase agreements. Home buyers, who prefer owning to renting, are often faced with the problem of building up a down-payment, because mortgages finance only a fraction of the cost of the property. The higher the down-payment, the stronger the effort for young households to accumulate capital in order to meet the payment at the time of purchase. Obviously, in economies where down-payments are reduced due to a developed mortgage market, liquidity constraints are less severe, and the saving rate may be low as a result. Similarly, access to new consumer-credit facilities softens the effects of constraints on durable-goods purchases. According to the way in which credit markets and mortgage markets solve problems of informational asymmetries, market imperfections may take different forms. In perfectly-functioning markets, consumers can borrow and lend the desired amounts at the prevailing interest rate. An extreme form of market failure takes place when asset or insurance markets are completely missing or very thin (see annuity markets in Europe). However, it is more common to observe imperfect credit and mortgage markets where some rationing takes place, or where interest rates are characterized by a wedge (spread) between the borrowing rate and the prevailing market rate.

The financial liberalization experienced in Europe throughout the recent decade has, by and large, lifted existing credit restrictions (e.g. in housing markets, see Chiuri and Jappelli, 2000). It is not obvious how this will affect aggregate saving. If consumers were "saving too much", in a life-cycle sense, because of liquidity constraints, there might be a negative effect on saving due to the introduction of new financial products.[12] On this point, it is important to stress that liquidity constraints do not imply that individuals consume a fixed fraction of their disposable income, only, if the borrowing constraint is binding, consumers are forced into a corner solution, hence the implied reduction in saving might be much lower than projected if only a few households are at the corner. Furthermore, credit restrictions often apply to the purchase of durables, hence increased access to credit may simply encourage durables purchases, and, under certain conditions, this may be viewed as substituting forced liquid savings for durable goods purchases.

The empirical implications of the variety of arrangements and the severity of the credit constraints are complex. In countries where the required down-payment is low the saving rate tends to be low.[13] The down-payment may affect the home-ownership decision and the timing of purchase: the higher the down-payment, the smaller the probability of a household making a purchase, and the longer the waiting period before the purchase takes place. Given the age structure existing in a given economy and the propensity to save of different cohorts, this may have sizeable and conflicting effects on aggregate saving (Hayashi et al., 1988). Note that the same arguments based on constrained consumers can be put forward for other durable goods and, although the stock value of durable goods may be considerably smaller than the stock value of homes in the economy, consumer durables represent an important part of the personal sector, and of households' budgets. Hence, new financial products, when lifting credit constraints, may greatly affect the relationship between saving and durables consumption, in some cases, non-durables consumption may even be unaffected.

Finally, both for home purchases and for credit provision in general, an important role is played in Europe by "informal markets": it is often maintained that households and families substitute for missing credit or insurance markets.[14] Intergenerational transfers (whether truly altruistic or strategically motivated)

---

12 As argued in Hayashi (1987), the existence of future-expected liquidity constraints may also induce sub-optimal behaviour in a life-cycle sense ("too much" saving). Future constraints may be important for young households, and yet hard to pick up in empirical analysis, and hence hard to quantify.

13 This is shown by Jappelli and Pagano (1994). In particular, they show that in periods where the down-payment in Finland, Sweden, Ireland and the UK was signficantly lower than in other countries (i.e. during the 1970s), the saving rate was also significantly lower. Further evidence on this can be found in Chiuri and Jappelli (2000).

14 See the seminal paper by Kotlikoff and Spivak (1981).

could partly explain differences in saving behaviour across countries as well as differences in the age-saving profile. What is relevant in this context is that some *inter vivos* transfers or bequests take place within the household or between related households (e.g. from parents to children) to compensate for the existence of various forms of liquidity constraints. For example, it is estimated that, in Italy, more than a third of homeowners obtained the property directly as a gift or bequests from relatives, or received financial support for the purchase (Guiso and Jappelli, 2000).

This discussion shows that it is hard to draw a conclusion on whether changes in credit markets have a sizeable impact on saving, the existence of informal markets may further contribute to blurring the picture. The empirical work is faced with the difficulty of carefully describing market arrangements and pinpointing, amongst the several key aspects affecting saving decisions, the effects directly related to improved market conditions. An example of these difficulties is summarized in Banks and Tanner (1999), who review the debate concerning the fall in saving experienced during the late 1980s in the UK. While Muellbauer and Murphy (1990) argue that this was due to the consumption boom following the euphoria of rising housing prices and relaxed credit arrangements (hence, a direct consequence of housing-market distortions coupled with improved credit availability), Attanasio and Weber (1994) argue that the growth in consumption experienced by UK households is simply the result of permanent changes in the lifetime resources of younger cohorts. In the latter case, changes in saving behaviour are not directly linked to changes in credit markets, but to growth in human capital.

4 Substitutability in old-age provision, risks and tax incentives

I have argued that the adequacy of saving for retirement may be the key aspect to worry about in analysing the effects of ageing on saving and I have described the variety and complexity of the existing pension and credit arrangements in Europe. A first conclusion is that it is very hard to impute a change in saving behaviour to a specific policy change or institutional change. However one area of research which is promising in this respect, precisely because of the institutional variability, is the issue of substitutability between different forms of provision for consumption in retirement.

4.1 Substitutability between pension wealth and private wealth: an overview

If saving for retirement is one major motive for saving, and if some form of mandatory public pension provision is in place, which cannot be easily reversed (i.e. annuities markets are imperfect or do not exist), then permanent changes in pension provisions should be reflected in changes in private saving. One way to pose the question is to ask whether a reduction of the first pillar will create new saving, or whether some other saving may be "crowded out". The extent of crowding out is the subject of the famous debate between Feldstein (1974) and Barro (1974). Auerbach and Kotlikoff (1983) pioneered the discussion on the difficulties, with the empirical identification of the substitution parameters both in time series and cross-sectional studies for the US. The structural instability of the parameters is due to the changes in population structure and to the fluctuations in interest rates, as well as in contribution rates, during the transition after the introduction of a new pension system.[15] For Europe, the empirical evidence is scanty, making it particularly hard to reach any conclusion. In this paper, I will briefly review the empirical literature and focus on examples where some studies do exist and represent very diverse realities: the Netherlands, the UK and Italy.

Related to the issue of crowding-out is the effect of savings incentives. In most pension-system transitions that have occurred in recent history, the transition was supposed to be speeded up by tax relief for private old-age provision. Economic theory is ambiguous on the effect of tax subsidies on the volume of private savings because the substitution effect (higher after tax returns make saving more attractive than consumption) has an impact opposing that of the income effect - the subsidy increases total income, which increases consumption in all periods) - see Bernheim, (1999) for a survey. Moreover, many have contributed to the debate by providing empirical evidence on the effects of tax subsidies on American households' retirement saving, and discussing the "creation" of new saving: Venti and Wise (1990), Gale and Scholz (1994), Attanasio and De Leire (1994). The issues are summarized by Skinner and Hubbard (1996); Engen, Gale and Scholz (1996); Poterba, Venti and Wise (1996); Attanasio and Banks (1998); Bernheim (1999) and Wise (2001).

Despite the amount of evidence existing in the US, it is unclear whether saving-related tax relief or similar subsidies increase total saving. Subsidies may increase saving in the specific form that is subsidized, possibly to the detriment of other saving forms. One has to bear in mind that only if the major motive for saving is saving for retirement, then households would value forms of savings which are a close substitute for pension-assets, but which are typically illiquid. It is to be decided whether tax relief has an impact: tax relief is not everything. The great

---

15 The argument is clearly reversed for cases of retrenchment of public schemes.

success of individual retirement accounts (so-called "IRAs") and employer-sponsored retirement-saving plans (so-called "401(k) plans") also seems to be due to information and advertising, and consistent capital-market regulation that reduced the uncertainty of investors. This is in great contrast to the UK experience, summarized by Disney (1996), where the lack of regulation and information led to the so-called mis-selling scandals that undermined investment in private accounts. In this scandal, inappropriate financial products were sold to households, often by door-to-door salespersons, resulting in huge financial losses to many families. State intervention seems necessary to provide for the smooth working of the capital market. Also, we learn from the US experience, that tax relief only works for the upper two-thirds of the income distribution. Neither IRAs nor 401(k) plans will have a noticeable impact on the future retirement income of the lower third of the income distribution. In Germany and in Italy, these observations have sparked an extended discussion on whether to make private provision mandatory. The argument in favour of obligatory savings is mainly the need to cover all persons, in order to avoid pockets of poverty and to prevent moral hazard. The arguments against mandatory savings essentially rest on the well-known economic efficiency result against taxation.

In Europe, there exists one further complication: the sheer variety of tax treatment of private pensions across countries totally blurs the picture on which type of taxes may encourage or discourage different types of assets.[16]

This brief overview suggests that one important aspect in relating saving to ageing is to assess the empirical relevance of substitutability between forms of saving. Much of the controversy on the correct way to approach this problem may rest on two issues: the issue of definition of pension assets *versus* non-pension assets and the issue of informational content of the available data. The former requires, in practice, to judge where to draw the line between the two forms of wealth, but it involves a number of important considerations and it may carry with it very different implications. The latter is a basic identification problem: there is a difference between ideal data and actual data.

4.2 Substitutability: theoretical and methodological issues

Saving can be defined as disposable income minus consumption. Given the obvious relationship between saving and wealth, an alternative approach is to take the difference between the wealth levels corresponding to two points (ages) in time. In the latter case, the definition of private wealth becomes important: the problem of drawing a line between "pension" assets and other assets is not confined to an accounting exercise. It is related to the institutional setting and to

---

16 See Dilnot (1992), World Bank (1999) and Börsch-Supan and Lührmann (2000) on the variety of pension-taxation rules in Europe and elsewhere.

the insurance-markets configuration, i.e. to the types of risks that consumers face and to whether some assets are tradable or non-tradable. Hence, in practice, this choice of definition may turn out to be country-specific. The problem of definition and measurement of pension assets is also different, according to whether one is looking at national-accounts data or survey data.

At one extreme, some authors question the basic distinction between pension wealth and private wealth, and argue that, since the major motive for saving is provision for retirement, we should think in terms of total wealth, including public social security wealth (Jappelli and Modigliani, 1998). Similarly, according to this view, one should not think of payroll taxes as taxes, but as forced saving. This is an extreme view, because it maintains the assumption that the only motive for saving is to protect consumption during retirement, and that public pension wealth is a perfect substitute for private wealth.

If one takes a more general approach, in order to analyse this question it is useful to consider the different features of saving instruments available to consumers. Bernheim (1987b) distinguishes between bequeathable wealth and non-bequeathable wealth (corresponding to "pension wealth"). In fact, the provision of old-age insurance is the response to the risk faced by individuals of surviving to a given age without adequate resources to finance consumption. The basic insurance feature of annuity contracts is that the annuity income is paid contingent upon survival, and risk-pooling allows for a higher return than financial notes provide. Indeed, the traditional argument in favour of public pension programmes is that annuity markets, tailored to insure against longevity risks, are often missing or underdeveloped. Hence the basic nature of pension assets, whether private or public, is that they are non-bequeathable.[17] Annuity and pension contracts are in principle irreversible investments, and this has important implications for the optimal timing of purchase and for the portfolio mix of pension and non-pension wealth (Brugiavini, 1993). There are different degrees of non-tradability and different features, which contribute to making the contract irreversible. In some cases, workers can borrow back a certain fraction of the annuity premium (usually at high borrowing rates and in the presence of a collateral). Occupational pension schemes and personal pension schemes (second pillar and third pillar) envisage transferability rules between schemes, but, given the privileged tax treatment, they do not normally allow for premature withdrawal from the scheme.[18] There may be considerable losses even in transferring rights

17 Some annuity contracts may actually be in the form of joint life-insurance and annuity contracts. This has the advantage, for the insurance company, of beating the adverse selection in both directions. However, these contracts are not very popular in Europe, probably also because they are as overpriced as traditional annuities.

18 Workers may leave the scheme prematurely and, if the value of the fund to which they are entitled is not transferred to another pension plan, they are normally entitled to a deferred annuity payable when they become eligible for retirement.

from one scheme to another, according to vesting and portability rules adopted by the different pension plans. Hence, by and large, we could conclude that both individual contracts (annuities, personal pensions and individual pension plans) and group contracts (occupational pensions) are typically ear-marked to retirement provision - i.e. non-fungible - and illiquid. However, there is a third aspect emerging, for pragmatic reasons, in the accounting exercise, i.e. the problem of valuing pension assets. At the aggregate level, the definition of disposable income has only recently been revised to include private pension benefits while excluding private pension contributions.[19] However, for some countries, this revision does not apply to mandatory occupational pensions (e.g. the Netherlands in Table 1 above). The problem is even more serious when estimating wealth and saving on the basis of survey data at individual or household level. In this case, a distinction is drawn between defined-benefit (DB) plans and defined-contribution (DC) plans, including money-purchase schemes: while DB plans do not have a market value, DC plans can be evaluated at market prices.[20] Hence the convention is to include DC plans in the definition of private wealth and to exclude DB plans, although both types of plans normally share the feature of being earmarked to retirement and being rather illiquid (see for example Wolff, 1998).

A fourth criterion often adopted rests on the distinction between mandatory and non-mandatory saving decisions. Estimating "non-bequeathable wealth" on the basis of this criterion is rather complex. In fact, mandatory wealth may include private pension wealth when this is mandated. For example, in the UK, workers must have a second-tier provision, though this can take many forms and the actual contribution can also vary, within given limits.

This discussion also hints at the potential substitutability between specific forms of wealth. Because of the different degrees of fungibility and liquidity that pension assets possess, it may easily be that pension assets are perceived as close substitutes for long-term financial assets, but not for wealth in real estate, say. Or it may be that the only sensible question to ask is what is the degree of substitutability between different forms of non-bequeathable wealth (e.g. social security and occupational pensions).

However, if consumers apply optimal portfolio rules, the covariance structure of returns on different assets should also matter. In fact, many risks, beyond longevity risks, potentially affect retirement-income provision and returns in general. PAYG systems and most occupational pension and collective pension contracts automatically pool longevity risks, as benefits are normally independent of an individual's survival probabilities, hence longevity risks are well covered by

---

19 See, for example, OECD (2000b), Economic Outlook, Sources and methods, where the System of National Accounts SNA68 has been updated to the system SNA93.

20 In both cases, the exercise may be non-trivial in survey data because of the existence of different eligibility conditions etc.. However, for DB plans, also the value of pension rights has to be imputed on the basis of a number of assumptions.

these schemes and attention is mostly shifted to other sources of uncertainty, particularly "macro" risks, which cannot be pooled within a given pension scheme or pensions system. Demographic risks are associated with both PAYG and funded systems. The demographic risk of a PAYG system has been documented extensively (OECD, 1988; World Bank, 1994). While most demographic trends can roughly be foreseen, residual uncertainty remains. Most importantly, it is far from clear how pension policy will respond to the demographic changes. Thus, from an individual's point of view, the consequences of demographic changes to her/his pension are uncertain. This can be expressed in the variance of predicted internal rates of return in PAYG systems. Political risks refer to the risk that claims on retirement income are altered by political action. In the case of funded pensions, funds may be used for purposes other than those originally designated, or, in the worst case, simply be appropriated, as in some developing countries (Börsch-Supan, Palacios and Tumbarello, 1999). PAYG pension systems face the risk of legislative changes. German and Italian workers, for example, have experienced a flurry of pension reforms since 1990, and further reforms are expected. While these reforms may address the demographic changes, they are superimposed by short-term considerations. The uncertainty facing individual workers with respect to their future PAYG pension claims has frequently been documented in polls and surveys.

Capital-market risks are intrinsic to funded pension systems. They include variation in the real rate of return, the risk of inflation if future pension claims are expressed in nominal terms, and the risk of exchange-rate fluctuations if claims are denominated in a different currency from consumption. While PAYG systems are exposed to demographic risk, they are immune to inflation risks because claims are typically indexed by wages and/or purchasing power. The risks for funds based on stocks and bonds are reasonably well understood. Less understood are the risks associated with investment in housing and in the book reserves popular in some countries. The wealth of most households, notably in the middle class, is concentrated in the own family home. While housing wealth may serve as a shield against inflation, it is exposed to its own variant of "capital-market risk" (in the sense that its price may fall relative to consumption) and a potentially large demographic risk (Mankiw and Weil, 1989; Hendershott, 1991).

This long list of potential risks suggests that the scope for diversification, particularly when a large fraction of wealth is already in illiquid form (e.g. social security and housing), goes beyond the basic distinction between bequeathable and non-bequeathable, wealth and involves different levels of substitutability. It should be added that differential taxation of assets may induce substitutability between assets. For example, if, at the initial equilibrium, investing in housing attracts tax privileges, introducing substantial tax relief on private-pension assets may lead to a new equilibrium, characterized by a shift away from housing into pension assets.[21] The final implication of this discussion is that, even within a common framework of analysis, where comparable estimates of wealth holdings have been computed (Börsch-Supan, 2001), and definitional issues seem settled, it is hard to address the substitutability problem.

4.3 Substitutability: empirical issues

I am not attempting to survey, in this paper, the huge empirical literature existing in the US on this topic: I just confine attention to the ongoing debate in Europe. However, recent papers by Gustman and Steinmeier (1998) and Gale (1998) have revived attention on estimation of the substitutability parameter stemming from the life-cycle model. As I mentioned, lack of evidence based on micro data is due partly to the inherent difficulty in measuring public pension (or social security) wealth and partly to the fact that observed cross-sectional differences in saving behaviour can be due to a variety of factors observed and unobserved. The presence of unobserved heterogeneity makes it difficult to identify the effect that different endowments of social security wealth have on different individuals' saving behaviour. In other words, besides the definition of private wealth versus pension wealth, the major problem is data quality. In order to translate the problem of estimating substitutability between pension wealth and private wealth into an empirical exercise, it is useful to refer to an ideal experiment (Bernheim, 1999). The ideal data should be characterized by a large random sample of the population, randomly partitioned into the "control" group and the "treatment" group, or into more treatment groups with different levels of treatment. The two sub-samples should be identically random in all respects, but for the treatment. I.e. one needs exogenous variation in participation in the public scheme or eligibility to the tax-favoured scheme under investigation.[22] This is a well-known empirical problem as unobserved heterogeneity, which correlates with saving choices, would bias the results.

Having described the ideal setting one could compare the existing literature with this framework. However, in Europe the question has been addressed only recently, and there exist very few examples to build on. Early contributions include Rossi and Visco (1995)[23] for Italy, who attribute a large part of the decline that occurred during the 1980s in the saving rate to the increased generosity of social security in the 1970s. In other countries, the relationship between pension

---

[21] Variations in risk aversion and liquidity constraints may justify differences in the sensitivity of saving to tax relief when applied to different assets. There could also be "less motivated occasional" savers (Bernheim, 1999), who are less sensitive (or more sensitive) to tax incentives of one type.

[22] For the US, this means exogenous variation in eligibility to IRA accounts.

[23] The studies are based on long-run relationships between, saving, growth and changes in public pension wealth.

wealth and private fungible wealth emerging from time-series data is poorly determined.[24] Cross-sectional analysis based on Italian micro data (Brugiavini, 1987; Jappelli, 1995) obtains low estimates of the substitution parameter. These studies are all subject to the Auerbach-Kotlikoff (1983) critique, and are quite far from the ideal setting. In Continental Europe, this debate is complicated by the fact that in some cases households seem to be "overannuitized". Evidence by Börsch-Supan and Stahl (1991) and Börsch-Supan (1992, 1994) shows that the annuity income of a majority of pensioners (mostly PAYG income, provided through the public system) far exceeds consumption expenditures. The next generation of pensioners appears to be aware of this fact, since saving rates in Germany drop for the now middle-aged cohorts. Given this trend, a reduction in public pension wealth may not necessarily be channelled into new savings, or only after a transition period of new, costly, learning. At the same time, these households, if left to their own devices, may also shun pension assets because they do not want to lock their wealth into long-term saving for retirement.

Turning to more recent studies, three countries provide useful examples: the Netherlands, the UK and Italy. However, while for the Netherlands there exist studies addressing directly the problem of substitutability, in the UK there are many related studies which hint at the problem without providing direct evidence. Looking at the recent Italian pension reforms is a useful exercise, which I will describe in more detail.

4.4 Substitutability. A case study: The Netherlands

The question of substitutability has been addressed directly on micro-level data in the Netherlands. I can summarize the evidence by referring to two studies: Alessie, Kapteyn and Klijn (1997) and Euwals (2000). Both papers investigate the question by distinguishing the two mandatory parts of the Dutch pension system: a public social security component (AOW, PAYG financed) and a private occupational-pension-funded component. The definition of private household wealth includes bank accounts, insurance, stocks, bonds, real estate, mortgages and debts. The computed household social security wealth and pension wealth depends on the age, marital status and employment status of the two spouses in the household, plus the rules in place for the social security system and the occupational pension (normally, final salary scheme). The major difference between the two papers is given by the sample used in the application and, partly, in the results. Euwals makes use of a sample drawn from the CentER Savings Survey (DCSS), and focuses attention on the household-wealth holdings of the 1994 wave. The author finds that, though important in some cases, the impact of social security is not well identified, due to a substantial lack of variability. The author himself admits to a number of shortcomings, particularly the typical biases

---

[24] See Kohl and O'Brien (1998) for a detailed survey.

pointed out by Gale (1998), which I will discuss later in this section. By making use of the Dutch Socio-Economic Panel (SEP), Alessie-Kapteyn-Klijn find slightly different results. By and large, one could conclude that private wealth and mandatory wealth show some degree of substitutability in the Netherlands. More than for the final result, the Dutch example is illuminating in two respects: (1) it is hard to reach one simple conclusion and (2) data problems and identification problems are at the heart of the matter.

5 Recent empirical work on "substitutability" in Italy

Clear-cut "experiments" of the type advocated by Bernheim (1999), in which the pension system changes but everything else remains constant, are hard to come by. Attanasio and Brugiavini (2000) rely directly on a test of how changes in saving rates are related to changes in pension wealth, following the 1992 Reform of the Italian Social Security System. Before describing the methodology and the results of this paper, it is useful to provide a general overview of saving in Italy.

Figure 2 plots the cohort-adjusted profile of median discretionary and mandatory saving, based on the Bank of Italy sample (SHIW), for repeated cross-sections in the years 1989, 1991, 1993, 1995 and 1998. This is a nation-wide cross-sectional sample of the Italian population. Discretionary saving is defined as disposable income minus total consumption. Mandatory saving is social security payroll taxes, minus social security benefits.[25] All values are expressed in 1998' thousand Euro. The smoothed line in the figure is the fitted line of a regression on a third-order age polynomial, a full set of cohort dummies and a set of restricted time dummies. This methodology for smoothing the profile deals with cohort effects, which are very important in this framework, and provides a cleaner picture of households' age-saving profiles. Figure 3 plots the cohort-adjusted profile of mean and median discretionary saving rate. Discretionary saving rate is computed as 1 minus the ratio of total consumption to disposable income. It is clear that Italian households have high saving rates (see Börsch-Supan, 2001 for a comparison with other countries): there is a smooth and small decline in saving only after age 60. If we take the extreme view of Jappelli and Modigliani (1998), and assume that households save only to protect retirement and social security wealth and private wealth are perfect substitutes, then it is interesting to contrast the discretionary saving profile, which is almost flat, and the mandatory social security-saving profile, which is hump-shaped. The latter is obviously so by construction, as pension rights accumulate up to retirement age, and decline once retirees start drawing their benefits. These figures suggest that social security

[25] Figures 2 and 3 are drawn from Brugiavini and Padula (2001). Details of the calculations are provided in that paper.

wealth in Italy is a large fraction of wealth and that mandatory saving varies with age, more than is the case for private discretionary saving.

5.1 Exploiting pension reforms

In Attanasio and Brugiavini (2000), changes in public pension wealth are used to identify the effect of social security entitlements on household savings. For such a purpose, two Italian cross-sections are used, one collected just before, and the other just after, a major reform of the social security system. The Italian social security system, started in the early 1970s, became progressively more generous and, by the end of the 1980s, was clearly unsustainable. In September 1992, the Amato government introduced the first of a series of important reforms. By changing retirement age, eligibility, early retirement arrangements and so on, the Amato reform induced a substantial reduction in the pension wealth of many Italian households. Furthermore, the Amato Reform received much attention from the press and was much debated on TV and radio programmes: the importance of the changes brought about by the reform could not go unnoticed. What is more relevant to the issue of substitutability, however, is that the reform changed the pension wealth of observable groups of households in different ways. Typically, public sector, younger employees were affected more than relatively older, private sector employees. In what follows, we try to identify the elasticity of personal savings to public pension wealth by comparing the changes in these different groups.

In 1993, the year following the Amato Reform, the private saving rate as recorded by survey data, increased.[26] Obviously it is not possible to attribute the change entirely to the reform of the social security system. Many things happened between 1992 and 1993, including the fact that the economy entered a substantial recession. By the same Act by which it changed the social security system, the Government introduced a number of measures aimed at re-equilibrating the Italian fiscal situation.

The approach adopted by Attanasio-Brugiavini (2000) is close to an ideal experiment. First, it is based on the relationship between public pension wealth, as measured by the present discounted value of future entitlements, and the *saving rate* of individual households. Second, it makes use the variation in pension wealth induced by a large legislative change that occurred in Italy in 1992 to identify the parameters of interest. This allows control for the possible presence of fixed effects at the group level. Third, groups (mainly public sector and private sector employees) are "exogenous" to the reform, and the reform applied a differential treatment to the different groups.

---

[26] From 24% to above 25%, note, however, the discrepancy between this result and the saving rate from national accounts data (Brugiavini and Padula, 2001).

It is useful, first, to sketch the basic idea that helps in discussing the empirical specification of the paper.

Figure 2. Italy. Cohort-adjusted age-profile of saving
(Thousands, 1998 Euro)

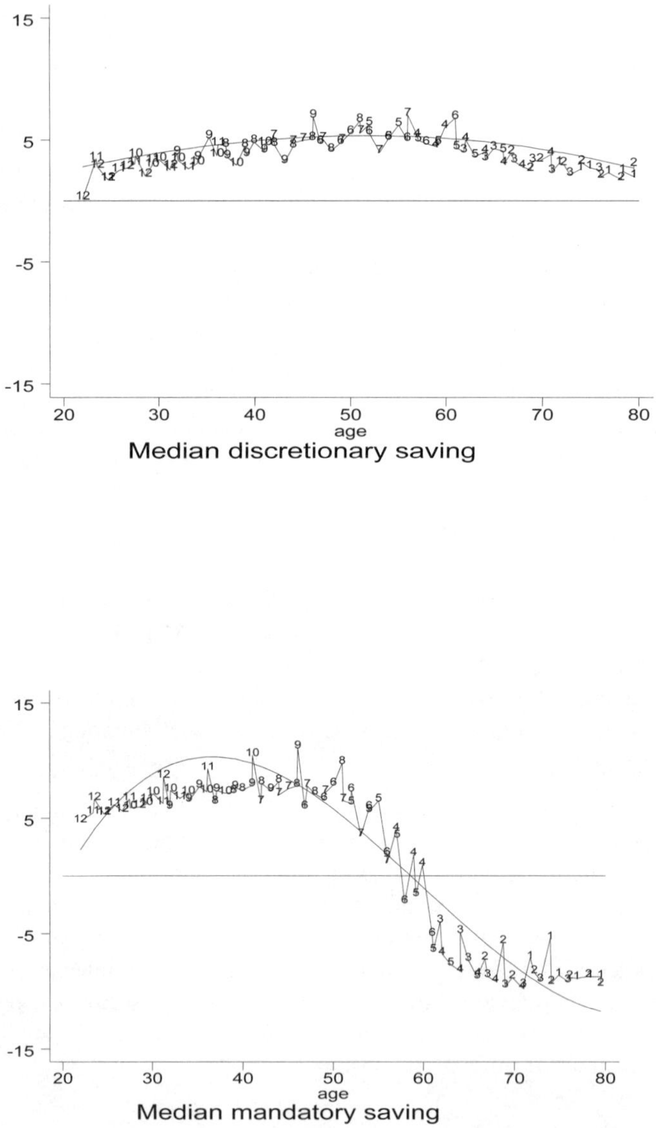

Figure 3. Italy. Cohort-adjusted age-profile of discretionary saving rate

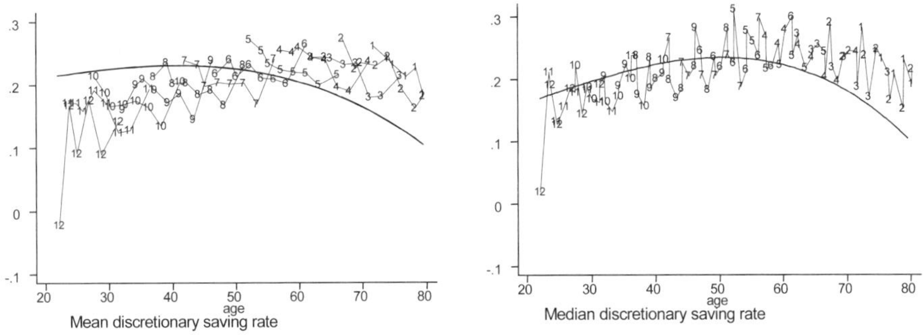

5.2 The basic idea

I have discussed the relationship between social security wealth and private wealth in the framework of the life-cycle model. In Italy, one could safely assume that public pension systems cannot be "undone" through reverse annuities. A permanent and unexpected reduction in an individual's social security wealth implies an increase in her savings and a reduction in her current consumption. More generally, for individuals who are not at a corner, changes in their pension wealth would induce a change in their consumption and saving behaviour. The link is particularly obvious in an un-funded Pay-As-You-Go system, in which, contributions are often perceived as taxes, and benefits are only conditional on meeting eligibility requirements.

Attanasio and Brugiavini (2000) consider a version of the model without uncertainty. This choice, which simplifies the discussion considerably, is not fully consistent, however, with the consideration of an 'unexpected' change in pension wealth at some point in the life cycle, and it reduces the motive to save *vis-à-vis* other risks. However, a complete treatment of uncertainty would not add much to the basic idea. The basic ingredients of the model are as follows: let's assume that people work during the first two periods of their lives and then retire. Labour supply is exogenous and is compensated with a deterministic level of earning. When people retire, they receive a level of benefits b. For simplicity, we assume the utility function is isoelastic. A generic individual maximizes her intertemporal utility, subject to the usual intertemporal budget constraint. There exists a constant interest rate r, β is the discount factor and $1/\gamma$ the elasticity of intertemporal substitution. The resulting age-1 saving rate and the *planned* age-2 saving rate depend on the value of human wealth according to the parameters of the model. To be more precise, the saving rate at each age depends on the ratio of the presented discounted value of pension wealth to the current level of income. These results can be used to derive the effect on second-period consumption (and saving) of an anticipated change in pension wealth. However, in order to consider the effect of an unanticipated change in pension wealth on consumption and wealth for a consumer aged 2, *at the current age*, what is relevant is the re-optimization problem of this individual, who has saved a certain amount in period 1. The re-optimization will yield saving rates which look quite different from the *planned* saving rates.

The final result is that the coefficient of pension wealth, in its relationship with current saving rates, varies with age and depends on the discount factor β. It can be shown that, in a general model with N periods of work before retirement, the coefficient of the ratio of the present discounted value of benefits to current income is equal to a factor *k(a,d)*, defined as follows:

$$(1) \qquad k(a,d) = \frac{1}{\sum_{j=0}^{ra-a} d^j} = \frac{1-d}{1-d^{ra-a}};$$

where ***ra*** is retirement age, assumed to be fixed, ***a*** is the current age and *d* is a function of β and 1/γ.[27]

From equation (1) it is clear that the relationship between pension wealth and current saving rates depends on age. The change in saving generated by a change in pension wealth is greater, the closer the individual is to retirement age. This reflects the fact that younger individuals have a longer horizon over which to absorb the "unexpected" shock to pension wealth. A similar result is also discussed by Gale (1998), who argues that the existing studies induce a bias in the estimates of the substitution parameter, as these do not control for variation in age.

In this simple model, pension wealth accrues in one period and is, for all practical purposes, indistinguishable from private financial wealth. However, I have provided a number of reasons why public pension wealth is substantially less liquid than financial wealth. Furthermore, it accrues over the retirement period with a predetermined pattern, and, therefore, cannot be adjusted to particular changes in needs. It is therefore plausible to assume that public pension wealth is not a perfect substitute for private savings. For this reason, one can expect the coefficient of public pension wealth (adjusted for age and discount factor) to be negative, but different from minus one.

27 This result is derived for iso-elastic utility, and assuming that $d = \beta^{\frac{1}{\gamma}}(1+r)^{\frac{1-\gamma}{\gamma}} < 1$. Under log utility, for instance, this is obviously true, as in that case $d=\beta$.

5.3 The empirical specification for analysing substitutability around the time of the 1992-Italian reform.

Attanasio and Brugiavini (2000) use a flexible specification that tries to capture the main features of the simple model sketched above. In particular, they regress individual saving rates on the ratio of future pension wealth to current net income. To capture the fact that the effect of this ratio on saving rates depends on age, this is interacted with a polynomial in age. Furthermore, to capture the effect of remaining resources, the empirical specification includes the ratio of the present discounted value of future net earnings to current income. As for the ratio of pension wealth to earnings, this variable is also interacted with a polynomial in age.

To be more precise, the estimated equation is:

$$(2) \quad SR_{i,t} = X_{i,t}\beta + \phi(a_{i,t})FE_{i,t} + \widetilde{\theta}(a_{i,t})k(a_{i,t},d)\,PW_{i,t} + \varepsilon_{i,t}$$

where $SR_{i,t}$ is the saving rate out of current income for household i at time t, X is a vector of controls, $FE_{i,t}$ is the ratio of future to current earnings and $PW_{i,t}$ is the ratio of pension wealth to current earnings. The coefficient on PW is assumed to vary with age, as discussed above, both because of the way in which such a variable enters equation (1) above and because the degree of substitutability between future earnings and pension wealth could also vary with age. The separate terms allow these to be identified as separate effects, where $\widetilde{\theta}(a_{i,t})$ is the life-cycle-annuity factor and $k(a,d)$ is defined in equation (1). Under the hypothesis that financial and pension wealth are perfect substitutes $\widetilde{\theta}(a_{i,t})$ should be equal to -1 at all ages. Given assumptions on the discount factor, the interest rate, the elasticity of intertemporal substitution, and retirement age, it is possible to construct estimates of $k(a,d)$ and then of $\widetilde{\theta}(a_{i,t})$.

The model presented above excludes some important factors, such as uncertainty and changing needs linked to changes in family composition: controls for such variables are included in the specification. For this reason, one would like to add a flexible polynomial in age to equation (2) above. However, as discussed below, the econometric strategy employed calls for the introduction of cohort and time dummies that completely absorb the effect of age.

The estimation of equation (2) on individual data poses a number of problems. First, as pension wealth, that is the present discounted value of future benefits, is not typically observed, one has to estimate it. However precise the method used in estimating such a quantity, the resulting variable is likely to be affected by measurement error, which is likely to introduce a bias in the estimated coefficients. Moreover, an additional source of bias is introduced by the possibility that individual (and unobserved) heterogeneity in saving behaviour is related to the individual variables used to estimate pension wealth. To take this type of problems into account, one can use an Instrumental Variable technique, using as instruments group dummies interacted with year dummies. Groups are defined so as to capture systematic differences in social security wealth.

Another factor that one has to take into account in specifying a regression equation for individual saving rates at different points in time is the possibility that a macroeconomic shock, different from the pension reform, affects all individuals' saving rates. Such a shock could account for changes in saving behaviour across the periods used in estimation. Therefore, one should include time dummies in equation (2).

While the introduction of time dummies and the use of group/time dummies as instruments for pension wealth $PW_{i,t}$ deals with the presence of aggregate shocks, measurement error and pure idiosyncratic fixed effects, it does not remove the likely bias caused by the presence of group-specific fixed effects. That is, if individual fixed effects are given by the sum of a group-specific fixed effect and a pure idiosyncratic element, the instrumenting procedure described above will not remove the bias in the estimation of the coefficient of interest θ. To deal with this problem, therefore, group dummies have to be added to the empirical specification. All these considerations lead to a new equation where the error term is specified in terms of three components: time effect, group effects and an i.i.d. error.

This new equation is estimated by using as instruments for PW_{it} and FE_{it} the interaction of time and group dummies. The proposed strategy, therefore, is an application of a *difference-in-difference* estimator. Identification requires that PW_{it} (and FE_{it}) varies in a way that is not fully explained by group and time dummies. This hypothesis can easily be tested by checking the significance of interaction terms in a regression of PW_{it} on group, time dummies and their interaction.

The Amato reform of 1992 afforded the possibility of implementing such a technique as it induced changes in pension wealth that were systematically different across the groups considered. That is, the average change in saving rates and pension wealth is computed before and after the pension reform for a number of well-specified groups, and the former are regressed on the latter. It is crucial for the identification of the substitutability parameter, therefore, that the choice of groups is made in such a way that the changes in pension wealth induced by the reform are systematically different.

Estimating social security wealth is obviously quite hard. The procedure requires knowledge of a number of legislative details, as well as some assumptions on expected earning profiles, retirement age and so on. Furthermore, as social security wealth is used as a determinant of individual saving behaviour, one would ideally require an estimate of social security wealth as perceived by the individual. If the social security system before (or after, for that matter) the Amato reform was widely perceived to be unsustainable, our estimates of social security wealth might deviate substantially from perceived wealth. This argument constitutes probably the most convincing reason for using average changes in estimated social security wealth to identify the relationship of such a variable with the personal saving rate. In 1991, the Italian social security system was one of the most generous in the world, and clearly unsustainable in the long run. As these issues were widely debated in the run-up to the reform of 1992, it is not unreasonable to think that a change in the legislation was widely expected. However, it should be stressed that the type of reform that would eventually be implemented was not obvious. In particular, it was not obvious who would be affected more by the reform, and by how much.

5.4 The 1992 pension reform and its differential effects on social security wealth

The Italian social security system was at the time (and still is) a patchwork of different public funds, each characterized by its own rules in terms of eligibility and benefit payments, and it is obviously based on a Pay-As-You-Go financing system.

The changes introduced by the 1992 reform mainly concerned the criteria of eligibility for a public pension and the size of the benefit outlays. The reform envisaged a gradual increase in retirement age from 60 to 65 for men, and from 55 to 60 for women in the private sector, and a gradual increase in the minimum number of years of social-security-tax payments from 15 to 20. Eligibility requirements for *early retirement* were tightened. The reduction in benefits was achieved by replacing a pension calculation, which was basically of a final salary type (average of last five years' gross earnings), by a career average earnings calculation (the benefits are a fraction of the average earnings over the entire working life of an individual). The reform envisaged a transitional period which left the rules in place for workers on the verge of retirement almost unaffected, while greatly affecting younger workers. While the system remained extremely generous overall, younger workers were potentially losing a substantial share of their pension wealth from the reform, particularly if their age-earnings profile was sufficiently steep. It is relevant to note that the seniority criterion adopted by the reform did not affect only those entering the labour market in 1993. A worker in the public sector, say, who had already paid contributions for 14 years (i.e. he is about 35-40 years old in 1993, according to when he entered the labour market) and who was planning to retire at age 65, was bound to lose substantially in terms of pension wealth.

Each component of the 1992 reform has a different effect on each of the groups we consider. In many cases, computing the net combined effect of the many changes induced by the reform on pension wealth is not obvious. Some broad patterns, however, are quite clear and have been discussed in the literature, particularly the 'wealth tax' imposed by the reform on the younger active population due to the change in the benefit-computation method, which affects mainly younger generations and public sector employees.

It is clear that expectation formation is one aspect of the exercise which requires further attention. Attanasio and Brugiavini assume that the specific features of the reform are essentially unexpected. At the same time, once the reform takes place, individuals make use of the new rules for future eligibility and for the calculation of the future stream of benefits and contributions (from the current age to retirement age). However, it is assumed that individuals cannot forecast exactly how the rest of the tax system implicitly changes as a result.[28]

In other words, Attanasio and Brugiavini emphasize the between-groups redistributional features of the system: it can be argued that public-sector employees were the group that most enjoyed the generosity of the Italian social security system under the pre-reform regime, and were presumably most affected by the changes. It is this type of variability that allows the authors to perform their tests. The research strategy used is based on dividing the sample into a number of groups chosen to maximize the variation, across groups, in pension-wealth changes induced by the reform. The institutional details discussed above were therefore crucial in informing the definition of the groups.

The hypothesis that group membership is constant over time is particularly problematic for young and old households, as household headship might change endogenously with age at the beginning and at the end of the life cycle, as household formation and dissolution are endogenous to saving decisions. This consideration should make us cautious in interpreting the results we present below.

There is a substantial decline, both in absolute and in relative terms, in pension wealth brought about by the reform. For males, young cohorts are those that experience the largest losses in terms of accrual rate. For females, the losses seems particularly large for the youngest cohort in the public sector. The household-accrual rate, which measures the loss in household-pension wealth brought about at the margin by the reform, ranges between -34% for households whose head is younger than 32 and works in the public sector and -14% for households whose head is on the verge of retirement. Interestingly enough, a major role in this result is played by the changes in pension wealth for spouses (mostly wives), as there is a very high correlation between the occupations of the two spouses.

---

[28] I wish to thank Alan Auerbach for raising this point. Basically, Attanasio and Brugiavini do not model the entire implicit intergenerational redistribution mechanism going via the tax system, but rather focus attention on the rules of the public pension system.

5.5 The effects of pension wealth on saving rates

The regression considered in Attanasio-Brugiavini relates, for each household in the sample, saving rates to a variable that depends on pension wealth and one that reflects the effects of future earnings. These two variables are interacted with a polynomial in age. In addition, a year dummy captures the aggregate effect of 1993, a term in age captures differences in age within each cohort and, in some of the specifications, group dummies control for fixed-group effects.

Two sub-samples of the active population are selected: (1) private-sector and public-sector employees only and (2) all workers, including the self-employed. Results are reported in Table 3. The left-hand side of all the equations is the saving rate, defined as income minus consumption, divided by income. In the left panel of the Table, results obtained in the smaller sample are shown, while those for the whole sample can be found in the panel on the right.

In each of the two panels, equation (2) is estimated, using two different methodologies.[29] First, standard IV is implemented, then group effects are accounted for by using a "diff.-in-diff." estimator. In the Table the coefficients on the interaction of pension wealth and earning with a polynomial in age, the coefficient on the 1993 dummy, the Sargan test of over-identifying restriction (when relevant) are all reported. To facilitate the interpretation of the results, the average of the coefficient on pension wealth over the 25-64 age interval is computed. In the first column of the two panels, the interaction of time and group dummies is used to instrument for the pension wealth and the future earning variables. Notice that now the effect of public pension on saving rates is, on average, positive. However, as indicated by the rejection of the over-identifying restrictions, these estimates are likely to be mis-specified. In particular, the presence of group-specific fixed effects would make these estimates inconsistent.

When correcting for the factor k in equation (2), the figures obtained are economically significant. In the sample of private and public sector employees, the level of substitutability between pension wealth and financial wealth (that is, the coefficient on pension wealth divided by the factor k) averages –0.124 between ages 25 and 64 and averages –0.29 in the sample of individuals of these ages.[30] For the sample that includes all the active population, the same numbers are much larger in absolute value, being equal to –0.56 and –0.77. That means that, on average, in the larger sample, the coefficient on pension wealth is not far away from the value of –1.

To give an idea of the way in which the coefficients on pension wealth corrected for the factor k vary with age, the two panels of Figure 4 plot the coefficient by

29 In Attanasio and Brugiavini (2000), results for OLS estimates are also presented.

30 This reflects the fact that there are more household heads aged between 35 and 45 in our sample. The coefficient is lowest around these ages.

age. These show the shape of the age profile for the coefficients implied by the estimates in Table 3. For both sub-samples, the estimates imply that the coefficient is positive in the first part of the life-cycle, and turns negative around age 30 (just after in the smaller sample, and just before in the larger one). It is lowest just past 40 and then increases again, reaching zero around retirement age. The figures, however, suggest caution in interpreting these age profiles too literally. It should be remembered that the coefficients are obtained by interacting pension wealth with a quadratic in age. Given the limited amount of variability available to us (remember, we are using differences *across groups* in the effects of the 1992 reform), we were not able to fit more flexible functional forms. It might be that the rather extreme values obtained by the coefficients, especially at the beginning of the life cycle are a consequence of the rather restrictive assumption we impose.

Figure 5 provides a kernel-estimated density of the estimated coefficient from the "diff.-in-diff." procedure, after trimming, for the case of "all the active population". It is interesting to see that the bulk of the distribution is concentrated around high negative values.

Table 3 Italy. Estimating the effect of pension wealth

	Sub-sample of private-sector and public-sector employees (N=6603)		Sub-sample of active population (N=9366)	
	(1)	(2)	(1)	(2)
	IV	Diff. in diff.	IV	Diff. in diff.
PW_{it}	0.0076	-0.0312	0.0184	-0.0642
	(0.0034)	(0.0217)	(0.0029)	(0.0207)
$Age\ PW_{it}$	0.0004	-0.0184	0.0051	-0.0269
	(0.0015)	(0.0097)	(0.0013)	(0.0088)
$Age^2\ PW_{it}$	-0.0011	0.0221	0.0002	0.0273
	(0.0005)	(0.0116)	(0.0006)	(0.0120)
FE_{it}	-0.0123	0.0920	-0.0178	0.1750
	(0.0043)	(0.0691)	(0.0050)	(0.0577)
$Age\ FE_{it}$	0.0018	0.0448	0.0073	0.0621
	(0.0025)	(0.0303)	(0.0025)	(0.0262)
$Age^2\ FE_{it}$	-0.0015	-0.1103	-0.0054	-0.0911
	(0.0016)	(0.0463)	(0.0018)	(0.0488)
1993 dummy	-0.0279	0.0815	-0.0585	0.1730
Group dummies	*No*	*Yes*	*No*	*Yes*
Number of IV	19	19	27	27
Sargan test	18.34	1.187	50.53	10.41
(P-value)	(0.049)	(0.76)	(0.0001)	(0.17)
Avg coeff on PW (25-64)/	0.0060	-0.0056	0.0210	-0.0344
Adjusted	0.1034	-0.1241	0.2896	-0.5580
Avg coeff on PW (sample 25-64)/	0.0067	-0.0170	0.0206	-0.0469
Adjusted [a]	0.1170	-0.2908	0.3146	-0.7739

Notes: Standard errors in parentheses. All specifications include a constant and a term in age to capture within cohort differences. "Groups" are defined by cohort and occupation interactions. PW is the ratio of pension wealth to current income. FE is the ratio of future human wealth to current income ; a) The adjustment is performed correcting the age dependent coefficient on PW by a factor k computed as in equation (1) with $\gamma=1$ and $\beta=0.98$.

Source: Attanasio and Brugiavini, (2000).

Figure 4. Adjusted coefficient on pension wealth as function of age

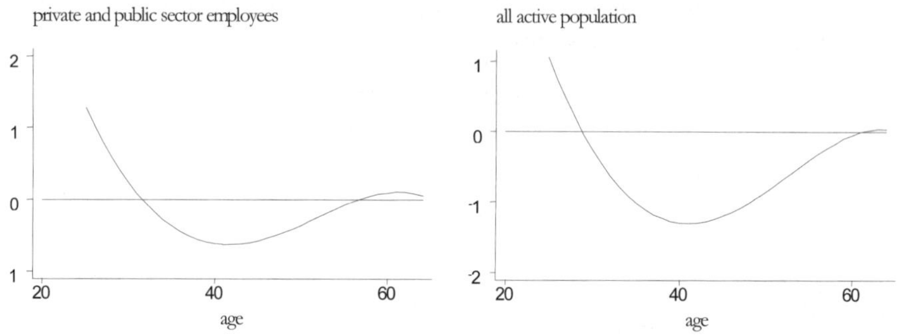

Figure 5. Kernel density for the estimated coefficient
(all active population)

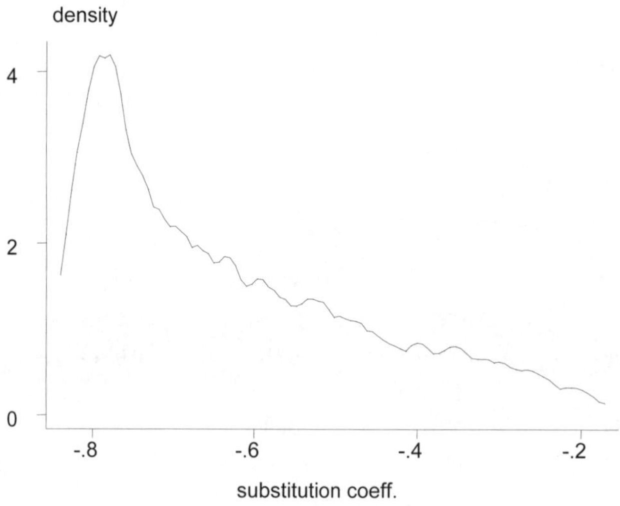

These results imply that public pension wealth is likely to be a good substitute for private savings and that people "react" to pension reforms. If this effect prevails in most European countries then, even if the stripped-down version of the life-cycle theory suggests that ageing could imply a reduction in aggregate saving. The public pension reforms that Europe is most likely to experience, could actually produce the opposite effects, either via the creation of funded private pensions or via private saving.

6 Conclusions

In this paper I have provided the motivation for relating saving to ageing by first looking at some stylized facts in Europe. European countries are ageing fast and pension reforms have already been enacted or rank high in the policy agenda. There are many channels through which ageing could affect saving: pension arrangements and the likely changes in pension programmes are one obvious link. However compositional effects and macro feedback effects make it hard to assess how much variation in private saving can be imputed to variation in "pension accumulation". In Europe, there are other important considerations to take account of: changes in financial markets, the advent of institutional investors and important changes in informal credit arrangements. All these may have contributed (and may contribute in the future) to explaining trends in saving behaviour. In the second part of the paper, I focus attention on empirical work, looking at substitution between private saving and pension assets. While the empirical evidence is scanty, and often leads to opposite conclusions, precisely because Europe is ageing fast and because of the different institutional arrangements, it may be the ideal ground to analyse this issue. I describe in detail an exercise based on Italian micro-data recorded at household level, which makes use of the 1992 Italian pension reform as a "quasi-experiment". This episode is particularly useful for several reasons. First of all, one can use two large and consistent household surveys that immediately precede and follow the reform. Second, the reform did not change the *nature* of the pension system, in that the Italian system remained an un-funded defined-benefit system. The 1992 reform, however, substantially changed the presented discounted wealth of a large majority of Italian households. Furthermore, the reduction in public pension wealth was far from uniform across households. A first specification relates individual saving rates to the level of (log) pension wealth, and accounts for the possibility of unobserved heterogeneity and measurement error in the estimate of pension wealth by using as instruments cohort and occupation dummies, interacted with year dummies. This strategy ignores the possibility of group-specific fixed effects. To take this into account we use an alternative strategy that is equivalent to using a "difference-in-difference" estimator. The results indicate that pension wealth is a relatively good substitute for private financial wealth.

More interesting: estimates of the degree of substitutability depend on age. The degree of substitutability between public pension wealth and private wealth obtained in the Italian example suggests that Europe may actually experience a growth in saving and capital accumulation, as a result of public pension reforms either through the creation of funded pensions or through an increase in private household saving.

References

Alessie, R., A. Kapteyn and F. Klijn. 1997. Mandatory Pensions and Personal Savings in the Netherlands, *De-Economist*, 145(3), pp. 291-324.

Alessie, R., S. Hochguertel and A. van Soest. 2001. Household Portfolios in the Netherlands. In: L. Guiso, M. Haliassos and T. Jappelli eds., *Household Portfolios*, Cambridge, Mass.: MIT-Press, in press.

Attanasio, O. 1997. Consumption and Saving Behaviour: Modelling Recent Trends, *Fiscal studies,* 18, No.1, pp. 23-47.

Attanasio, O. 1998. Consumption Demand, *NBER Working Paper No. 6466.*

Attanasio, O., J. Banks, C. Meghir and G. Weber. 1999. Humps and Bumps in Lifetime Consumption, *Journal of Business & Economic Statistics*, 17(1), pp. 22-35.

Attanasio, O. and T. De Leire. 1994. IRAS and Household Saving Revisited: Some new Evidence, *NBER Working* Paper 4900.

Attanasio, O. and G. Weber. 1994. The UK Consumption Boom: Aggregate Implications of Microeconomic Evidence, *Economic Journal*, 104.

Attanasio, O. and J. Banks. 1998. Trends in Household Saving don't Justify Tax Incentives to Boost Saving, *Economic Policy*, October.

Attanasio, O. and A. Brugiavini. 2000. Household Saving in Italy, mimeo, University of Venice, Italy.

Auerbach, A. J. and L. J. Kotlikoff. 1983. An Examination of Empirical Tests of Social Security and Savings, in: E. Helpman et al., eds., *Social Policy evaluation: an Economic Perspective,* Academic Press, New York, pp. 161-179.

Banks, J. and S. Tanner. 1999. *Household Saving in the UK*, Institute for Fiscal Studies, London.

Barro, R. J. 1974. Are Government Bonds Net Wealth?, *Journal of Political Economy*, 82,5, pp. 1095-1117.

Bernheim, B. D. 1987a. The Economic Effects of Social Security: Toward a Reconciliation of Theory and Measurement, *Journal of Public Economics*, 33, pp. 273-304.

Bernheim, B. D. 1987b. Dissaving After Retirement: Testing the Pure Life Cycle Hypothesis, in Z. Bodie, J. Shoven and D. Wise, eds., *Issue in Pension Economics,* NBER, The University of Chicago Press, Chicago.

Bernheim, B. D. 1999. Taxation and Saving, mimeo, Stanford University.

Blommestein, H. J. and N. Funke, eds., 1998. *Institutional Investors and the new Financial Landscape*, OECD, Paris.

Börsch-Supan, A. and K. Stahl. 1991. Life-cycle Savings and Consumption Constraints. *Journal of Population Economics*, 4, pp. 233-255.

Börsch-Supan, A. 1992. Saving and Consumption Patterns of the Elderly: the German Case. *Journal of Population Economics, 5*, pp. 289-303.

Börsch-Supan, A. 1994. Savings in Germany - Part II: Behaviour. In: Poterba, J. (ed.), *International Comparisons of Household Savings*. Chicago, London: University of Chicago Press, pp. 207-236.

Börsch-Supan, A., R. Palacios and P. Tumbarello 1999. Pension Systems in the Middle East and North Africa: A Window of Opportunity, *The World Bank und SFB Arbeitspapier,* Nr. 99-44, Universität Mannheim.

Börsch-Supan, A. and M. Lührmann. 2000. *Prinzipien der Renten- und Pensionsbesteuerung*, Frankfurter Institut: Bad Homburg.

Börsch-Supan, A. and M. Miegel, eds., 2001. *Pension Reform in Six Countries: What can We Learn from Each Other?*, Springer-Verlag: Berlin, Heidelberg, New York.

Börsch-Supan, A. and A. Brugiavini. 2001. Savings: the Policy Debate in Europe, *Oxford Review of Economic Policy*, 17 (1), pp. 116-143.

Börsch-Supan, A., ed. 2001. *International Comparisons of Household Saving: A Study of Life-Cycle Savings in Seven Countries*, New York: Academic Press.

Browning, M. and A. Lusardi. 1996. Household Saving: Micro Theory and Micro Facts, *Journal of Economic Literature*, 34, pp. 1797-1855.

Brugiavini, A. 1987. Empirical Evidence on Wealth Accumulation and the Effects of Pension Wealth: an Application to Italian Cross Section Data, 1987, *Financial Markets Group,* D.P. 20, LSE, London.

Brugiavini, A. 1993. Uncertainty Resolution and the Timing of Annuity Purchases, *Journal of Public Economics*, 50, pp. 31-62.

Brugiavini A. and M. Padula. 2001. Too much for Retirement? Saving in Italy, *Research in Economics*, 55, pp. 39-60

Carroll, C. D. 1992. Buffer-stick Saving and the Life-cycle/Permanent Income Hypothesis, *Quarterly Journal of Economics*, 112, pp. 1-56.

Chiuri M.C. and T. Jappelli. 2000. Financial Markest Imperfections and Homeownership: an International Comparison, *CSEF Working Paper*, Salerno.

Deaton, A. 1991. Saving and Liquidity Constraints, *Econometrica*, 59, pp. 1221-48.

Deaton, A. 1995. *Understanding Consumption*, Clarendon Press.

Davies, E. 1998. Pension Fund Reform and European Financial Markets, *Financial Markets Group, LSE*, Special Paper No. 107.

Dilnot, A. W. 1992. Taxation of Private Pensions: Costs and Consequences. In OECD, *Private Pensions and Public Policy*, Paris: OECD.

Disney, R. 1996. *Can We Afford to Grow Older?*, The MIT Press, Cambridge.

Disney, R., M. Mira D'Ercole and P. Scherer. 1998. Resources During Retirement, OECD, *Ageing Working Paper*, AWP 4.3, Paris.

Disney, R. 2000. Crises in OECD Public Pension Programmes: What are the Reform Alternatives?, *Economic Journal Features*, 110, pp. F1-F23.

Engen, E., W. Gale and J. Scholz. 1996. The Illusory Effects of Saving Incentives on Saving, *Journal of Economic Perspectives* 10(4), pp. 113-138.

Engen, E., W. Gale and C. Uccello. 1999. The Adequacy of Household Saving, *Brooking Papers on Economic Activity*, 2, pp. 65-187.

European Commission. 1999. *Towards a Single Market for Supplementary Pensions*. Results of the Consultation on the Green Paper on Supplementary Pensions in the Single Market, COM(1999), 134.

Euwals, R. 2000. Do Mandatory Pensions Decrease Household Savings: Evidence for the Netherlands. *De Economist*, 148, pp. 643-670.

Eymann, A. and A. Börsch-Supan. 2001. Household Portfolios in Germany. In: L. Guiso, M. Haliassos and T. Jappelli, eds., *Household Portfolios*, Cambridge, Mass.: MIT-Press.

Feldstein, M. 1974. Social Security, Induced Retirement and Aggregate capital accumulation, *Journal of Political Economy*, 82,5, pp. 905-926.

Gale W. and J. Scholz. 1994. IRAs and Household Saving, *American Economic Review*, 84, pp. 1233-1260

Gale W. 1998. The Effects of Pensions on Household Wealth: a Reevaluation of Theory and Evidence, *Journal of Political Economy*, 106, 4, pp. 706-723.

Gruber, J. and D. Wise, eds. 1999. *Social Security and Retirement Around the World*, The University of Chicago Press: Chicago.

Guiso, L. and T. Jappelli. 1994. Intergenerational Transfers and Capital Market Imperfections, Evidence from a Cross-section of Italian Households, in A. Ando, L. Guiso and I. Visco, eds., *Saving and the Accumulation of Wealth*, Cambridge University Press, Cambridge.

Guiso, L. and T. Jappelli. 2000. Private Transfers, Borrowing Constraints and the Timing of Homeownership, *CFES Working Paper No. 17*, Salerno.

Gustman, A. L. and T. L. Steinmeier. 1998. Effects of Pensions on Savings: Analysis with data from the Health and Retirement Study, *NBER Working Paper No. 6681.*

Hamermesh, D. S. 1982. Consumption During Retirement: the Missing Link in the Life Cycle, NBER Working Paper No. 930.

Hayashi, F. 1987. Tests for Liquidity Constraints: A Critical Survey and some new Observations. In: T. F. Bewley, ed., *Advances in Econometrics, Fifth World Congress*, 2, Cambridge University Press.

Hayashi, F., T. Ito and J. Slemrod. 1988. Housing Finance Imperfections, Taxation and Private Saving: a Comparative Simulation Analysis of the United States and Japan, *Journal of Japanese and International Economies*, 2, pp. 215-238.

Hendershott, P. 1991. Are Real House Prices Likely to Decline by 47 Percent? *Regional Science and Urban Economics* 21(4), pp. 553-63.

Hubbard, G., J. Skinner and S. Zeldes. 1995. Precautionary Savings and Social Insurance, *Journal of Political Economy*, 103, pp. 360-99.

Jappelli, T. and M. Pagano. 1994. Saving Growth and Liquidity Constraints, *Quarterly Journal of Economics*, 109, pp. 83-109.

Jappelli, T. 1995. Does Social Security Wealth Reduce the Accumulation of Private Wealth? Evidence from Italian Survey Data, *Ricerche Economiche*, 49, pp. 1-31.

Jappelli, T. and F. Modigliani. 1998. The Age-Saving Profile and the Life-cycle Hypothesis, University of Salerno, *CSEF Working Paper No. 4*.

Kohl, R. and P. O'Brien. 1998. *The Macroeconomics of Ageing, Pensions and Savings: A Survey*, OECD Working Paper, AWP1.1, Paris.

Kotlikoff, L. J. and A. Spivak. 1981. The Family as an Incomplete Annuity Market, *Journal of Political Economy*, 89.

Kotlikoff, L. J., A. Spivak and L. H. Summers. 1982. The Adequacy of Savings, *American Economic Review*, 72, pp. 1056-1079.

Laibson, D., A. Repetto and J. Tobacman. 1998. Self-control and Saving for Retirement, *Brooking Papers in Economic Activity*, 1-98, pp. 91-196.

Mankiw, N. G. and D. N. Weil. 1989. The Baby Boom, the Baby Bust, and the Housing Market, *Regional Science and Urban Economics* 19(2), pp. 235-58.

Mansori, K. S. 2001. *Economic Liberalization and Savings Rates*, CESifo Working Paper No. 418.

Modigliani, F. and R. Brumberg. 1954. Utility Analysis and the Consumption Function: An Interpretation of Cross-section Data. in: Kurihara, K., ed., *Post-Keynesian Economics*. New Brunswick, NJ: Rutgers University Press.

Muellbauer, J. and A. Murphy. 1990. Is the UK Balance of Payments Sustainable?, *Economic Policy*, 11, pp.345-83.

OECD. 2000a. *Reforms for an Ageing Society*, Paris.

OECD. 2000b. *Economic Outlook,* 67, Paris.

Pagano, M. 1993. The Flotation of Companies on the Stock Market: a Coordination Failure Model, *European Economic Review*, 37.

Poterba, J. 1994. *International Comparisons of Household Saving*. Chicago, Chicago University Press.

Poterba, J., S. Venti and D. Wise. 1996. How Retirement Saving Programs Increase Saving, *Journal of Economic Perspectives*, 10(4), pp. 91-112.

Rossi, N. and I. Visco. 1995. National Saving and Social Security in Italy, *Ricerche Economiche*, 49, pp. 329-356.

Saint-Paul, G. 1992. Technological Dualism, Incomplete Financial Markets, and Economic Development, *Journal of International Trade and Economic Development,* 1, pp. 13-26.

Skinner, J. und R. G. Hubbard. 1996. Assessing the Effectiveness of Saving Incentives, *Journal of Economic Perspectives* 10(4), pp. 73-90.

Venti S. and D. Wise. 1990. Have IRAs Increased U.S. Savings? Evidence from Consumer Expenditure Surveys, *Quarterly Journal of Economics*, 105, pp. 661-98.

Wise, D. A. 2001. United States Support in Retirement: Where We Are and Where We Are Going, in: Börsch-Supan, A. and M. Miegel, eds., 2001. *Pension Reform in Six Countries*, Springer: Berlin, Heidelberg, New York.

Wolff, E. N. 1998. Recent Trends in the Size Distribution of Household Wealth, *Journal of Economic Perspectives*, 12,3, pp. 131-150.

World Bank. 1994. *Averting the Old Age Crisis; Policies to Protect the Old and Promote Growth*, The World Bank, Washington.

World Bank. 1999. The Tax Treatment of Funded Pensions, *World Bank Pension Reform Primer* 2.3, Washington, D.C.

Discussion

Reinhold Schnabel

The paper by Agar Brugiavini addresses the relation between population ageing and saving in Europe. Population ageing will eventually lead to major reductions in benefit levels in European social security systems, which are usually set up as pay-as-you-go systems. The main focus of the paper is the impact of these pension cuts on household saving. The first, and larger, part of the paper (sections 1 to 4) consists of a survey of the literature on ageing, saving and the provision of income during retirement. In this part, Brugiavini discusses stylized facts on saving, the structure of the financial and insurance markets, theoretical issues of substitution between different forms of retirement income or wealth and empirical work on substitutability. The second part of the paper summarizes the methodology and main results of a paper by Attanasio and Brugiavini (2000) on the saving behaviour of Italian households, which makes use of institutional variation induced by a recent and major pension reform. This comment follows the set-up of the paper, and discusses in some detail the relation between population ageing and saving behaviour. In doing so, I try to add some aspects which may be of importance in assessing the potential impact of population ageing on saving. In particular, the ageing process and the induced pension reforms are likely to have strong effects on other economic decisions, like labour supply and intergenerational transfers, that mitigate saving reactions.

The first section ("Ageing and saving") presents some basic stylized facts on saving and population ageing in Europe. It is stressed that saving rates have declined in almost all OECD countries, with a few exceptions, notably France and Belgium. While there are several explanations, empirical analysis has not yet been able to pin down the reasons for the aggregate decline in saving rates – nor has empirical analysis been able to explain the cross-sectional variation of saving. This is especially true of European countries. One of the main reasons for this situation is the lack of reliable data on saving at the household level. In many countries, data on consumption, saving and wealth are not available at all. Often, data that have been sampled (usually by governmental authorities) are not available to the scientific community. In Germany, for instance, the income and expenditure survey (EVS) of the Federal Statistical Office, which is conducted every five years, has not been made accessible until very recently. Even if data on saving are available (as in the German EVS), questionnaires and sample designs are usually inadequate for economic research. To give some examples: questionnaires lack important information on the education and labour-market participation of the respondents. Panel designs, which would enable the researcher to trace the behaviour of individuals or households over time, are generally not available with adequate data on consumption and saving.

The basic problem considered by Brugiavini is the dramatic process of double ageing of European economies. It has to be stressed that the demographic changes are driven by rising life-expectancies and, at the same time, fertility rates that are far below the reproduction level. The latter fact distinguishes the European ageing process from the US-American. The pure demographic old-age dependency ratio – i.e., the ratio of persons aged 65 and above to potential workers aged 20 to 64 – will increase dramatically over the next 30 to 40 years. In countries like Germany and Italy this ratio will double. In the absence of behavioural reactions (that is, if labour supply and retirement stayed at their current levels) the ratio of pensioners to workers would increase in proportion to demographics. It is worth noting that this ageing process creates considerable problems not only for pay-as-you-go pension systems but also for funded pension systems, since the capital-labour ratio of the economy is bound to rise and interest rates tend to be lowered. This in turn has a depressing impact on funded retirement incomes.

There are many possible ways of coping with the problem of population ageing, both politically and individually. Only one response among many others is an increase in the saving rate (or a policy directed towards increasing it). In view of the inevitable ageing of European economies, some countries have already started to reform their pension systems. There have been major reforms in Italy (1992) and Germany (1992 and 2001). Among the features of those recent reforms are: increases in official early retirement and regular retirement ages, introduction of actuarial adjustments for early retirement, changes in indexation of benefits (e.g., moving from gross wage to net wage indexation or to inflation indexation), and a lowering of replacement rates. Those reform steps aim at mitigating the burden of population ageing by lowering the long-run expenditure of the pay-as-you-go pension systems. This can be done in two ways: by lowering the number of pensioners relative to workers and/or by lowering the replacement rate (that is, expenditure per pensioner).

The first is accomplished through an increase in the labour supply of the elderly and postponement of retirement. Currently, average retirement ages are below 60 years for many European countries. Therefore, raising the retirement age has a considerable impact on the budget of the pay-as-you-go pension systems. One instrument is the introduction of an actuarial adjustment that reduces the pension benefits by a certain factor per year of early retirement. This has a direct mechanical effect on the budget. A second instrument is a modification of eligibility rules, which may simply preclude early retirement or make it contingent on fulfilment of some condition (e.g., number of service years, disability tests). Note that if an individual postpones retirement to the normal retirement age (i.e. changes the labour supply) he or she can completely avoid a lower benefit level. The extent of that behavioural reaction has important implications for the level of disposable retirement income, and thus for the potential saving gap. If the introduction of actuarial adjustment is fully accommodated by a shift in retirement age, this policy does not induce any need for higher saving. Thus, labour supply and saving decisions are inevitably and directly linked.

However, policies like the introduction of actuarial adjustments cannot completely offset the effect of population ageing, since it is to a large extent driven by the long-run decline in fertility rates. Moreover, the rise in retirement age can offset only a part of the increase in life-expectancy, insofar as morbidity does not fall as fast as mortality. It has been shown that, in order to keep the contribution rates and taxes at the current levels, it will be necessary to lower the replacement rate in addition to raising the retirement age. This also has an immediate mechanical effect on the budget of the public pension system even if retirement behaviour does not respond. In the case of Germany, the reform in the year 2001 will eventually reduce the replacement rate from 71 to 64 percentage points, which amounts to a benefit cut of 10 per cent. This reduction in benefit levels is – assuming a balanced budget – accompanied by an equally large reduction in contribution rates. Clearly, the aggregate *net* wealth effect of reducing the replacement rate must be zero in a pay-as-you-go pension system. However, there are two important aspects of such a benefit cut: a redistribution effect and a disposable income effect. While the present discounted values of pension benefits are lower for all generations, the young generations will have to pay lower payroll contributions (or taxes) such that the net present discounted value of their pension wealth will increase. The intermediate generations will suffer a loss in terms of net present discounted values (Schnabel, 1998).

The cut in the benefit level lowers future retirement income and increases current labour income. One possible behavioural reaction is to increase saving out of current income in order to compensate for the lower future benefit level. Again, the benefit level may also be augmented through postponed retirement, which yields higher pension benefits in most European pension schemes. In Germany, for example, the 10 per cent pension cut induced by the 2001 reform could be offset by working 18 months longer. A third way to compensate for the lower pension level is to reduce private transfers. This is a feasible option for households with sufficiently high income and wealth. In Germany, retirement income on average used to be higher than consumption, yielding positive saving in old age and considerable bequests (intended or unintended). In addition to that, private inter vivos transfers are very high. Schnabel (1999a) shows with data of the EVS that the elderly have transferred on average 7 per cent of their disposable income per year. In total, they use only about 86 per cent of their disposable income for their own consumption. Moreover, the average private wealth of German households that reach retirement age exceeds € 150,000; the median wealth reaches almost € 150,000 at the beginning of retirement (Schnabel, 1999b). The level of private saving is on average of the same order of magnitude as the social security contributions. Given the high saving rates at working ages, combined with the positive saving rates and the high levels of private transfers in old age, it seems to be feasible for average German households to adjust to considerable pension cuts without additional saving and without reductions of consumption during retirement.

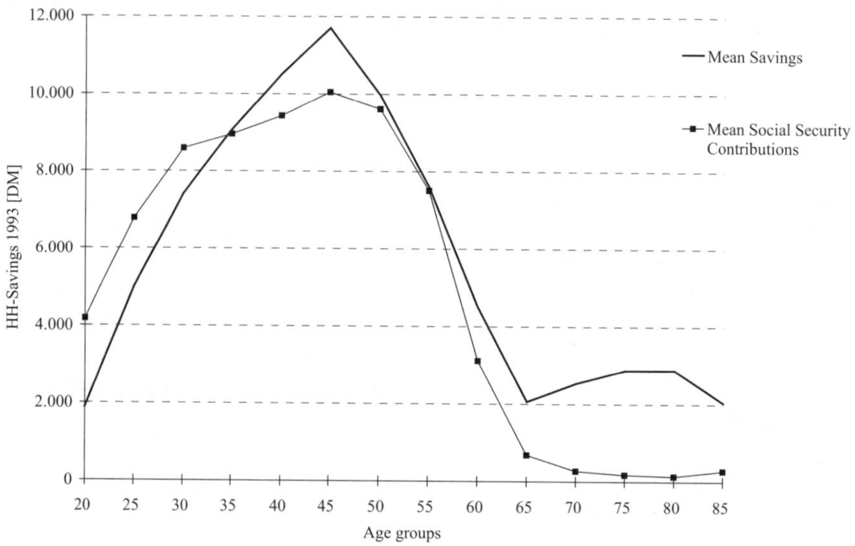

Figure: Social Security Contributions and Private Saving. Own calculations based on the EVS 1993.

Given the wide range of potential behavioural responses, the reaction of saving behaviour to population ageing and pension reforms is theoretically undetermined. It is therefore necessary to conduct empirical analyses of the substitution process between social security wealth and private wealth. To this end, Brugiavini uses the Italian pension reform of 1992 in order to study the effects of a pension reform on private saving decisions. The reform reduces the social security wealth of different cohorts and different groups of the working population in different ways. This heterogeneity can be used to identify behavioural responses. The main assumption of the empirical analysis is that the reform can be interpreted as a quasi-experiment. The differential treatment of cohorts and of groups within cohorts should ideally be random in order to estimate the parameters of the econometric model consistently. If the reform has been anticipated, the reactions to the pension cuts may have taken place before the enactment. In this case, the difference-in-differences estimator underestimates the true behavioural response. Thus, the parameter estimate yields a lower bound. Brugiavini and Attanasio use social security wealth as the main explanatory variable in a difference-in-differences approach. What Brugiavini calls social security wealth is rather the (expected) present discounted value of pension benefits, and does not include the value of contributions. The social security wealth variable indicates the level of disposable income during retirement. The lower the value of this wealth variable, the larger is the necessity of performing additional private saving in order to smooth consumption. The estimation yields mixed results. The difference–in-differences estimate of the wealth variable (PW) is insignificant in the smaller sample of employees. The estimate becomes significant in the larger sample of the whole active population. The adjusted coefficients of the PW-variable suggest a relatively high degree of substitutability between pension wealth and private saving, but the variance of those coefficients is not reported, and is probably high. This empirical work is a very interesting first step; however, further evidence (e.g., using pension reforms in other countries) is needed in order to confirm the results, and to draw strong policy conclusions.

References

Attanasio, O. and A. Brugiavini. 2000. Household Saving in Italy, mimeo, University of Venice, Italy.

Schnabel, R. 1998. Rates of Return of German Pay-as-you-go Pension System, Finanzarchiv.

Schnabel, R. 1999a. The Golden Years of Social Security – Life-cycle Income, Pensions and Saving in Germany, SFB-504 Discussion Paper 99-40, University of Mannheim.

Schnabel, R. 1999b. Vermögen und Ersparnis im Lebenszyklus in Westdeutschland, SFB-504 Discussion Paper 99-43, University of Mannheim.

Ageing and International Capital Flows

Axel Börsch-Supan, Alexander Ludwig and Joachim Winter*

1 Introduction

In the vast majority of countries, populations are ageing, and demographic change will continue well into the 21st century. While the patterns of population ageing are similar in most countries, timing differs substantially, in particular between industrialized and less developed countries. It is well known that, within each country, demographic change alters the time path of aggregate savings, even more so in countries where fundamental pension reforms – that is, a shift towards more pre-funding – are implemented (Börsch-Supan, 1995a and b; Reisen, 2000). To the extent that capital is internationally mobile, population ageing will also induce capital flows between countries.

In this paper, we present a quantitative analysis of the capital flows induced by differential ageing processes across countries and by pension reforms. We develop a stylized multi-country overlapping-generations model which is, to our knowledge, new to the literature, and we use long-term demographic projections for different sets of countries to project international capital flows over a 50-year horizon. For tractability, we focus on Germany as a country with one of the most severe ageing problems in the world, and with a public pay-as-you-go (PAYG) pension system in an ongoing reform process (e.g. Birg and Börsch-Supan, 1999; Börsch-Supan, 2000). We expect results for countries such as France and Italy to be very similar in patterns and quantity - Italy is a bit smaller than Germany but ages more, while France most likely will see fewer capital flows since the French ageing process is less dramatic. To separate the direct effect of population ageing on capital markets and potential feedback effects from pension reform, we present our projections for both of two counterfactual scenarios: (a) maintaining Germany's current generous pension system, and (b) introducing a one-third transition to a funded pension system as described by Börsch-Supan (2000). The most likely development will be in between these two scenarios.

---

* We would like to thank Alan Auerbach, Ulrich Grosch, Florian Heiss, Heinz Herrmann, Gary Hufbauer, Ulf von Kalkreuth, Melanie Lührmann and Howard Rosen for their helpful comments. This research project was supported by the Deutsche Forschungsgemeinschaft, the VW-Foundation, and the Gesamtverband der Deutschen Versicherungswirtschaft.

Our simulations predict substantial capital flows due to population ageing. Our results confirm that population ageing results, at least initially, in a higher capital stock, but when the baby-boom generations begin to consume their retirement savings, the capital stock will decrease. International capital flows follow this trend. At the peak of savings around the year 2020, German capital exports to all other OECD countries will exceed 7% of GNP. Most of the capital flows will actually stay within the EU. 15 years later, capital flows will briefly decline to 2%, but remain positive through the ageing process.

Moreover, our simulations show that a transition to a partially-funded system does not crowd out existing savings totally. Capital exports from Germany to the OECD countries - again, mainly to EU-countries - will be higher in this case, never falling below 2% after a peak of about 9% in the year 2020. The decrease in the rate of return, which results from both population ageing and pre-funded pensions, is modest: approximately one percentage point if we assume a closed economy. The return on capital can be improved substantially by international diversification, that is, by investing pension savings in countries with a more favourable demographic transition path than Germany.

The effects of international diversification on savings behaviour and its interaction with pension reforms receive rapidly-increasing attention as the pension-reform debate progresses. Deardorff (1985) contains an early analysis, and Reisen (2000) provides a comprehensive overview of these issues. Reisen argues strongly that there are pension-improving benefits of global asset diversification. In a theoretical paper, Pemberton (1999) highlights the importance of international externalities caused by the effects of national pension and savings policies on the world-interest rate. More recently, Pemberton (2000) goes a step further and shows that - while the switch from a pay-as-you-go system to a fully funded pension system implies that (at least) one generation necessarily loses - in a world where pension reform takes place in many small, open economies, an intergenerational Pareto improvement is possible (for some production technologies). Pemberton supports this finding by numerical simulations of a stylized model for the OECD countries. However, Pemberton's extremely stylized overlapping-generations model cannot account for realistic paths of demographic change within different regions. Our model represents a significant improvement in the direction of a more realistic quantitative projection.

The remainder of this paper is structured as follows. Section 2 presents some facts - empirical evidence and theoretical explanations - on the effects of population ageing on international capital flows. In section 3, we present a stylized overlapping-generations model that can be used to evaluate these effects quantitatively. Section 4 contains our simulation results for two different pension policies and three different capital-mobility scenarios. Section 5 concludes.

2 Some facts about population ageing and international capital flows

At mid-1998, world population stood at 5.9 billion. While the world population has constantly grown, its annual growth rate has decreased from 2.04 percent during the period from 1965 to 1970 to 1.33 percent between 1995 and 2000. It is expected that this decrease in world population growth will continue. In the medium variant of the United Nations' current world-population projections, the growth rate is projected to decrease to 0.3 percent by 2050. By then, world population will have increased to 8.9 billion. 97 percent of this increase takes place in less developed regions (United Nations, 2000).

These demographic changes - the so-called *demographic transition* (e.g. Birg, 1996) - are characterized by falling mortality rates, followed by a decline in birth rates, resulting in population ageing. While the patterns of population ageing are similar in most countries, timing differs substantially, in particular between industrialized and less developed countries. Europe has almost passed the closing stages of the demographic transition process. It is now, and is projected to remain, the geographic region that is most affected by ageing. By 2005, population growth is projected to be negative in Europe. The median age in Europe is projected to increase from 37.1 years in 1998 to about 47 years by 2050. The proportion of children is projected to decline from 18 percent to 14 percent while the fraction of older persons will increase from 20 percent to 35 percent by 2050. Other regions of the world that are substantially affected by ageing are Northern America, Oceania, Asia, Latin America and the Caribbean (United Nations, 2000).

Asia and Latin America are only at the beginning stages of the demographic transition. So far, characteristics of a demographic-transition process cannot be identified in Africa – fertility is at the highest level worldwide, and even though child mortality is declining, life expectancy is still very low (Bloom and Williamson, 1998). The impact of AIDS is devastating: in a group of 29 African countries where the impact of AIDS has been studied by the United Nations, life expectancy is projected to decrease by seven years in the near future (United Nations, 2000).

The impact of population ageing can be expressed in the old-age dependency ratio, defined as the ratio of the number of pensioners to the number of workers. In Germany, this ratio will increase from about 60 percent in 2000 to 90 percent in 2050, according to Birg and Börsch-Supan (1999).[1] Analogous calculations for the rest of the European Union show an increase in the old-age dependency ratio from currently 45 percent to 60 percent in 2050 (Ludwig, 2001). The consequences of

1 These projections are for a medium scenario of demographic change, characterized by modest ageing, constant fertility and a modest increase in labour-force participation rates.

these increases are well-known and mirrored by the current debate on privatizing social security (e.g. Börsch-Supan and Brugiavini, 2001).

From a macroeconomic point of view, population ageing will change the balance between capital and labour, in particular in industrialized countries. Labour supply will be scarce whereas capital will be relatively abundant. This will drive up wages relative to the rate of return on capital, reducing households' incentive to save (if the interest elasticity of saving is positive). In addition, some fraction of the capital stock may become obsolete due to the shrinking labour force and diminishing returns to scale, making the accumulation of capital even less attractive. Developing countries are less affected since their population-age structure is younger; these countries are better characterized by a relatively low supply of capital and a relatively high supply of labour. As a result, the rate of return on capital is higher in developing countries. Capital exports to developing countries could therefore solve the ageing problems of industrialized countries by reducing the pressure on the interest rate and by shifting the production of consumption goods towards developing countries.

More generally, differences in the timing of demographic change across countries and regions induce international capital flows, and there is some empirical evidence that this mechanism is already at work (e.g. Higgins, 1998; Lührmann, 2001). Private net capital flows have increased remarkably during the past ten years. In 1996, the volume of these flows was six times higher than at the beginning of the nineties. Private capital flows make up around 80 percent of total world capital flows and clearly dominate public capital flows. 40 percent of private capital flows is foreign direct investment, another 40 percent is portfolio investment and around 20 percent is banking credits (which are becoming less and less important). Due to the increasing role of institutional investors such as pension funds, the share of portfolio investment is likely to increase in future (World Bank, 1997).

In a recent empirical study, Lührmann (2001) uses a broad panel of 141 countries that covers the period 1960-95 to investigate the effects of demographics on national saving and capital formation, and on international capital flows. She confirms that cross-country capital flows are indeed influenced by demographic variables. While this has been shown in other studies before, she can also show that, across countries, relative differences in the age structure are the most important determinants of capital flows, a finding that is even more important for the analysis of pension reform than the fact that the absolute age structure affects a country's capital balance. Moreover, as Lührmann (2001) shows, future changes in the age structure of countries are important determinants of current savings and investment decisions, a finding that confirms forward-looking household behaviour.

There are a number of theoretical arguments that establish a link between demographic change and international capital flows. The simulation model we present in this paper builds on the well-known life-cycle theory of consumption and savings by Modigliani, Ando and Brumberg (Modigliani and Brumberg, 1954; Ando and Modigliani, 1963). The aggregation of individual, cohort-specific life-cycle savings profiles leads to a decrease of national-saving rates in an ageing economy. Moreover, in a general equilibrium model of forward-looking individuals, it is not only the current demographic structure that alters the time path of aggregate savings, but also future demographic developments. Empirical evidence on how demographic change has affected savings behaviour across countries in the past is reviewed by Poterba (1998) and Brooks (2000).

In a general-equilibrium framework, there are two main channels for effects of demographic change on domestic capital formation. First, decreasing labour supply reduces demand for investment goods since less capital is needed. The magnitude of this effect depends on the elasticity of substitution between the production factors capital and labour. Börsch-Supan (1995a) estimates a CES production function and concludes that the elasticity of substitution between these two factors is close to one. This result indicates that production can be adjusted quite flexibly, which reduces the impact of demographic change on investment. Second, in a closed economy, a decline in national saving leads to a decline in investment, by definition. In an open economy, the link between these two aggregates is broken to the extent that capital is internationally mobile.

For quantitative projections of international capital flows induced by population ageing, the degree of capital mobility is crucial. This is essentially an empirical question, and there has been no shortage of research on this isseue since the famous puzzle of Feldstein and Horioka (1980).[2] In their original contribution, Feldstein and Horioka have shown that national saving rates are highly correlated among OECD countries. While the coefficient has fallen over time, it is still remarkably high. These findings have been interpreted as an indication that capital is imperfectly mobile. However, this interpretation has later been criticized, both because there are a number of alternative explanations for the observed correlation (a recent example is Obstfeld and Rogoff (2000) who focus on transport costs for goods) and because of econometric problems associated with simply regressing national saving rates on domestic investment rates (see, for example, Taylor, 1994). Even if capital is fully mobile, this does not necessarily imply that households do actually diversify their portfolios optimally. There is a large empirical literature on 'home-bias' in international portfolio choice (e.g. French and Poterba, 1991), and it is not yet fully understood why households do not optimally diversify their portfolios across countries. A recent empirical study by Portes and Rey (1999) suggests that information asymmetries across countries are a major source of home bias effects, and that capital flows are affected by both geographic and informational proximity. Applied to pension-reform policies, this

[2] See Obstfeld (1995), Obstfeld and Rogoff (1996), and Coakley, Kulasi, and Smith (1998) for recent surveys.

literature suggests that households might be more willing to invest their retirement savings in 'similar' countries, such as the EU or OECD countries, than in, say, developing countries. Unfortunately, the latter are the countries where not only the highest returns are to be found over much of the next century, but which would also benefit most themselves from capital provided by the ageing industrialized nations. Blommestein (1998) and Holzmann (2000) discuss theses issues, both concluding that investments in emerging markets can help to solve the OECD countries' pension crisis at the margin, but are unable to solve the demographic problem alone, and stressing that additional reforms are needed. Our simulations will shed more light on the role of capital flows to developing countries.

In most of our simulations, we assume that capital is freely mobile only within industrialized countries. This contrasts with Fougere and Merette (1999) and Miles (1999) who state that modelling European countries as closed economies in general equilibrium models is closer to reality than modelling them as open economies. Certainly, the truth is somewhere in the middle, but we believe that allowing for free capital mobility in a multi-country model is a better approximation to reality, and warranted by the empirical evidence – at least when we restrict our model to perfect capital mobility in the OECD area.

3 Ageing and pension reform in a stylized overlapping-generations model

In this section, we present a dynamic macroeconomic model that allows us to analyse the effects of population ageing and of a shift from a pay-as-you-go system to a (partially) funded-pension system. The model is based on a version of the overlapping-generations model (Samuelson, 1958; Diamond, 1965) introduced by Auerbach and Kotlikoff (1987: chapter 3). Overlapping-generations models have been used extensively to study the effects of population ageing on social security systems, a purpose for which they are well suited, since they are based on households' and firms' optimal reactions to movements in the demographic structure and public policy measures. Recent examples include Kotlikoff, Smetters and Walliser (1999) and De Nardi, Imrohoroglu and Sargent (1999) for the United States, Miles (1999) for Great Britain, and Fehr (2000) and Börsch-Supan, Heiss and Winter (2000) for Germany. Miles and Iben (2000) present a comparative analysis of pension-reform schemes for the United Kingdom and Germany. Kotlikoff (1998) provides an overview of earlier applications of overlapping generations models.

To our knowledge, the multi-country version of the Auerbach-Kotlikoff model presented in this paper is new to the literature.[3] Our model builds on a closed-economy model for Germany developed by Börsch-Supan, Heiss and Winter (2000). In particular, we extend that model along three dimensions: (i) we consider several countries with differential ageing processes and assume perfect capital mobility between different regions; (ii) we implement technological progress; and (iii) we explicitly model variations of the planning horizon of different generations that are due to increasing life expectancy.[4]

Since the purpose of this paper is to study the macroeconomic effects of population ageing and of a fundamental pension reform, we restrict the analysis to a very stylized version of the standard overlapping-generations model that excludes many interesting aspects. However, we take great care to get the first-order effects of demographic change right, by using 75 cohorts and annual demographic projections. In our simulations, we use two data sources for the demographic projections: Birg and Börsch-Supan (1999) provide several demographic projections for Germany; we use the medium scenario (characterized by modest ageing, constant fertility and a modest increase in labour-force participation rates). For the other world regions, we use the medium variant of the United Nations' World Development Prospects (United Nations, 2000). Based on these demographic projections, we compute time-paths for the number of workers and pensioners for each of the countries and world regions in our model. These projections are described in detail by Ludwig (2001).

The most significant simplifications of our model relative to existing overlapping-generations models are as follows. (i) We do not explicitly consider taxes (other than the contributions to the pay-as-you-go pension system). (ii) We do not include labour supply in households' decision problem, but rather assume that all households supply one unit of labour until retirement.[5] (iii) We do not model intra-generational household heterogeneity and therefore cannot capture distributional effects. (iv) We assume perfect foresight. (v) The only factors of production are labour and real capital (i.e., we do not model human capital and therefore cannot account for endogenous growth). While these issues surely are important, especially if one wishes to analyse the effect of population ageing on labour supply in the presence of distorting taxes, we restrict our attention to

---

3 A formal description of the model presented in this paper can be found in Ludwig (2001).

4 Börsch-Supan, Heiss and Winter (2000) made the simplifying assumption of a constant planning horizon, and implemented changes in life expectancy by weighting cohort sizes accordingly.

5 However, we account for unemployment and labour-force-participation decisions since the aggregate workforce is adjusted according to the labour-market scenarios behind the demographic projections.

households' life-cycle savings decisions, as their primary means to prepare for demographic change and the decreasing generosity of public pensions.

The rationality of our agents – exemplified in their perfect foresight and their life-long planning horizon - is typical of economic models but also one of their main limitations. We are certainly aware that real-life saving and investment decisions, at the micro level, reveal many behavioural "anomalies". It is less clear, however, in how far these anomalies affect the aggregate behaviour of an economy. We have some confidence that our stylized model is sufficient to obtain the first-order effects of population ageing on domestic capital formation and international capital flows. This view is substantiated, for instance, by the results of Börsch-Supan (1995a), who uses a similar "ultra-rational" model to fit economic growth, saving and investment retrospectively for the 1960-1989 period and achieves a very satisfactory fit.

To keep the analysis tractable, our model focuses on Germany. We consider both the closed-economy case and alternative open-economy scenarios; the latter are different with respect to the regions within which capital can flow freely (within the EU, within the OECD, or across the whole world).

Our simulation model carries 75 overlapping cohorts. The economic life of a cohort begins at the age of twenty years and ends at a retirement age at which the model persons stop working, stop paying pension taxes, and begin to collect pension benefits from a pay-as-you-go system. We employ a flexible retirement pattern through age- and time-specific weights that represent the fraction of the population that is retired. This fraction increases from 0 to 1 over an extended retirement window from age 47 through 80. The time paths of these weights are cohort-specific, reflecting shifts in labour supply and retirement behaviour. Similarly, the life expectancy of each cohort is matched with the demographic projections.

The pension system of each country enters the model through a fixed time-path of the contribution rate, calculated from an exogenously-given time-specific replacement rate, defined as the ratio of the average net pension to the average net wage, by balancing the budget of the public pension system.

As a pension-reform scheme for Germany, we use the 'freezing' reform proposal by Birg and Börsch-Supan (1999) and Börsch-Supan (2000), but we should point out that the same mechanisms are at work in any scheme that involves the introduction of a funded component. The 'freezing' reform scheme assumes that the contribution rate to the pay-as-you-go pension system remains fixed - for Germany, at its current level of 21 percent. More specifically, we assume that the pension reform is publicly announced in the year 2001, and implemented by fixing the contribution rate in 2006. Thus, households that started their economic life before 2001 have a period of five years to adopt their life-time plans, while households that enter economic life after 2001 already face the new conditions. Since the pay-as-you-go pension system remains in place, freezing contribution rates results in lower public pension payments, given a rising old-age dependency ratio. This, in turn, results in lower replacement rates provided by the pay-as-you-go pillar of the pension system.

We do not explicitly model the funded component of the pension system. In our model, the funded component consists entirely of voluntary, private savings, as given by households' optimal life-cycle decisions. The rationality of our households implies that these voluntary savings increase exactly in proportion to the decrease in the pay-as-you-go pension replacement rate.

General equilibrium in this overlapping generations model is constructed via the production sector where, given factor inputs (capital and labour), output and factor prices are determined by a Cobb-Douglas production function with an exogenous growth rate of labour productivity that is constant for all countries and across time.[6] From static profit maximization, we obtain the interest rate which is identical for all countries due to our assumption of perfect capital mobility. In turn, the wage rate can be different since labour productivity might differ across countries.

In order to determine aggregate consumption, we derive optimal household behaviour from intertemporal utility maximization. Departing from a utility function with constant relative risk aversion, we compute the optimal path of saving and consumption. As in any life-cycle model, the trade-off between consuming today and consuming tomorrow is determined by the ratio of the interest rate and the time-preference rate, and by the degree of risk aversion.

Since factor prices (i.e., wage and interest rates) and both contribution rates to, and replacement rates of, the pay-as-you-go pension system are known, we can now determine the life-time consumption paths of all generations backwards, starting with zero wealth in the final period of life, and then iterating using the Euler equation and the budget constraint. The resulting time-paths of consumption determine aggregate saving and wealth in the household sector for each country. This yields total world-wealth holdings, as the sum of the wealth of all regions, which in turn is equal to the world-capital stock.

We carefully distinguish between the real capital located in each country, which might be owned by natives as well as foreigners, and the wealth owned by each country's inhabitants, which might be invested both domestically and abroad. The difference between the total wealth of a country's inhabitants and the capital stock located within this country's borders is foreign assets. Note that we assume that labour is not mobile, and therefore the only income from abroad is asset income.

Equilibrium of the model requires that the sum of all foreign assets across all countries is zero. It is computed numerically using a Gauss-Seidel algorithm (see

---

6 This implies that we abstract from any direct impact of demographic change on productivity; see Cutler et al. (1990) for a discussion.

Auerbach and Kotlikoff, 1987). Because we lack appropriate historical data on some exogenous variables, we start calculations in 1997, which we will further refer to as the base year of our calculations. This implies that we implicitly assume that the economy is in a steady state in 1997. The final steady state is reached after 150 years, when all the transitions in our model have settled. Below, we will report results only for the time period between 2000 and 2050 (even though the underlying simulations have a much longer horizon).

The parameter values used in the calibration of our model are as follows. The output share of capital in the production function (0.4099) is based on an estimation of the aggregate production function by Börsch-Supan (1995a). The annual rate of depreciation is 5 percent according to the German national accounts. We further assume an annual growth rate of labour productivity of 1.4 percent.[7] Base-year values of asset holdings are inferred from actual GDP data and inverting the production function. This implies that the initial weight of each country in our simulations is determined by its relative share in current world GDP. Households' preference parameters (the rate of time preference equals 8 percent, the coefficient of relative risk aversion equals 3) are chosen such that the calculations of the model match the empirical counterparts of the base year's aggregate households saving rate and net aggregate capital stock.

It is not easy to fit such a simplified model to historical data. In our model, the common conceptual difficulties of calibrating macroeconomic models are amplified by the fact that the 'world' we model changes with the number of regions that we consider in our capital-mobility scenarios. On the one hand, it would therefore make sense to adjust the calibration parameters each time we change the number of regions that we consider. On the other hand, this would change households' reactions to changes in policy, and it would therefore be more difficult to interpret our results with respect to a reform of the public pension system. For that reason, we keep the parameters' values constant across the capital-mobility scenarios. As a base scenario for calibration, we assume perfect capital-mobility within the European Union. It turned out that the saving rate deviates only marginally from the observed saving rate when we extend the capital-mobility region to the OECD, using the same parameters, but the deviation is larger when we apply these parameters to the closed economy case. We interpret this finding as a confirmation of our earlier statement that perfect capital mobility within the EU or the OECD is a better approximation to reality than modelling Germany as a closed economy.

While our model is capable of matching the empirical counterpart of Germany's net capital stock in our base year, the calculated rate of return on capital is well

---

7 See e.g. Maddison (1987) and Cutler et al. (1990).

above our target value of 7.5 per cent.[8] For a better interpretation of our results, we normalize the following values of the rate of return on capital to 7.5 per cent in 2000. We carried out an extensive sensitivity analysis with respect to the values of households' preference parameters and the annual growth rate of labour-force productivity which confirmed that the findings reported below do not change qualitatively under alternative parameter values; detailed results are available upon request.

4 Simulation results for alternative pension and capital-mobility scenarios

We now present the results of our macroeconomic simulation model. For tractability, we focus on Germany as a country with one of the most severe ageing problems in the world and with a pension system in an ongoing reform process (e.g. Birg and Börsch-Supan, 1999; Börsch-Supan, 2000). We expect results for countries such as France and Italy to be very similar in pattern and quantity - Italy is a bit smaller than Germany but ages more while France most likely will see fewer capital flows since the French ageing process is less dramatic.

To separate the direct effect of population ageing on capital markets and potential feedback effects from pension reform, we present our projections for both of two pension-policy scenarios: (a) maintaining Germany's current generous pension system, and (b) introducing a one-third transition to a funded pension system as described by Börsch-Supan (2000). These are two extreme cases, and they are both counterfactual: The current system is politically unsustainable and cannot survive, while the German pension reform that was passed in February 2001 is by no means as fundamental as the one we consider in our simulations. So the most likely scenario for Germany's future pension system is somewhere in between our extreme scenarios. However, by comparing two polar scenarios, we can show that a good portion of the capital-market effects of population ageing arise even without a fundamental pension reform. Hence, all figures have two panels: In panel (a), the forces of population ageing are shown under the assumption that the current pay-as-you-go pension systems stay as they are. In panel (b), we always show the *differential* effect if Germany implements a fundamental pension reform of the "freezing" type as outlined in the earlier sections.

---

8 For the twenty years between 1974 and 1993, the annualized aggregate rate of return on capital was 9.1 percent in the United States and 7.4 (7.1) percent in Germany (Japan) (McKinsey Global Institute, 1996; Börsch-Supan, 1999). We therefore define a rate of return of 7.5 percent as the target value for calibration. This is also in the range of reasonable values suggested by Pemberton (1999).

In addition to our pension policy scenarios, we consider three alternative capital-mobility scenarios: investment only within Germany (the closed-economy case), investment in the EU countries, and investment in the OECD countries. There are two reasons for choosing these rather modest capital-mobility scenarios: first, as already noted in section 2, there is a broad consensus that capital is quite mobile among OECD countries, while this is much less clear for developing countries. Second, as we will show below, beneficial effects of capital-mobility already show up when capital is freely mobile among countries of the European Union, and including more countries does not change our results substantially. Finally, while we initially assume that a fundamental pension reform is implemented only in Germany, we end this section with a brief analysis of simultaneous pension reforms in other countries of the world.

4.1 Savings and the capital stock

We begin with by looking at aggregate savings rates. Figure 1 shows that projected aggregate savings rates under a fundamental pension reform would be substantially higher than under the present system. For example, in the year 2035, when the peak of the ageing problem occurs, savings rates are projected to be very low under the current pay-as-you-go system. Depending on the capital mobility scenario, the aggregate savings rate declines from currently around 12.1 percent (1998) to between 8.3 and 8.8 percent. This is the pure effect of population ageing under the current system. In contrast, under a fundamental pension reform, the aggregate saving rate settles at around 10.4 percent on the assumption of perfect capital mobility within the EU. These projections show that optimal life-cycle behaviour generates additional saving under a fundamental pension reform – in our model, it is not the case that additional retirement saving induced by a pension reform crowds out other saving totally, as is often claimed. Our projections indicate a substitution of about one third, leaving two-thirds to new saving. Note that all variations of the aggregate saving rate shown in Figure 1 are in the range of historical variations in German saving rates.

Figure 1. Projections of the German aggregate saving rate under alternative pension systems and capital-mobility scenarios.

Figure 1a. Current pension systems

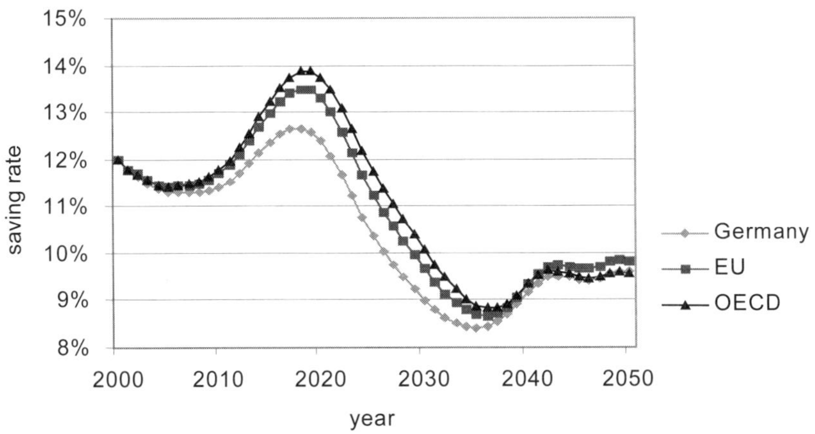

Figure 1b. Differential impact of fundamental pension reform

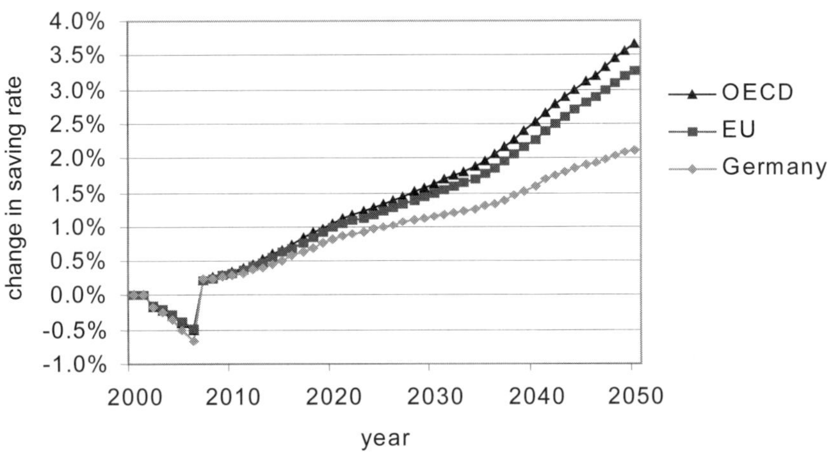

Notes: This figure shows projections of the aggregate saving rate of German households as a percentage of GNP. Pension reform only in Germany. Germany: Germany as a closed economy, EU: perfect capital mobility in the EU area, OECD: perfect capital mobility in the OECD area.
Source: Own calculations, based on demographic projections by Birg and Börsch-Supan (1999) and the United Nations (2000).

Figure 1b shows the difference in the savings rate, in percentage points, that is due to a fundamental pension reform. One important result is that considerable adjustments occur both when a pension reform is announced and when it is implemented. These adjustments reflect rational behaviour of households who react instantaneously to changes in their economic environment (and a fundamental pension reform is of course a major change). The first adjustment occurs in 2002, in the year after the announcement of the pension reform, the second in 2007, in the year after the reform itself. This latter adjustment is an upward jump, as expected, since the pension reform induces additional private saving. The reason for the first adjustment, a decrease in savings after the announcement of the fundamental pension reform, is that we aggregate across households which react very differently to this announcement. Such a reform induces additional retirement savings for all households. But contrary to older households, young households have the prospect of higher net wages after the reform is implemented. This future income effect dominates, and younger households therefore decrease savings during the period from 2002 to 2006. In aggregation, the weight of these young households is higher than the weight of older households, which have less time to exploit higher net wages, or are already dissaving.

Next, we aggregate savings to obtain Germany's foreign position and capital stock. Figure 2a shows projections of the total capital stock under the current pension system. A first observation is that movements in the aggregate capital stock are by far less pronounced in the open economy. These movements are caused by the alternating dominance of demographic effects and of growth in labour productivity. The economy gradually accumulates capital until the peak of the ageing process is reached in 2030. After 2030, when the ageing process has almost reached its peak, the capital stock decreases. In the open economy scenarios, this decrease in the domestic German capital stock is far less pronounced. The German capital stock increases to about 114 or 109 percent of its current value if capital is freely mobile within the EU or within the OECD, respectively, compared with 128 percent if Germany is assumed to be a closed economy. Under a fundamental pension reform, the decrease in the capital stock in the closed economy scenario, caused by ageing, is less pronounced since more capital is accumulated as a result of the pension reform (figure 2b). The increase in the aggregate capital stock is now higher than in both open-economy scenarios. This result confirms that under a pension reform, relatively more capital is invested abroad.

Finally, under a fundamental pension reform, there is a small downward blip in the aggregate German capital stock between 2000 and 2010 (figure 2b). This blip results from a complicated interaction between the adjustments made by different cohorts during the announcement and implementation phase of a fundamental pension reform: as discussed above, dissaving of younger cohorts dominates for a

Figure 2. Projections of the aggregate German capital stock under alternative pension systems and capital-mobility scenarios (Index, 2000=100)

Figure 2a. Current pension systems

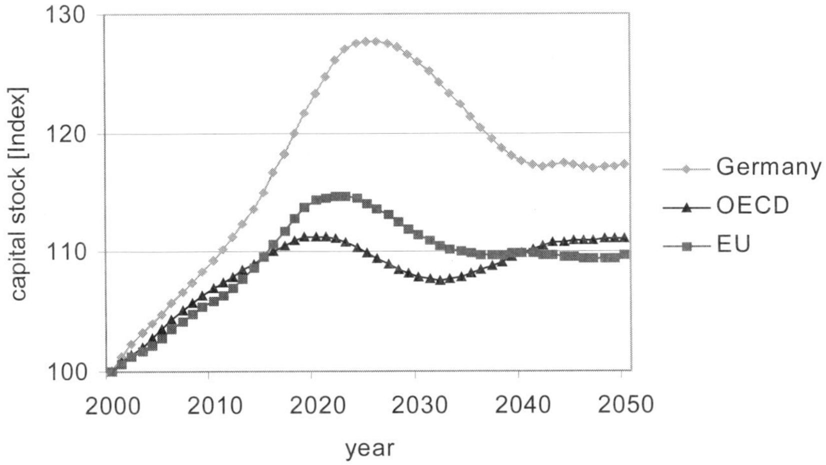

Figure 2b. Differential impact of fundamental pension reform

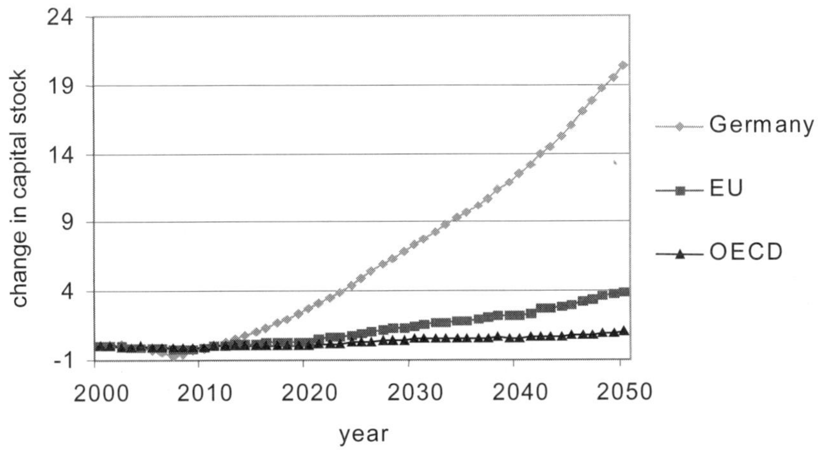

Notes: This figure shows projections of the aggregate German capital stock. Pension reform only in Germany. Germany: Germany as a closed economy, EU: perfect capital mobility in the EU area, OECD: perfect capital mobility in the OECD area.

Source: Own calculations, based on demographic projections by Birg and Börsch-Supan (1999) and the United Nations (2000).

Figure 3. Projections of aggregate German foreign assets under alternative pension systems and capital mobility scenarios (Index, 2000=100)

Figure 3a. Current pension systems

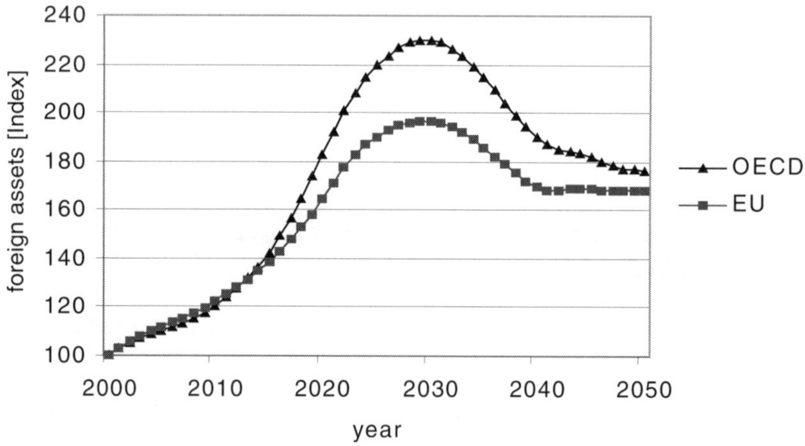

Figure 3b. Differential impact of fundamental pension reform

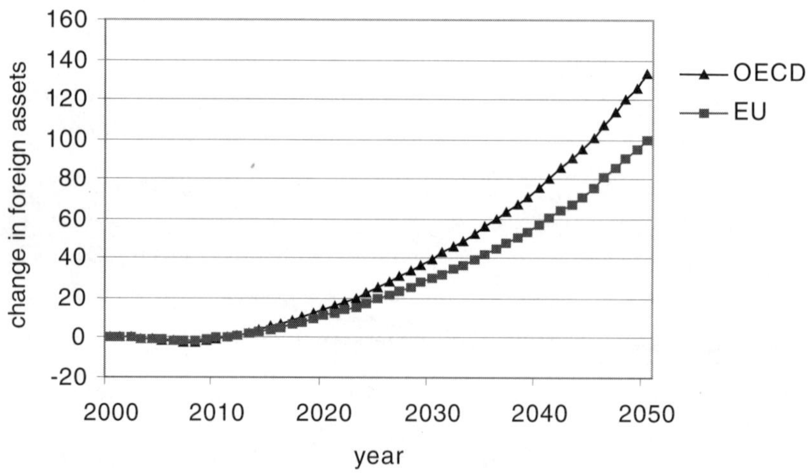

Notes: This figure shows projections of aggregate German foreign assets. Pension reform only in Germany. Germany: Germany as a closed economy, EU: perfect capital mobility in the EU area, OECD: perfect capital mobility in the OECD area.

Source: Own calculations, based on demographic projections by Birg and Börsch-Supan (1999) and the United Nations (2000).

Figure 4. Projections of German net capital exports under alternative pension systems and capital-mobility scenarios

Figure 4a. Current pension systems

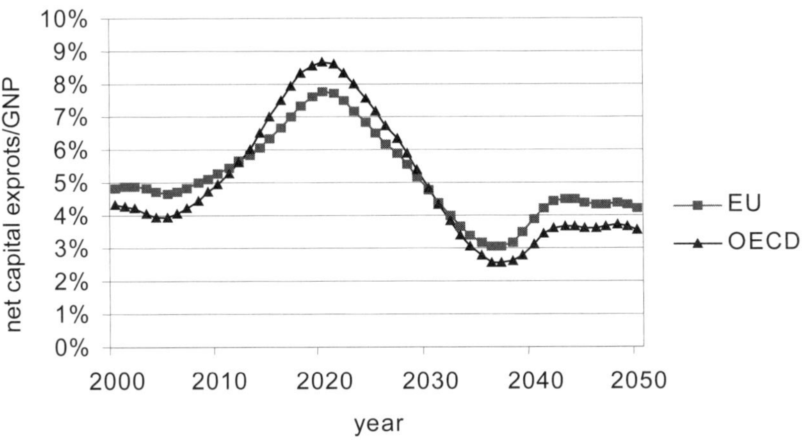

Figure 4b. Differential impact of fundamental pension reform

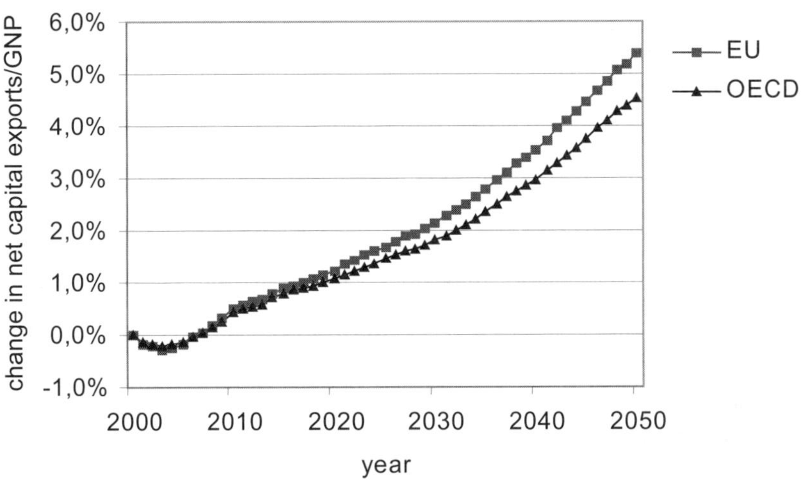

Notes: This figure shows projections of German net capital exports as a percentage of German GNP towards the EU and the OECD respectively. Pension reform only in Germany. EU: Net German capital exports to the other countries of the European Union when there's capital mobility only within the European Union, OECD: Net German capital exports to the other countries of the OECD when there's capital mobility only within the OECD.

Source: Own calculations, based on demographic projections by Birg and Börsch-Supan (1999) and the United Nations (2000).

brief period. This effect carries through into all figures shown below, where similar blips can be found.

4.2 International capital flows

We now turn to the main topic of this paper, international capital flows. The effect of ageing on German capital exports is shown in figure 3. Under the current pension system, foreign asset holdings of German households first increase and then, after a peak is reached in 2030, decrease again. This decrease in foreign asset holdings is less pronounced under a fundamental pension reform. Until 2050, German foreign asset holdings are projected to more than double. Net capital flows from Germany to other regions are shown, as a percentage of GNP, in figure 4. When the ageing process peaks, Germany's capital exports are reduced substantially, even though there are more attractive investment opportunities abroad. The reason for this is that, in this period, the saving rate is at its lowest. During our simulation period, Germany never becomes a net capital importer.

Our model tends to overestimate German foreign asset holdings since the only driving forces of international capital flows are the return differentials between world regions that are caused by the demographic transitions. In reality, many other factors generate and, most importantly, restrict international capital flows. These factors are discussed in our conclusions.

Next, we take a closer look at net capital flows in the OECD scenario. Figure 5 shows net capital exports of different regions within the OECD, as a percentage of total capital flows. The region named "OECD 12" includes all OECD countries except Japan, the United States and the countries of the European Union. Like Germany, Japan is projected to be a net capital exporter due to the effects of ageing. In Japan, the ratio of the number of persons aged over 65 to the number of workers is expected to increase from currently slightly above 20 percent to more than 50 percent by 2050. At the same time, Japan has implemented a social security reform programme, that, among other things, intends to increase retirement age by five years by the year 2050.[1] We implement this reform programme in our simulations and thus the increase in the old-age dependency ratio is far less pronounced than the direct effect of population ageing. Therefore, our model predicts decreasing net capital exports of Japan.

Figure 5 also shows that the main capital import region is the European Union except Germany, denoted as "EU 14". We further assume that the net exports of region *i* to region *j* are equal to the product of the export share of a region *i*, expressed as net capital exports as a percentage of total 'world' exports, and net

---

9 See Leibfritz, Roseveare, Fore and Wurzel (1996).

Figure 5. Projections of net capital exports of the OECD area upon the assumption of perfect capital mobility within the OECD

Figure 5a: Current pension systems

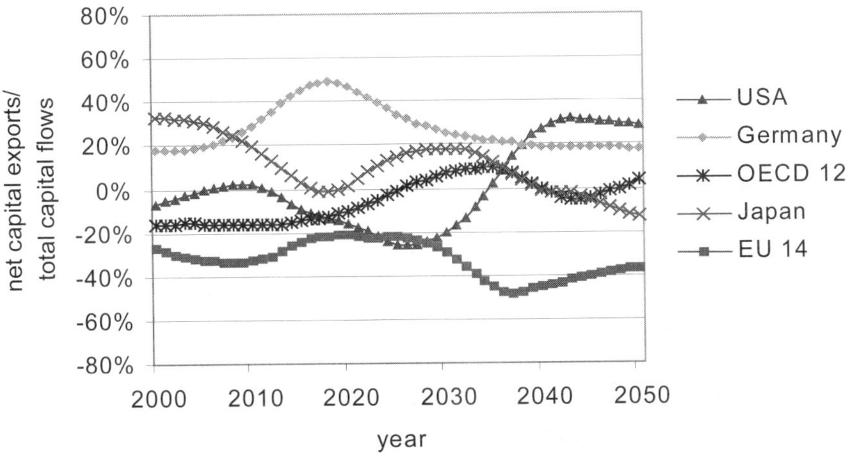

Figure 5b. Differential impact of fundamental pension reform

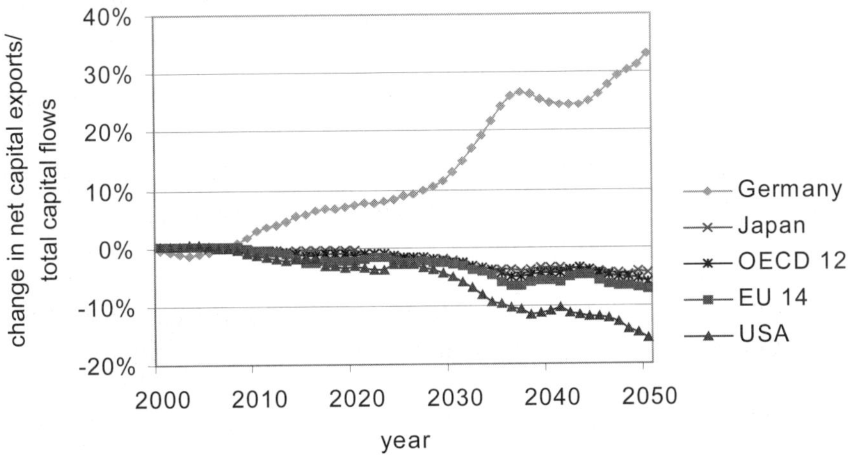

Notes: This figure shows projections of net capital exports of the respective region as a percentage of total capital flows under the assumption of perfect capital mobility within the OECD. Pension reform only in Germany. EU 14: All countries of the European Union except Germany. OECD 12: All OECD countries except the countries of the European Union, Japan and the United States.

Source: Own calculations, based on demographic projections by Birg and Börsch-Supan (1999) and the United Nations (2000).

imports of region *j*. This assumption is consistent with our model, since households are indifferent between regions with respect to their portfolio choice. Under this assumption, the region EU 14 absorbs about 36 percent of total German exports until 2008. Then, the United States becomes the most important import destination of German capital exports. In 2026, the US itself imports around 26 percent of total German exports. After 2038, the region EU 14 again takes over this position. The United States is less affected by the ageing process than are Germany and Japan. As for Japan, we implement reform proposals aimed to increasing retirement age by 2 years in our simulations. Therefore, the United States is a capital importer initially, but takes over the role of a capital exporter when Japan becomes an import country due to the reform proposal. As Figure 5b indicates, a fundamental pension reform in Germany would lead to an enormous increase in Germany's export share. By 2050, it is projected to have increased from 19 percent in 2001 to above 30 percent, at the expense of the export shares of all other countries.

4.3 Rates of return on capital

Population ageing will affect the return on capital because labour will become scarcer relative to capital, thereby decreasing the price of capital relative to labour. As can be seen from figure 6a, the return on capital in the closed economy scenario decreases slightly more than 0.7 percentage points between the years 2010 and 2026 - this is the direct effect caused by ageing.[10] This decrease is only around 0.2 percentage points when capital is freely mobile within the European Union, while the return does not decrease at all in the OECD scenario. A fundamental pension reform leads to an additional reduction in the rate of return on capital, caused by the increasing supply of capital and diminishing returns. In the closed-economy scenario, the rate of return is reduced by roughly 1 percentage point in 2050 relative to the rate of return under the current pension system. This decrease is much less than is often claimed in the public debate, and similar in magnitude as in the closed-economy model by Börsch-Supan, Heiss and Winter (2000). Moreover, the decrease in the rate of return on capital reduces to only 0.2 percentage points if capital is freely mobile within the EU. In the OECD scenario, the yield difference almost disappears.

10 The general upward tendency in the rate of return results from the exogenous growth in labour productivity. It is amplified in the open-economy scenarios since other world regions are generally younger than Germany. When the process of demographic change ends, and as the economies approach the steady state, the return on capital decreases again (this effect cannot be seen in our figures which have a shorter time horizon).

Figure 6: Projections of the rate of return to capital under alternative pension systems and capital mobility scenarios

Figure 6a: Current pension systems

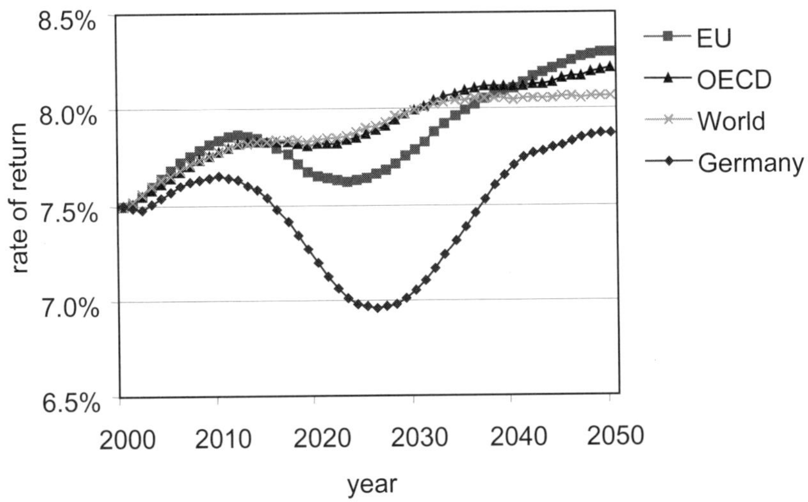

Figure 6b: Differential impact of fundamental pension reform

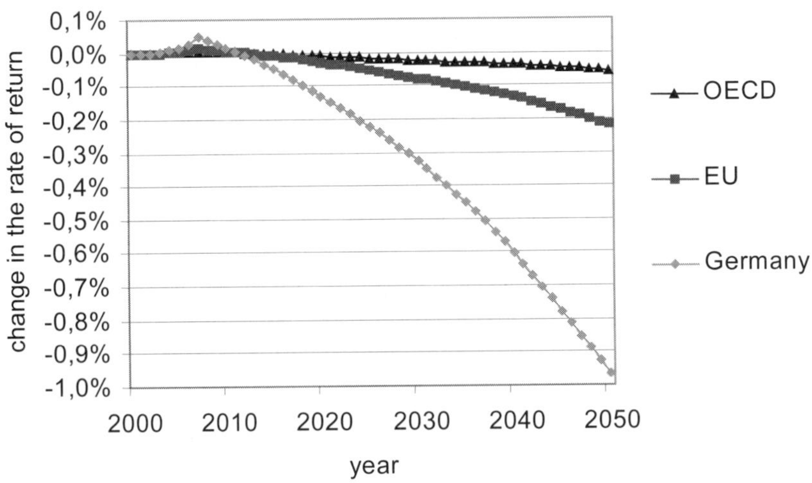

Notes: This figure shows projections of the rate of return to capital. Pension reform only in Germany. Germany: Germany as a closed economy, EU: perfect capital mobility in the EU area, OECD: perfect capital mobility in the OECD area.

Source: Own calculations, based on demographic projections by Birg and Börsch-Supan (1999) and the United Nations (2000).

These results suggest that household savings induced by a fundamental pension reform should be invested internationally, not only for reasons of risk diversification (which are of course not present in our deterministic model), but also for the sake of higher returns, which are available in other countries with different ageing processes and more favourable capital-labour ratios. Our results also confirm our earlier claim that the most important beneficial effects of capital mobility already show up under very modest capital mobility scenarios. Indeed, there is almost no difference between the OECD scenario and a scenario where we allow for perfect capital mobility in the entire world, including developing countries, as can be seen from figure 6a, where we add the entire world as a forth capital-mobility scenario.

4.4 Welfare aspects of population ageing, pension reform and capital mobility

The final step of our analysis focuses on welfare aspects of population ageing, pension reform, and capital mobility. Here, we focus only on the effects of pension reform and international capital flows on aggregate consumption. For ease of presentation, we restrict the analysis to a comparison between the closed-economy scenario and free capital mobility among OECD countries. Figure 7a shows that aggregate consumption in the open-economy scenario exceeds aggregate consumption in the closed-economy scenario from the year 2030 on. These differences in aggregate consumption are higher under a fundamental pension- reform scheme. Moreover, in the long run, consumption gains due to a fundamental pension reform are higher in the open economy case (figure 7b).

We conclude this section with a brief discussion of the effects of simultaneous pension reforms in several countries. We constrain the analysis to the effects of stylized pension reforms in other countries of the European Union. For simplicity, we assume that all pension systems are simultaneously reformed in the same manner, by freezing contribution rates to the public pay-as-you-go pension system, as in the reform scenario for Germany. We further assume that capital is perfectly mobile within the European Union. Here, we concentrate on the effects on the rate of return on capital. Recall that when Germany was assumed to be the only country that implements a pension reform, the rate of return on capital decreased by 0.2 percentage points in the EU scenario. As can be seen from figure 8, the decrease in the rate of return on capital is slightly larger in magnitude (0.3 percentage points) when all European economies simultaneously reform their pension systems. This effect is small, and we therefore conclude that the international capital market is strong enough to absorb additional capital that is generated by pension reforms throughout the world, the main reason for this strength being differences in the timing of population ageing across countries.

Figure 7: Projections of aggregate consumption of German households under alternative pension systems and capital-mobility scenarios

Figure 7a. Current pension systems

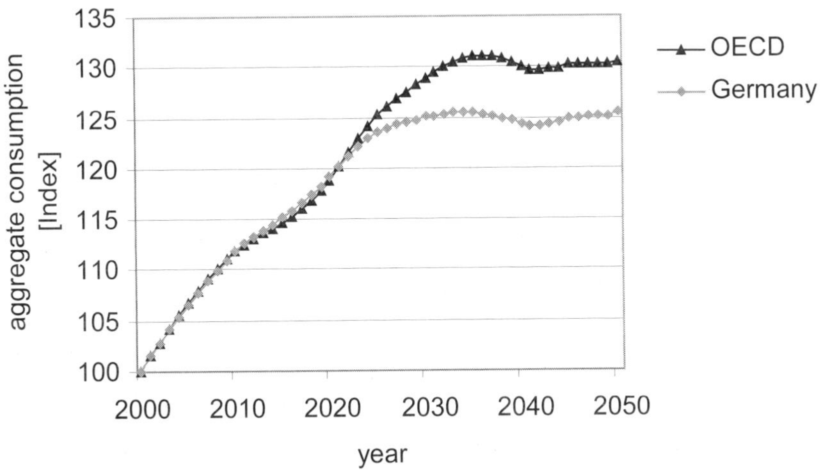

Figure 7b. Differential impact of fundamental pension reform

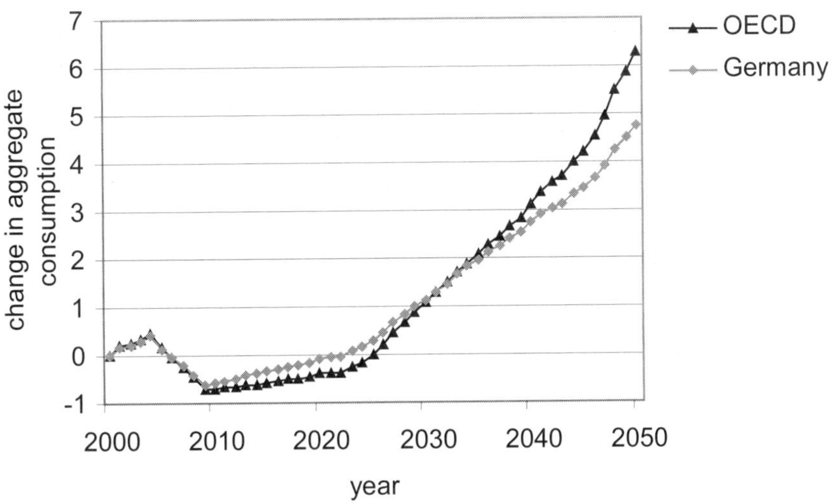

Notes: This figure shows projections of aggregate consumption of German households. Pension reform only in Germany. Germany: Germany as a closed economy, OECD: perfect capital mobility in the OECD area.

Source: Own calculations, based on demographic projections by Birg and Börsch-Supan (1999) and the United Nations (2000).

Figure 8: Projections of the rate of return to capital under alternative pension system scenarios and perfect capital mobility in the EU (Differential impact of fundamental pension reform)

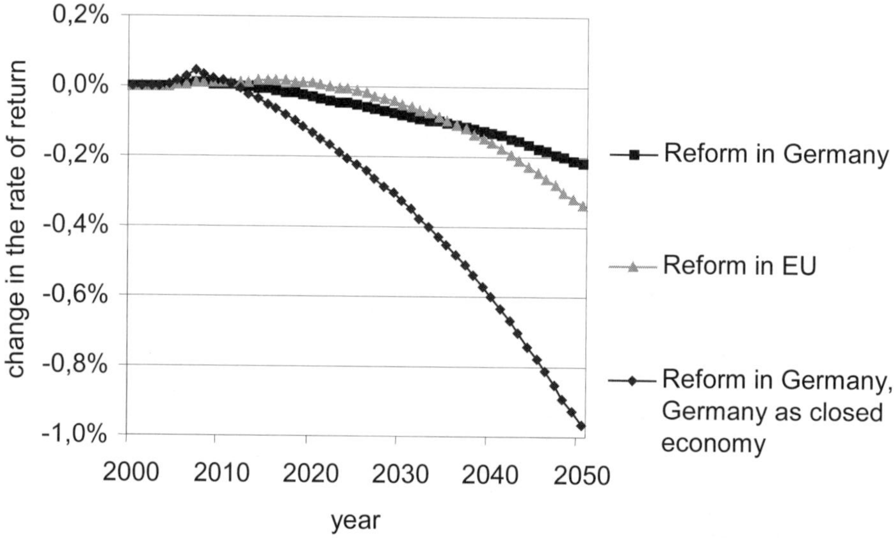

Notes: This figure shows the differential impact of pension reforms on the rate of return on capital. Pension reform in Germany or simultaneous pension reforms in all other countries of the European Union respectively. Perfect capital mobility in the EU. Current Systems: current PAYG pension systems, Reform in Germany: pension reform only in Germany, Reform in EU: simultaneous pension reforms in the entire EU.

Source: Own calculations, based on demographic projections by Birg and Börsch-Supan (1999) and the United Nations (2000)

5 Conclusions

In this paper, we have analysed the consequences of population ageing and a fundamental pension reform - that is, a shift towards more pre-funding - for capital markets in Germany and for international capital flows. We developed a stylized overlapping-generations model to predict capital formation and movements over a long horizon, taking demographic projections as given. Our results confirm that population ageing results, at least initially, in a higher capital stock, but when the baby boom generations begin to consume their retirement savings, the capital stock will decrease after 2030. Our simulations suggest that the decrease in the rate of return on capital, which results from secular shifts in the capital-labour ratio associated with an ageing population and retirement saving, is about 1 percentage points less by 2050, and only if all capital is invested exclusively in Germany.

However, capital markets these days are anything but closed national markets, and the return on capital can be improved substantially by international diversification. International capital flows induced by population ageing and pension reform follow the time-path of saving just described. At the peak of savings around the year 2020, German capital exports to all other OECD countries are projected to exceed 7% of GNP. 15 years later, they will briefly decline to below 2% but remain positive throughout the ageing process. Our simulations show that a transition to a partially-funded system does not crowd out existing savings totally. Capital exports from Germany to the OECD countries will be higher in this case, never falling below 2% after a peak of about 9% in the year 2021.

Our analysis has shown that the decrease in the rate of return on capital due to ageing between 2010 and 2026 is moderated substantially when capital is invested abroad. Moreover, the decrease in the return on capital due to a fundamental pension reform is only 0.2 percentage points if capital is freely mobile within the countries of the European Union versus 1 percentage point if Germany is modelled as a closed-economy. This suggests that closed economy overlapping-generations models overestimate the transitional burden of such a fundamental pension reform.

A few remarks on the economic model we used to simulate the macroeconomic effects of a fundamental pension reform are in order. We have already mentioned that our overlapping-generations model is very stylized and some important economic mechanisms are not taken into account, most importantly, endogenous labour-supply decisions and taxation. While it would certainly be interesting to explore these issues in our model, we do not anticipate that they would change the basic message of our analysis.

An important feature which is not reflected by the overlapping-generations model of sections 3 and 4 is financial-market risk. Our analysis concentrated on the long-term path of the rate of return on capital, in a model with no stochastic aggregate fluctuations, so there was no role for risk. However, real-world investments are risky, and, in their savings and portfolio decisions, households are concerned not only about the (expected) rate of return, but also about its variance, that is, about portfolio risk. This raises the question of whether countries such as Germany are really willing to invest substantial fractions of their retirement wealth abroad.

Finally, our multi-country model operates using the explicit assumptions of (i) perfect capital mobility, and (ii) a world-wide monetary union (we have no exchange rates in our model). First, the fact that we have free capital mobility implies that there are no institutional restrictions on capital exports, and that households are willing to hold unlimited foreign assets. Both assumptions might not be warranted in the real world (we only mention the literature on home-bias in international capital holdings; for example, French and Poterba, 1991). Second, while a monetary union is a realistic assumption for capital flows within the EU, exchange-rate reactions between major regions of the world are an important aspect that we did not address. They have, indeed, no role in these one-good economies. However, a distinction between home and tradable goods, demanded differentially by the various age-groups, would imply real exchange-rate effects due to population ageing.

While more research on these issues is certainly warranted, we are confident that our simulations are informative about the main effects of population ageing on international capital flows. Our simulations suggest that significant effects of capital mobility arise even if capital flows are restricted to Europe or the OECD, and this does not appear to be an unrealistic scenario.

References

Ando, A. and F. Modigliani. 1963. The 'Life-Cycle' Hypothesis of Saving: Aggregate Implications and Tests. *The American Economic Review,* 89(3), pp. 605-618.

Auerbach, A.J. and L.J. Kotlikoff. 1987. *Dynamic Fiscal Policy.* Cambridge, MA: Cambridge University Press.

Birg, H. 1996. *Die Weltbevölkerung – Dynamik und Gefahren.* Beck'sche Reihe Wissen, Nr. 2050. München: C.H. Beck Verlag.

Birg, H. and A. Börsch-Supan. 1999. *Für eine neue Aufgabenteilung zwischen gesetzlicher und privater Altersversorgung.* Berlin: Gesamtverband der Versicherungswirtschaft.

Blommestein, H. 1998. Ageing-induced Capital Flows to Emerging Markets do not Solve OECD's Basic Pension Problem, in Blommestein, H. J. and N. Funke, eds. *Institutional Investors in the New Financial Landscape.* Paris: OECD, pp. 349-362.

Bloom, D.E. and J.G. Williamson. 1998. Demographic Transitions and Economic Miracles in Emerging Asia. *World Bank Economic Review*, 12(3), pp. 419-455.

Börsch-Supan, A. 1995a. Die Rolle von Direktinvestitionen bei der regionalen Entwicklung eines alternden Europas, in Gahlen, B., H. Hesse and H.-J. Ramser, eds. *Standort und Region: Neue Ansätze zur Regionalökonomik* (Wirtschaftswissenschaftliches Seminar Ottobeuren, 24). Tübingen: J.C.B. Mohr (Paul Siebeck).

Börsch-Supan, A. 1995b. The Impact of Population Ageing on Savings, Investment and Growth in the OECD Area, in OECD, ed. *Future Global Capital Shortages: Real Threat or Pure Fiction*? Paris: OECD, pp. 103-141.

Börsch-Supan, A. 1999. Capital Productivity and the Nature of Competition. *Brookings Papers on Economic Activity, Microeconomics*, pp. 205-248.

Börsch-Supan, A. 2000. A Blueprint for Germany's Pension Reform. Paper Presented at the Workshop *Reforming Old-Age Pension Systems*, Magdeburg, May 25-26, 2000.

Börsch-Supan, A. and A. Brugiavini. 2001. Savings: The Policy Debate in Europe. *Oxford Review of Economic Policy*, 17(1), pp. 116-143.

Börsch-Supan, A., F. Heiss and J. Winter. 2000. *Pension Reform, Capital Markets, and the Rate of Return*. Institut für Volkswirtschaftslehre und Statistik, University of Mannheim, Discussion Paper No. 589-00.

Brooks, R. 2000. *What will Happen to Financial Markets when the Baby BoomersRetire?* International Monetary Fund, Working Paper No. 00/18.

Coakley, J., F. Kulasi and R. Smith. 1998. The Feldstein-Horioka Puzzle and Capital Mobility: A Review. *International Journal of Finance & Economics*, 3(2), pp. 169-188.

Cutler, D.M., J.M. Poterba, L.M. Sheiner and L.H. Summers. 1990. An Ageing Society: Opportunity or Challenge? *Brookings Papers on Economic Activity*, 1, pp.1-73.

De Nardi, M., S. Imrohoroglu and T.J. Sargent. 1999. Projected U.S. Demographics and Social Security. *Review of Economic Dynamics*, 2, pp. 575-615.

Deardorff, A.V. 1985. Trade and Capital Mobility in a World of Diverging Populations, in Johnson, D.G., and R.D. Lee, ed. *Population Growth and Economic Development: Issues and Evidence.* Madison, WI: University of Wisconsin Press.

Diamond, P.A. 1965. National Debt in a Neoclassical Growth Model. *American Economic Review*, 55, pp. 1126-1150.

Fehr, H. 2000. Pension Reform During the Demographic Transition. *Scandinavian Journal of Economics,* 102(3), pp. 419-443.

Feldstein, M. and C. Horioka. 1980. Domestic Saving and International Capital Flows. *Economic Journal*, 90, pp. 314-329.

Fougere, M. and M. Merette. 1999. Population Ageing and Economic Growth in Seven OECD Countries. *Economic Modelling*, 16, pp. 411-427.

French, K.R. and J.M. Poterba. 1991. Investor Diversification and International Equity Markets. *American Economic Review, Papers and Proceedings*, 81, pp. 222-226.

Higgins, M. 1998. Demography, National Savings, and International Capital Flows. *International Economic Review*, 39, pp. 343-369.

Holzmann, R. 2000. *Can Investments in Emerging Markets Help to Solve the Ageing Problem?* World Bank, Social Protection Discussion Paper No. 00/010.

Kotlikoff, L.J. 1998. *The A-K Model: Its Past, Present and Future.* National Bureau of Economic Research, Working Paper No. 6684.

Kotlikoff, L.J., K. Smetters and J. Walliser. 1999. Privatizing Social Security in the United States: Comparing the Options. *Review of Economic Dynamics*, 2, pp. 532-574.

Ludwig, A. 2001. Rentenreform im Globalen Kontext. Forthcoming. Frankfurt: Peter Lang Verlag.

Lührmann, M. 2001. *The Role of Demographic Change in Explaining International Capital Flows.* Mimeo, University of Mannheim.

Leibfritz, W., D. Roseveare, D. Fore and E. Wurzel. 1996. Ageing Populations, Pension Systems and Government Budgets – How do they Affect Savings? In OECD ed. *Future Global Capital Shortages – Real Threat or Pure Fiction?* Paris: OECD, pp. 47-102.

Maddison, A. 1987. Growth and Slowdown in Advanced Capitalist Economies: Techniques of Quantitative Assessment. *Journal of Economic Literature,* 25, pp. 649-698.

McKinsey Global Institute. 1996. *Capital Productivity,* Washington, D.C.

Miles, D. 1999. Modeling the Impact of Demographic Change upon the Economy. *Economic Journal*, 109, pp. 1–36.

Miles, D. and A. Iben. 2000. The Reform of Pension Systems: Winners and Losers across Generations in the United Kingdom and Germany. *Economica*, 67, pp. 203-228.

Miles, D. and A. Timmermann. 1999. Risk Sharing and Transition Costs in the Reform of Pension Systems in Europe. *Economic Policy*, 29, pp. 253-286.

Modigliani, F. and R. Brumberg. 1954. Utility Analysis and the Consumption Function: An Interpretation of Cross-Section Data, in Kurihara, K.K., ed. *Post-Keynesian Economics*. New Brunswick: Rutgers University Press, pp. 388-436.

Obstfeld, M. 1995. International Capital Mobility in the 1990s, in Kenen, P.B., ed. *Understanding Interdependence: The Macroeconomics of the Open Economy.* Princeton, NJ: Princeton University Press, pp. 201-261.

Obstfeld, M. and K. Rogoff. 1996. *Foundations of International Macroeconomics.* Cambridge, Massachusetts: The MIT Press.

Obstfeld, M. and K. Rogoff. 2000. *The Six Major Puzzles in International Macroeconomics: Is there a Common Cause?* National Bureau of Economic Research Working Paper No. 7777.

Pemberton, J. 1999. Social Security: National Policies with International Implications. *Economic Journal*, 109, pp. 492-508.

Pemberton, J. 2000. *National and International Privatization of Pension.* Mimeo, University of Reading.

Portes, R. and H. Rey. 1999. *The Determinants of Cross-Border Equity Flows.* National Bureau of Economic Research, Working Paper No. 7336.

Poterba, J.M. 1998. *Population Age Structure and Assets Returns: An Empirical Investigation.* Mimeo, Massachusetts Institute of Technology.

Reisen, H.. 2000. *Pensions, Savings and Capital Flows. From Ageing to Emerging Markets.* Cheltenham, E. Elgar.

Samuelson, P.A. 1958. An Exact Consumption-Loan Model of Interest with or without Social Contrivance of Money. *Journal of Political Economy*, 66, pp. 467–482.

Taylor, A.M. 1994. *Domestic Saving and International Capital Flows Reconsidered.* National Bureau of Economic Research, Working Paper No. 4892.

United Nations Population Division. 2000. *World Population Prospects: The 1998 Revision.* United Nations, Washington, D.C.

World Bank. 1997. *Private Capital Flows to Developing Countries – The Road to Financial Integration.* World Bank Policy Research Report. New York: Oxford University Press.

Discussion

Gary Hufbauer

In this path-breaking paper, the authors apply a sophisticated multi-country, multi-cohort model to evaluate the impact of ageing on international capital flows and rates of return. They focus on Germany, under scenarios with and without pension reform, and with and without capital mobility. Pension reform is defined as capping the contribution rate at 21 percent of wages, the current level. Without reform, the contribution rate rises as necessary to fund pensions at the current wage replacement rate. Capital mobility is defined as flows sufficiently large to equate rates of return within the European Union, or within the entire OECD. In addition to their German results, the authors provide supplementary calculations on net capital flows to and from the United States, Japan, the OECD and the EU.

My comments on this excellent paper are divided into four short parts: main results; an interesting puzzle; alternative assumptions; and policies for capital mobility.

1 Main results

The main results for Germany may be summarized for the year 2038, when the ageing process peaks. The household savings rate (now about 12% of German GDP) drops to 8.7% under the current pension system, but only falls to 9.5% with pension reform. With pension reform, *and no capital mobility*, the German-owned capital stock located in Germany is about 5% higher than it would otherwise be. However, assuming capital mobility within the EU or the entire OECD area, the German-owned capital stock located in Germany remains about the same as it would be without pension reform. In other words, the combination of pension reform and capital mobility mainly adds to German-owned assets held abroad. This is where international capital flows come into play.

Assuming capital mobility in the OECD, German net capital exports follow a sinuous trajectory. Under the current pension system, starting at 3% of GNP in 2000, net capital exports fall to 1.5% in 2008, then rise to 5.3% in 2020, then fall nearly to zero in 2038. With pension reform, net capital exports are 0.5% higher in 2020 (5.8% rather than 5.3%), and 1.5% higher in 2038 (1.5% rather than zero).

In a world of OECD capital mobility, but without pension reform, German-owned assets held abroad rise to an index value of 180 in 2030 (2000 = 100), before falling to an index of 120 in 2038. With pension reform, German-owned assets held abroad rise to an index of 200 in 2030, and then fall to an index of 160 in 2038. Capital mobility virtually stabilizes the rate of return to German-owned capital (at home or abroad); but without capital mobility, the rate of return drops by about 0.3% by 2038.

2 An interesting puzzle

From this summary of main results, I draw two conclusions about the relationship between ageing and international capital flows depicted by Börsch-Supan and his colleagues. First, capital mobility matters *a lot*; second, pension reform also matters, *but less*. The tenor of these conclusions comports with my own rough, back-of-the-envelope calculations of potential capital flows from OECD nations to emerging markets (Dobson and Hufbauer, 2001). However, I would have expected public pension reform - especially the contribution cap modelled by Börsch-Supan, Ludwig and Winter - to make a much bigger difference to household savings rates in Germany, and hence to outward capital flows in a world of capital mobility.

Why do I think public pension reform would make a bigger difference to household savings rates? My stylized model, shaped by the empirical work of Martin Feldstein and his NBER associates,[1] and influenced by my own casual observation of American retirees, goes something as follows. In the absence of pension reform, most people will just "leave it to luck and Social Security". Despite all the good advice from banks, mutual funds, and financial columnists, they won't save regularly for their retirement years. Instead, they will vaguely hope for a high-paying job tomorrow, a windfall inheritance, or the lottery. In realistic moments, they reckon there's always Social Security. Good times and a booming stock market, such as the United States enjoyed in the 1990s, reinforce these tendencies. The only regular saving done by most American households is to pay down the home mortgage. As a result, most couples retire with liquid financial assets of less than $50,000.

---

1 For a bibliography of NBER scholarship on the relation between U.S. private saving and Social Security, see www.nber.org/reporter. For a global perspective, see MacKenzie, Gerson and Cuevas (1997).

By contrast, Börsch-Supan, Ludwig and Winter project German household savings behaviour using a ultra-rationalistic lifetime utility maximization model. I am not persuaded that households determine their savings on the basis of lifetime utility maximization. If the authors can provide persuasive empirical support for this proposition, I would like to read it. Maybe German savings genes are different from American savings genes; maybe I don't understand why the U.S. household savings rate (as conventionally measured) has fallen nearly to zero. I remain to be convinced about the proposition of lifetime utility maximization.

Until persuaded by contrary evidence, I will cling to the forecast that the combination of German ageing and the current pension system will sharply depress the household savings rate in 2038. The floor will not be the relatively benign rate of 8.7% modelled by Borsch-Supan and his colleagues - but a much, much lower level.[2] Conversely, pension reform will "shake up" the population. Younger generations may still count on "lady luck", but they will no longer believe that "public pensions will always be there".[3]

If I am right that pension reform makes a bigger difference, then the international capital consequences are also larger. In the absence of pension reform, Germany could well become a net capital importer on a large scale after 2030 (like the United States today), selling off assets wholesale, not only to other OECD countries, but also to successful emerging countries.

3 Alternative assumptions

The model ingeniously developed by Borsch-Supan, Ludwig and Winters provides a rich playground for "what if" calculations. Like other readers of this paper, I have my own wish list of alternative assumptions and extensions.

Capital-intensity of demand. As specified, the model does not make the composition of demand depend on ageing. Intuitively, it seems that an ageing society, with flat or falling population, will require far less new housing and infrastructure (roads, schools, etc.) than a younger society with rising population. Hence, the demand for capital-intensive output (at a given level of per capita income) should be significantly lower for an ageing society. Peterson (2000)

---

2 Based on econometric evidence, Higgins (1998) estimates that German savings will fall by 9.0% of GDP between the level achieved in 1985-89 and 2025. Thirteen years later, in 2038, when German ageing reaches its peak, an additional fall could be expected. Among OECD countries examined by Higgins, Germany and Japan experience the worst age-induced drop in national savings as a percentage of GDP.

3 I believe the same forces are at work in the United States and other OECD countries. My simplistic explanation for the persistently high household savings rate in Japan is that public pensions barely exist.

thinks this could be an important feature of the European economy in the next thirty years. Peterson is not an economist – but he is a very savvy investor, which could be a better credential. Börsch-Supan and his colleagues could put Peterson's hypothesis to the test with an equation that models the composition of demand as a function of ageing.[4]

Productivity growth. Börsch-Supan and his colleagues project German technological progress (which I translate into productivity growth) at a rate of 1.4% per year. While I don't count myself among super-enthusiasts for the new economy, I am a believer. I think the United States is capable of 2% annual productivity growth - on the basis of technology already developed - for at least another decade. Particularly critical, I think, are downstream user industries that are building information and communications technology into their production and distribution systems. Moreover, I think that significant new technologies in energy, biology, materials, and other fields will reach commercial application in a few years. Again, their adoption will depend far more on economic systems than on the technology availability. Based on experience over the last 20 years, Anglo-Saxon capitalism is the best system known today for translating better technology into faster economic growth. Not everyone loves Anglo-Saxon capitalism - certainly not in Germany. Flexible labor practices, wholesale privatization, hostile takeovers, industrial consolidation, a dominant role for capital-market finance rather than bank finance - these are all ingredients of the Anglo-Saxon model. Many of them are unpopular in Europe. Yet it is possible that significant features of the Anglo-Saxon model will find their way into the German economy, and boost productivity growth at least to 2% annually. This possibility should play a role in the pension debate: a larger dose of Anglo-Saxon capitalism might enable Germany to afford a slower transition to pension reform. Börsch-Supan and his colleagues could contribute to the policy debate by analyzing the trade-off between more rapid productivity growth and the requisite speed of pension reform.

Share and property values. Finally, I am curious how stock-market and real estate valuations affect the results. As the experience of the 1990s teaches us, Tobin's q value (the ratio between the capital-market value of an asset and its replacement cost) can exceed 1.0 by substantial magnitudes for extended periods. The q value for the S&P 500 is now above 2.5, and similar high values prevail in Europe. If I understand their model correctly, Börsch-Supan and his colleagues implicitly assume that q is 1.0. What happens to savings and international capital flows when q is allowed to have higher values, and indeed fluctuate?

---

4 Higgins (1998) - who is an economist - estimates a fairly dramatic fall of 10.3% in German investment as a percentage of GDP, between the 1985-89 level and 2025, as a consequence of ageing.

Likewise, what are the consequences when home and apartment prices rise and fall relative to household income? The residential construction market is sufficiently perfect that new homes and apartments normally spring up whenever construction and land costs are slightly below the selling price for the existing housing stock (i.e., when Tobin's q is slightly above 1.0). However, residential land prices can rise and fall sharply-depending on population and income pressures-adding or subtracting enormous household wealth. Does this affect national savings and international capital flows?

4 Policies for capital mobility

Capital mobility is essential for maintaining rates of return in the multi-country model developed by Börsch-Supan, Ludwig and Winters. It could be even more important than they have suggested if pension reform actually happens, or if ageing reduces the capital-intensity of demand.

In their paper, the authors assumed two scenarios for capital mobility - mobility within the EU alone and mobility within the entire OECD. In these political spaces, institutional frameworks respect the rights of investors and ensure reasonable supervision of major financial institutions.

However nearly all OECD countries face the same ageing and pension-reform problems that confront Germany. If they behave responsibly, these countries collectively should become net capital exporters to the younger emerging markets. Institutional frameworks are weak for investment in many emerging markets, both on the demand side (local banking standards, corporate governance, etc.) and on the supply side (surveillance of international banks and portfolio managers based in the G-10 countries). In the wake of the Asian crisis, and fresh episodes in Russia, Brazil, Turkey, and Argentina, much has been written about the demand side of international capital markets. The IMF and other international institutions are beginning to improve local institutions - but only beginning. Wendy Dobson and I (2001) have written about severe problems on the supply side of international capital markets. From a policy standpoint, the Bundesbank and other G-10 financial regulators can make an enormous difference. Nearly all the capital invested in emerging markets travels under the flag of a few hundred large corporations, banks, and financial houses. In our view, G-10 financial regulators can do much more to improve the quality of bank lending and portfolio investment in emerging markets.

References

Dobson, W.W., and G. C. Hufbauer. 2001. *World Capital Markets: Challenge to the G-10.* Washington DC: Institute for International Economics.

Higgins, M. 1998. "Demography, National Savings, and International Capital Flows." *International Economic Review*, 39(2), pp 343-69.

MacKenzie, G.A., P. Gerson and A. Cuevas. 1997. *Pension Regimes and Saving.* Occasional Paper No. 153. Washington DC: International Monetary Fund.

Peterson, P. 1999. *Gray Dawn: How the Coming Age Wave will Transform America – and the World.* New York: Random House.

Discussion

Ulrich Grosch

In their paper, Börsch-Supan, Ludwig and Winter focus on ageing-induced capital flows, choosing Germany as an example. They derive their capital-flow forecasts using a multi-country overlapping generations growth model, and combine it with long-term demographic projections for several world regions. Moreover, they distinguish three different capital-mobility scenarios (closed economy, liberalized capital flows among EU countries, or liberalized capital flows among OECD countries) and two different pension systems (Germany's current pay-as-you-go pension system and a one-third transition to a funded system). The strength of this approach is - in my opinion - the very detailed modelling of the future development of the age structure. Return differentials caused by differences in the ageing process between countries are therefore the driving forces of the capital flows in this model.

The simulation results which I consider to be most important are as follows:

- There are some fluctuations in the German savings rate over the simulation period, with a strong decline between 2020 and 2035, when the ageing problem peaks. A fundamental pension reform would moderate this decline, and it would do so even more under the assumption of perfect capital mobility. The implication is - as Börsch-Supan, Ludwig and Winter stress - that the additional retirement savings induced by a pension reform would not totally crowd out other savings.
- The German capital stock is projected to increase until the second half of the 2020s and to decrease during the following 15 years (closed economy scenario). These fluctuations - caused by the interaction of demographic effects and growth in labour productivity - have been dampened by capital mobility. A fundamental pension reform would increase the German capital stock significantly if Germany were regarded as a closed economy.
- In the open economy model, German households accumulate foreign assets until about 2030 and draw on these assets thereafter. This decrease would be less pronounced in the case of a fundamental pension reform. In both scenarios, Germany would end up in 2050 with significantly more foreign assets than in 2000.
- Germany's financial account is characterized by net capital exports over the whole period under review, with strong fluctuations in the magnitudes, starting

from 4 % to 5 % of GNP in 2000 and reaching nearly 9 % of GNP in 2020 before decreasing to about 4 % of GNP at the end of the simulation period.[1] The profile of net capital exports reflects the movements of the savings rate discussed above. A fundamental pension reform would lead to even more net capital exports in nearly all years.

- From a detailed analysis of the capital flows among the major OECD regions, Börsch-Supan, Ludwig and Winter conclude that it is the other EU countries that mainly absorb Germany's capital outflows. Net capital flows of the United States and Japan switch directions several times during the simulation period, with Japan being the major net capital exporter at the beginning of the period under review and the US at the end. A fundamental pension reform in Germany alone would substantially increase Germany's share in "world" capital exports.
- The evolution of the rate of return on capital reflects the pattern of the ageing German population, decreasing strongly between 2010 and 2026 and increasing afterwards. The fluctuations are much less pronounced if capital is mobile internationally. A fundamental pension reform would reduce the rate of return on capital by about 1 percentage point in the closed-economy scenario and only marginally in an open economy.
- Börsch-Supan, Ludwig and Winter close their analysis by discussing welfare considerations and the political feasibility of pension reform.

My further comments focus on those aspects of the paper that have to do with international capital flows - leaving aside other important questions, particularly those concerning the necessity and feasibility of pension reform in Germany (and elsewhere). Hence, I will first compare the results of the model with those of some other studies. Second, I will say a few words on the potential political implications of the capital movements in the model. And, last but not least, I will comment on some limitations of the specific type of model to analyse the impact of ageing on international capital flows.

---

1 It is rather difficult, however, to reconcile the developments of external stocks and flows. In the model, Germany, on the one hand, records net capital exports over the whole reference period, while, on the other, foreign assets are decreasing as from 2030. To bring these diverging trends together, the depreciation of foreign assets (5 %) plays a role. But taking this effect into account leads to highly implausible levels of GNP and foreign assets. It should be noted that in 2000 Germany's net foreign assets were only about 3 ½ % of GNP. For net capital exports to be totally absorbed by the depreciation on foreign assets, net foreign assets would have to make up about 100 % of GNP in 2030, which seems to be incompatible with figure 3.

Comparison with other studies

There are only a few studies that deal with the effects of ageing on international capital flows and foreign asset holdings in a comprehensive (multi-region, general-equilibrium) manner. Furthermore, it is rather difficult to compare these studies, as they are different in their design and in their specific assumptions, (including overlapping-generations models, optimal-growth models and general-equilibrium macro-econometric models); different assumptions, relate to the population trends and the country selection. In particular, there is – as far as I know – no other comprehensive study besides the one we are discussing here that focuses especially on Germany.

Despite these limitations, there are striking similarities to, but also some differences from, the results of the Börsch-Supan, Ludwig and Winter model:

- In a model similar to that of Börsch-Supan, Ludwig and Winter, but including non-OECD countries, Brooks (2000) foresees current account surpluses for the EU, Japan and North America at the beginning of this century (2010) and deficits for the EU and North America from around 2030 onwards; in Japan the swing will take place somewhat later.
- Cutler et al. (1990), using an optimal growth model, project swings in the US current account from surplus to deficit in (around) 2007 and in the opposite direction after 2020.
- Turner et al. (1998), using OECD's MINILINK model, forecast initial current account surpluses of the US, Japan and the EU, all switching to deficits after 2020, yet at different points in time. Consequently, net foreign assets first increase and subsequently decrease when the ageing pressures become more and more severe.

In summary, the results of Börsch-Supan, Ludwig and Winter's simulations of the profile of the capital flows of the US and Japan are broadly in line with some of the projections of these models. Current trends and the most recent forecasts of the US b.o.p. by international organizations, however, show large and persistent current account deficits and corresponding net capital imports. The swing to net capital exports projected by Börsch-Supan, Ludwig and Winter and other authors has not yet materialized. The same is also true of Germany, which – contrary to the simulations – recorded net capital exports and, correspondingly, a current account deficit in 2000.

The results for the EU are more difficult to compare, as Börsch-Supan, Ludwig and Winter have singled out Germany and do not present aggregate capital flows for the EU-15. Moreover, one has to bear in mind that the results of the model relate mostly to intra-OECD capital flows, whereas other models have a broader country coverage.[2]

Taking this feature into account, the magnitudes of the capital flows might be considered surprisingly large. Peaking at over 8 % of GNP (even without a pension reform), German net capital exports seem to be significantly higher than the figures published by many other authors for the various other regions they focus on. On the other hand, Higgins (1998) estimates the demographic effects on past current account balances to exceed 6 % and expects even larger balances for the future. However b.o.p balances of such a magnitude will inevitably have political and economic consequences, that will be addressed below.

Apart from these quantitative differences, Börsch-Supan, Ludwig and Winter seem to give international capital flows greater importance in solving the ageing problems of industrial countries than other authors. Their position is summarized in the following sentence:

> *"Capital exports to developing countries could therefore solve the ageing problems of industrialized countries by reducing the pressure on the interest rate and by shifting the production of consumption goods towards developing countries." (p. 58)*

This contrasts with the assessment of Turner et al. (1998), who conclude that *"... age-related pressures could be alleviated to only a modest extent by OECD investment in the non-OECD world."* Similarly cautious conclusions are drawn by Cutler et. al. (1990, pp. 36-7) and GoldmanSachs (2001).

Policy implications of the capital flows

From a political point of view, Germany's projected very large net capital exports (or, equivalently, the current account surplus) over extended periods might become a major political issue. Germany indeed recorded net capital exports of between 4 % and 5 % of GNP at the end of the eighties, but only for four years. The model simulations show German net capital exports of more than 4 % for about 30 years or more (figure 4). Even the past situation, however, caused severe tensions in the international economic and political dialogue.

To assess the potentially "explosive effects" of such large external imbalances, one does not need to look back that far, however. After the Asian crisis, the question of "fair burden sharing" was discussed intensively, and the external positions of major world regions played a significant role in the debate, too.

---

2 Börsch-Supan, Ludwig and Winter mainly focus on capital flows within the EU and the OECD; only on one occasion, related to the impact of capital mobility on the return to capital, do they present results for a comprehensive country coverage.

Therefore, I am inclined to think that prolonged capitalaccount or current account balances of the magnitudes suggested by the simulations of Börsch-Supan, Ludwig and Winter might be politically unsustainable. The same might be the case for large holdings of cross-border assets, particularly in the case of foreign direct investment, which occasionally encounters some resistance among the population of the host country.

Limitations of the model

Taking due account of the restrictive assumptions of the underlying model, one could argue, however, that one should not worry too much about the magnitude of the capital flows in the simulations. The authors themselves point to the fact that their model might overestimate German foreign asset holdings, as, in reality several factors (one example being, restrictions on international capital flows) tend to dampen international capital flows.

Moreover, in the deterministic, perfect-foresight setting, the capital flows are induced only by return differentials between world regions that are caused by different demographic trends. There is therefore room neither for assetprice risk nor for exchange rate changes that - in reality - would have an impact on the direction and magnitude of capital flows. Some authors argue that incorporating risk in models like that presented by Börsch-Supan, Ludwig and Winter would reduce net capital flows compared with the deterministic case, if agents were risk-adverse. They therefore regard the results of deterministic models as an upper limit of potential capital flows (Brooks, 2000).

On the other hand, in a stochastic world, foreign assets offer diversification advantages if the returns on capital between different countries are not perfectly correlated. For this reason Reisen (2000) - among others - forcefully advocates an international diversification of pension assets for rapidly ageing countries.

To sum up, from a theoretical point of view, it is not clear what the effects of allowing for asset price risk on cross-border capital flows would be. Empirically, recent financial market crises have shown that the potential losses involved in such a diversification strategy may be substantial, particularly where investments in emerging markets (EM) are concerned (and even some of the OECD member countries have to be regarded as EM). In addition, world-wide capital-account liberalization has led to a significant increase in correlation among individual markets. International diversification might therefore offer fewer advantages than in the past. A recent study (Deutsche Bank Global Markets Research, 2001) shows that for Germany, returns per unit of risk of an internationally diversified equity portfolio would have been only marginally higher over the last five years than a purely domestic portfolio; for some other industrial countries, diversification would even have resulted in smaller returns per unit of risk. According to the cited study, only crisis-shaken EM could have achieved large diversification advantages during that period.

One last point is worth considering: exchange rates. The model does not allow for exchange-rate changes. However, the large and persistent b.o.p. (im)balances and the strong fluctuations in foreign assets imply strains in the ageing-induced adjustment process. This will inevitably lead to reactions of the exchange rate. Exchange-rate changes, for their part, might well reduce capital flows, as well as give rise to additional risk for pensioners investing abroad if protracted misalignments occur.[3]

All this is not to say that the overlapping-generations model is inadequate in analysing the consequences of ageing on capital flows. It does indeed have its merits and the results could serve as some sort of a benchmark. However, my conclusion would be that without explicitly incorporating financial-market risk and exchange rates, the results should be interpreted with care.

References

Börsch-Supan, A., A. Ludwig and, J. Winter. 2001. Ageing and International Capital Flows. *This Volume,* p. 55-83.

Brooks, R. 2000. *Population Ageing and Global Capital Flows in a parallel Universe.* IMF Working Paper WP/00/151.

Cutler, D. M., J. M. Poterba, L. M. Sheiner and L. H. Summers. 1990. An Ageing Society: Opportunity or Challenge. *Brookings Papers on Economic Activity*, 1, pp. 1-73.

Higgins, M. 1998. Demography, National Savings, and International Capital Flows. *International Economic Review*, 39 (2), pp. 343-69.

Deutsche Bank Global Markets Research. 2001. *Exchange Rate Perspectives*. April 2001, pp. 17 and 32-3.

GoldmanSachs. 2001. Capital Market Implications of Global Ageing. *Global Economics Weekly* 01/08, 28 February 2001, pp. 3-12.

Reisen, H. 2000. *Pensions, Savings and Capital Flows. From Ageing to Emerging Markets.* Cheltenham, Edward Elgar.

Turner, D., C. Giorno, A. De Serres, A. Vourc'h and P. Richardson. 1998. *The Macroeconomic Implications of Ageing in a Global Context.* OECD Working Paper AWP 1.2.

---

3 In their conclusions, Börsch-Supan, Ludwig and Winter make some suggestions on how to incorporate real exchange rates in their model.

Developments in Decumulation: The Role of Annuity Products in Financing Retirement

Olivia S. Mitchell*

1 Introduction

Driven by the prospect of a global ageing boom, researchers and policy-makers throughout the developed world have devoted substantial effort to exploring what motivates household-asset *accumulation* patterns. It is during the accumulation phase that a society seeks to build up resources adequate to finance retirement. More remains to be learned about this process, but even less is known about what happens to people later in life, during which they are confronted with the need to *decumulate* their resources in an orderly manner in retirement. To understand this decumulation process better, along with the factors that influence it, this paper explores a particular financial tool that helps protect against key financial risks at the end of life. Specifically, I examine the role of the *life annuity*, a financial contract in which the purchaser exchanges a certain sum of money for a continuous stream of benefits until he dies. The insurer, or annuity provider, pools longevity risk across all contract-holders and is thus able to offer longevity insurance for the customer population.

Financial contracts that pay off in the event of survival have been in existence for centuries, with the *tontines* devised under Louis XIV's reign, and life annuities

* Research support was provided by the Pension Research Council at the Wharton School, the National Bureau of Economic Research, the Wharton-SMU Research Center at Singapore Management University and the Economic and Social Research Institute of the EPA of Japan. Helpful comments were provided by Alan Auerbach. Much of the research referenced draws on material appearing in Jeffrey Brown, Olivia S. Mitchell, James Poterba and Mark Warshawsky (2001). Opinions and errors are solely those of the author and not of the institutions with which she is affiliated.

appearing in Europe as early as the 1300s.[1] In modern times, life annuities have become a substantial component of the life-insurance market, but only a few of the products labeled annuities are, in fact, "true" life annuities. For instance, in the United States, purchases of life annuities totaled almost $120 billion in 1999, but the vast majority of the market consists of variable annuities rather than products paying out income for life (Brown, Mitchell, Poterba and Warshawsky, 2001, henceforth BMPW). In the UK and Australia, annuity markets also appear relatively small (Finkelstein and Poterba, 2000; Murthi, Orszag and Orszag, 1999; Doyle, Mitchell and Piggott, 2001). Recent interest in annuity markets has come from Latin America, resulting from more than a dozen countries having converted their pay-as-you-go national defined-benefit (DB) pensions into mandatory funded defined-contribution (DC) pensions (Mitchell and Barretto, 1997). Similarly, European pension reforms now under way are demanding that similar attention be paid to these insurance markets as well (Blake, 1999; Disney, 2000; Disney and Johnson, 2000; European Commission, 1997).

Under a DC pension plan, retirees are typically offered a range of choices regarding how to convert their accumulated assets into retirement payouts. Usually the plan offers a portion of the money as a lump-sum, and may also provide an option to disburse some funds as either a periodic payment or an annuity. Perhaps inevitably, whenever annuities come to play a key role in a nation's retirement system, questions come up regarding how these products should be structured and priced, and the extent to which governments should supervise and regulate the market for these products. For example, these debates surfaced in the most recent US presidential race, where advisers to President Bush formulated individual account-defined contribution proposals that included annuity promises as a cornerstone (Feldstein and Ranguelova, 2000).

In what follows, I explore lessons from research on the annuity market over the last decade and discuss its relevance for ageing, financial markets and government policy. In particular I pose and answer the following questions:

- How can annuities help retirees finance retirement?
- What problems arise in annuity markets?
- How could the functioning of these markets be enhanced?

The discussion proceeds in five sections. Section 1 summarizes the economics of annuities, while Section 2 sketches the methods commonly used to value annuities. Section 3 presents empirical results from a range of countries, and

---

[1] One famous German tontine, that of Gotha in 1752, has been called the "perfect illustration of rococo finance". In exchange for a purchase price, the plan promised surviviors a fixed income of 3% per year plus a variable amount determined by lottery and mortality outcomes (Jennings and Trout, 1982).

Section 4 addresses the question of how the functioning of these markets might be enhanced. A final section concludes.

2 Handling Retirement-Wealth Decumulation

Many critics have decried Westerners' inability or unwillingness to save for their own retirement, with the US often identified as the most spendthrift country due to its low and oftentimes negative saving rate. Nonetheless the strong US economy over the past 25 years has provided older generations with substantial wealth on hand as they look to retirement. Thus pre-retirement wealth for US households in their 50'-s amounts to $360,000 exclusive of Social Security (in 1992$), with the median at $192,000 (see Table 1). If these households retire at age 62, their non-Social Security assets would be anticipated to amount to $438,000 on average ($239,000 at the median). Of this total, at least $100,000 of retiree assets and perhaps more will have to be managed during retirement.

Table 1. Components of Household Wealth on the Verge of Retirement: US Health and Retirement Study ($1992)

Wealth Decile	Current Wealth	Projected Wealth at Age 62	Projected Wealth at Age 65
Mean	**$478,313**	**$566,431**	**$625,066**
-Housing	65,940	76,410	80,507
-Financial	175,974	205,653	228,133
-Social Security	119,793	128,712	142,018
-Pension	116,606	155,656	174,408
Median 10%	**$325,157**	**$382,678**	**$420,537**
-Housing	59,746	71,097	75,047
-Financial	66,530	71,004	71,175
-Social Security	133,606	143,864	160,824
-Pension	65,275	96,713	113,491

Source: Moore and Mitchell (2000).

All values weighted by relevant sample weights.

2.1 Assessing the Money's Worth of Life Annuities

Having found that many elderly people will have assets to spread over their retirement period, the next question is how people can manage their portfolios to ensure old-age economic security. One way to proceed is to simply divide available retirement wealth by one's life expectancy, a fractional consumption pattern known as "periodic withdrawal" sometimes found in pension payout schemes. For example, Table 2 indicates that a 65-year old man could anticipate living another 16.4 years on average, and a 65-year old woman 19.6 additional years. The problem is that were a retiree to divide his age-65 retirement wealth by the number of years remaining to him at age 65, he would be exposed to substantial risk of exhausting his funds before he died. Thus a 65-year old man who would anticipate dying at age 81 still faces a one-third chance of living to age 85, and almost a 20% chance of surviving to age 90. For his female counterpart whose expected age at death is age 85, there is almost a one-third chance of surviving to age 90. Hence the periodic withdrawal approach is likely to leave retirees short of funds. Clearly the periodic withdrawal approach does not insure a steady flow of retirement income until death.

Table 2. Remaining Life Expectancy and Probabilities of Survival to Selected Ages for 65-Year Old Men and Women in the Year 2000

	Men	Women
Remaining Life Expectancy (years):	16.4	19.6
Probability of Surviving to Age:		
70	88%	92%
75	74	82
80	56	69
85	36	51
90	18	31
95	6	14
100	1	4

Source: Brown (2000)

One way to enhance retirement security is to purchase a life annuity that guarantees an income stream until the insured's death. Such a financial product provides the retiree with an opportunity to insure against outliving his assets, in exchange for a premium paid to an insurer. The insurer pools assets across those with similar longevity expectations but who then experience different actual survival outcomes. In this way the life annuity provides income insurance to a homogeneous insured retiree population.

How valuable is a life annuity to the retiree? A financial answer to this question can be ascertained by computing the expected present discounted value (EPDV) of the life-income benefits paid to annuitants, in relation to the annuity-premium cost or purchase price. In the literature this is known as the "money's worth" of the life annuity. To illustrate the concept, we may assume that the product under examination is an immediate single-life annuity that pays out $A per period for life, in exchange for an initial purchase price (e.g. $100,000). The payout amount may vary across annuity policies on offer, and it may further be fixed in nominal terms for the life of the annuity contract, or it may be linked to some underlying real asset (which would then generate a real rather than a nominal annuity).

In the case of a nominal annuity, we follow Mitchell, Poterba, Warshawsky and Brown (1999: henceforth MPWB) in defining $q_{a,t}$ as the probability that an a-month-old individual who is alive at the beginning of month t will die during that month. Future mortality experience is expressed by letting Pj be the probability that someone who is 65 years old at the time when he purchases an annuity survives for at least j months: [2]

(1) $$P_j = (1 - q_{780,1}) * (1 - q_{781,2}) * \ldots * (1 - q_{780+j1,j}).$$

(To compute the EPDV of an annuity purchased by a 65-year-old, we need to forecast this individual's *future* mortality probabilities. This point is taken up in the next section.)

We denote the EPDV of a life annuity with monthly payout A purchased by an individual of age b as $V_b(A)$. This may depend on a term structure of interest rates, i_k, expressing the nominal short rate k periods into the future:

(2) $$V_b(A) = \sum_{j=1}^{600} \frac{A * P_j}{\prod_{k=1}^{j} (1 + i_k)}$$

---

2 In practice, our computations assume that no one lives beyond age 115 years; calculations are insensitive to this upper limit on lifespan.

In most of our research, we use the term structure of yields on Treasury bonds to estimate the time series of expected future nominal short-term interest rates,[3] though we have also explored the sensitivity of results to flat term structures, with the discount rate given variously by the ten-year Treasury bond yield, the thirty-year Treasury bond yield, and the BAA corporate bond yield. In addition, we have evaluated the sensitivity of results to taxes, since in the US at least the federal tax treatment of annuities is quite complex, governed by specialized rules described in BMPW (2001).[4]

In the case of real annuities, equation (2) must be adapted to reflect the fact that the amount of the payout is time-varying in nominal terms but fixed in real terms. As shown in BMP (forthcoming), the simplest approach is to let A_r denote the real annual payout, and to replace the nominal interest rates in the denominator of equation (2) with corresponding real interest rates r_k. Hence, to compute the EPDV of a real annuity the equation becomes:

$$(3) \qquad V_b(A_r) = \sum_{j=1}^{115-b} \frac{A_r * P_j}{\Pi_{k=1}^{j}(1+r_k)}.$$

The "money's worth" of an annuity may then be specified as the *ratio* of the EPDV of the annuity's payout stream to its purchase price. For a nominal annuity selling for \$100,000, for instance, the money's worth may be expressed as $V_b(A_n)/100{,}000$. These money worth ratios have proved useful in the literature, since they represent a currency-independent metric for comparing annuities across different groups, over time, and across countries.

2.2 Equivalent Wealth Valuation: Measuring the Utility of Longevity Insurance

Before reporting money's worth results for a range of countries and time periods, we note that risk-averse consumers will value annuities more highly than the simple financial money's worth approach implies. This is because people derive

3 Nominal yields on US Treasury bonds with fixed maturities of 1, 2, 3, 5, 7, 10, 20, and 30 years are used to estimate the term structure of expected short-term interest rates; the expected nominal short rate in each future period is computed as the nominal short rate that would satisfy the expectations theory of the term structure for the two adjacent long-term bonds.

4 Specifically, the US tax authority specifies the expected number of years over which the annuitant can expect to receive benefits, T', based on the annuitant's age at the time payouts begin. Tax law then defines an *inclusion ratio* that determines the share of annuity payments in each period that must be included in the recipient's taxable income until T'; after that all annuity income is subject to tax. The inclusion ratio is designed to measure the fraction of each annuity payout that reflects the capital income on the accumulating value of the annuity premium.

utility from insurance protection *per se*. In our research, we develop a metric for evaluating how a risk-averse consumer might value an annuity in utility terms, using the **equivalent wealth** concept (MPWB, 1999). Specifically, this represents the amount of wealth that a consumer would need *if he did not have access to an annuity market,* in order to achieve the same lifetime expected utility level that he could achieve by using that wealth to purchase a nominal annuity.

As developed in MPWB (1999) and summarized in Brown and Warshawsky (2000), the consumer is posited to maximize an expected utility function V by choosing an optimal consumption path $\{C_t\}$ from time 0 to time T (the maximum possible lifespan), given his rate of time preference ρ and a vector of cumulative survival probabilities $\{P_t\}$: $\underset{\{C_t\}}{Max}\sum_{t=0}^{T}\frac{P_t U(C_t)}{(1+\rho)^t}$. The budget constraint facing the individual depends on whether he has access to a fair annuity market. If not, the constraint is that the present value of future consumption, discounted using the riskless interest rate r, must be equal to his initial wealth, W_0: $W_0 = \sum_{t=0}^{T}\frac{C_t}{(1+r)^t}$.

If he could purchase an actuarially fair annuity, his budget constraint would become: $W_0 = \sum_{t=0}^{T}\frac{P_t C_t}{(1+r)^t}$. The difference between the two budget constraints is due to survival probabilities: without an annuity, the retiree's *actual* present value of future consumption must not exceed his initial wealth, but with an annuity, his *expected* present value of consumption cannot exceed his initial wealth. In other words, an annuity reduces the relative price of future consumption.

Evaluating how much a consumer values access to an annuity requires the researcher to assume a functional form for the utility function and to evaluate a range of risk-aversion parameters. In much of our research, we assume that the one-period utility function, $U(C_t)$, exhibits constant relative risk aversion: $U(C_t)=\frac{C_t^{1-\beta}}{1-\beta}$, where β is the Arrow-Pratt coefficient of relative risk aversion, and 1/β is the elasticity of intertemporal substitution in consumption. Using the indirect utility function V(.) corresponding to each budget constraint, we can then evaluate the maximum utility that the individual could achieve by following the optimal consumption path as a function of all parameters. Last, we can assess the consumer's utility gain in dollar terms, by computing how much additional wealth he would need if he lacked access to annuities, to make him as well-off as if he had annuities available. This is accomplished by finding α such that: $V_0(\alpha W_0)^{no\ annuities} = V_0(W_0)^{annuities}$. Here α is termed the "Annuity-equivalent Wealth " and deriving it requires dynamic programming algorithms described in

MPWB (1999) for the case of a single individual, Brown and Poterba (2000) for the case of couples, and BMP (forthcoming) for the case in which returns and/or inflation are uncertain.

3 Valuing Annuities in Practice

Moving from theory to practice, several types of data are required to value annuity products. These include information on actual annuity premiums and payouts, and on current as well as projected mortality risks. We investigate each of these in turn in this section.

3.1 Annuity Premiums and Payouts

To understand how annuity markets work, it is important to obtain data on actual payouts for life annuities and the premiums charged for these annuity promises. Information on average monthly nominal annuities for the US by sex over time appears in Table 3. Here we find, not surprisingly, that men can expect to receive higher payouts than women because of their shorter life expectancies conditional on purchase at age 65. The table also indicates that monthly payouts appear to be falling in the US market over time.

Table 3. Average Nominal Monthly Annuity Payouts on Single-Premium Immediate Annuities Offered to 65-Year-Old Men & Women: US Individual (Retail) Market 1995-1999

	Best's Survey				Annuity Shopper	
	1995	**1996**	**1997**	**1998**	**1998**	**1999**
Men	$7.94	$7.62	$7.72	$7.33	$7.32	$7.35
Women	7.17	6.86	7.00	6.62	6.59	6.67

Source: Derived from Brown, Mitchell and Poterba (forthcoming). Each entry indicates the monthly income per $1,000 based on the purchase of a $100,000 single-premium immediate life annuity policy.

Evidence from other countries is supportive of these findings. Information for the UK appears in Table 4, where we display data for different purchase ages as well as nominal and real annuities. Patterns are as expected. Benefit payouts for younger people are, logically, lower than for older people, due to the longer life expectancy for younger purchasers. Once again, payouts for women are below those for men. Payouts for a nominal annuity purchased at age 65 are similar in the UK and the US. The monthly payouts for the real benefit stream in the UK are lower than for the nominal annuity, again with women receiving less than men. Average nominal payouts for 55-year-olds in two other countries with good data, Singapore and Australia, appear in Table 5.

Table 4. Average Real and Nominal Monthly Annuity Payouts on Single-Premium Immediate Annuities Offered to Men and Women, by Age: UK Compulsory Pension Market 1998

	Age 60	Age 65	Age 70
A.Nominal Annuity Benefit:			
Men	£6.66	£7.55	£8.73
Women	6.03	6.67	7.61
B. Real Annuity Benefit:			
Men	4.76	5.63	6.80
Women	4.17	4.83	5.75

Source: Derived from Brown, Mitchell and Poterba (2000). Each entry indicates the monthly income per £1,000 premium, based on the purchase of a per £100,000 single-premium immediate life annuity policy. All annuity products analysed provide a 5-year guarantee period.

Table 5. Average Nominal Monthly Annuity Payouts on Single-Premium Immediate Annuities Offered to Men and Women at age 55: Singapore and Australia 2000

	Australia	Singapore
Men	\$8.46	\$7.98
Women	7.94	7.09

Source: Derived from Doyle, Mitchell and Piggott (2001). Assumed nominal interest rate 5 percent in Singapore and 6.96 percent in Australia, consistent with historical trend. Each entry indicates the monthly income per \$1,000 premium, based on the purchase of a per \$100,000 single-premium immediate life annuity policy. Values given in local currencies.

3.2 Mortality Processes[5]

To value an annuity payout for life, we must have data on the probability of the purchaser's survival into the future, yet it is not a simple matter to obtain projected mortality (or life) tables. This is because a mortality table is typically derived by obtaining actual data on the number of deaths by age and sex occurring in a given population over a specific period of time. The probability q_x that a group member of age x will die in the next year of life is then estimated by either fitting a hazard model to the empirical distribution of deaths in the population, or by applying a smoothing algorithm to the raw maximum-likelihood estimates of q_x. Then the smoothed estimates of q_x may be used to construct a complete mortality table. In the developed world, mortality is extremely rare at most ages, so a large number of lives must be observed in order to obtain reliable estimates.

A concern that arises in developing mortality projections for valuing annuities is that mortality processes may be heterogeneous across subgroups of the population. For example, insurers worry about *adverse selection*, which arises when annuity purchasers are more likely than average to live a long time in retirement. In this case, using mortality rates derived for the population as a whole would seriously overstate mortality among annuitants. One factor that may play a critical role in this context is the extent to which annuitization is optional or mandatory. For example, in the UK, occupational pension benefits must be partly annuitized on a mandatory basis, so dedicated annuitant-mortality tables exist in the UK for *voluntary* as well as *compulsory* annuitant populations (Finkelstein and Poterba, 1999; Murthi, Orszag and Orszag, 1999). In the US, federal Social Security payments are by law (real) life annuities, while defined-benefit company pensions are usually paid out as life annuities (though lump-sum cashouts are becoming more popular in recent years; cf Mitchell 1999). As a consequence, US mortality tables are available for group and individual (retail) annuitants, as well as for the general population (BMP 2000).

Not only do mortality tables differ across population subgroups; they also change over time. On the whole, mortality among older people has fallen steeply in developed countries over the last century, and this process seems likely to continue in the future.[6] The actuarial profession therefore projects future mortality trends by estimating so-called *period* mortality tables from past data, and then devising separate, forward-looking, *cohort* mortality tables by extrapolating trends in mortality. Of course, anticipated future declines in mortality built into cohort tables are only estimates, and could be wildly in error; nevertheless, future mortality estimates are essential in order to value annuity benefit flows for people alive today as they age into the future.

---

5 For additional, background see Mitchell and McCarthy (2001).

6 For more on changing mortality patterns over time, see Vaupel (http://www.demogr.mpg.de/).

There is no single universally-accepted method of comparing mortality tables across subgroups and over time, though it is clear that selecting the "right" mortality table has a profound effect on the value placed on a life annuity. One perspective is provided in Figure 1, illustrating the distribution of expected age at death for US males, conditional on surviving to age 65. The graph clearly illustrates that the population curve has greater mortality at younger ages, with the annuitant curve being shifted to the right. An obvious disadvantage of the graphical comparison is that it does not provide a measure for "how far apart" the two tables might be and what financial difference they would make in annuity valuation.

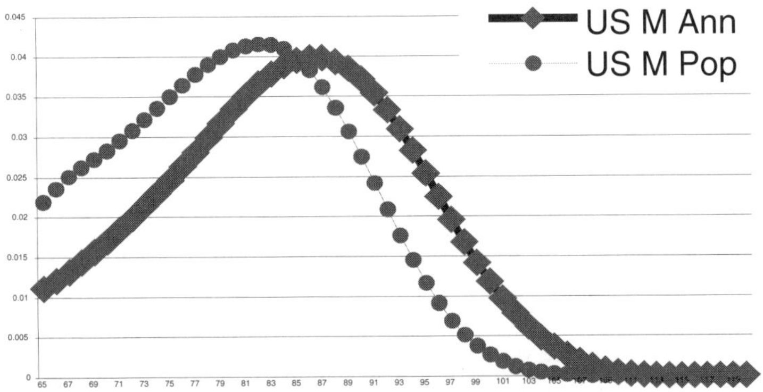

Source: Derived from Mitchell and McCarthy (2001).

Figure 1. Distribution of Age at Death: US Male Population vs Annuitant-Mortality Patterns

Other ways to compare mortality over time and across different groups include the A/E ("A over E") and the IRR measures. The first, A/E, expresses the number of deaths expected in a population with a given age structure using one table ("the benchmark") and then compares that with the expected number of deaths in a population of the same size in a second mortality table. When presented as a ratio multiplied by 100, the measure is equivalent to a ratio of the weighted average probabilities of death for the two mortality tables, using a specific population structure for the weights. In recent research (Mitchell and McCarthy, 2001) we use as the base the US Male period population table and compute the measure A/E = $\frac{\sum_x w_x q_x^*}{\sum_x w_x q_x} \times 100$ where q_x^* is the probability that an individual of age x dies according to the table in question, and q_x is the probability that an individual of age x dies according to the US Male period population table. The weights, w_x, are set so that $w_{65} = 100{,}000$, and $w_x = w_{x-1}(1 - q_{x-1})$.

Alternatively one could view the mortality process as akin to a mathematical discount rate, where if \$1 of today's money were to be divided in five years' time between survivors of a group of one million people alive today, each individual survivor's share would grow over time with mortality, just as it would with compound interest. To compare mortality tables in this perspective, we may use one mortality table to solve for the internal rate of return (IRR) required to equate the present value of a life annuity computed using a second mortality table (at some fixed interest rate). In other words, this approach solves for the r in this expression: $\sum_x \bar{a}^{r\%}_{\overline{x+1-65|}} \cdot {}_{x-65}p_{65} \cdot q_x = \sum_x \bar{a}^{5\%}_{\overline{x+1-65|}} \cdot {}_{x-65}p^*_{65} \cdot q^*_{65} = 10.18079$,

where ${}_{x-65}p_{65}$ is the probability that an individual alive at age 65 lives to at least age x according to the mortality table in question; q_x is the probability that an individual alive at age x dies before reaching age $x+1$; $\bar{a}^{r\%}_{\overline{x+1-65|}}$ is the present value at r percent per year of an annuity certain paid continuously for x-65+$^1/_2$ years; and ${}_{x-65}p^*_{65} \cdot q^*_x$ is the probability that an individual alive at age 65 dies aged x according to US Male population period mortality. This technique requires both a benchmark mortality table and an interest rate. Our analysis uses as a benchmark the US Male population period table and an interest rate of 5 percent per year, and solves for the interest rate required to equate the annuity in present value with some other mortality table (Mitchell and McCarthy, 2001).

3.3 International Comparisons

Comparisons of US and UK mortality tables are given in Table 6, with results for annuitants shown in the top panel and for the population as a whole in the bottom panel. Using the A/E metric, annuity mortality proves to be 10-15% below the base population group, with differences for men exceeding those for women. Also it is interesting to note that seemingly small mortality differences translate into relatively large implied internal rates of return (IRRs). Hence the figure of 6.55% associated with the US male *annuitant* cohort mortality table may be compared to the 5% return assumed in valuing a \$1 life annuity payable using the US male *population* period table.

In other words, the lower mortality for US male annuitants can be thought of as using a discount rate 155 basis points above the 5 percent base rate. In the UK, where mortality rates are higher for UK male compulsory annuitants, this is mathematically equivalent to discounting with an interest rate 100 basis points above the benchmark. In other words, a US insurer would have to earn approximately 100 basis points more on invested assets for men, and 59 basis points more for women, to provide the same payout as the UK compulsory annuity product.

Table 6. Comparing Mortality Patterns Across Countries: Cohort Results for the US & the UK

	A/E(%)	A/E(%)	IRR(%)	IRR(%)
	1	2	3	4
	Men	Women	Men	Women
A. Annuitants conditional on attaining age 65				
US	69.0	55.3	6.55%	7.20%
UK	82.3	55.9	5.92%	7.23%
% (US-UK)/US	(19.28)	(3.2)		(3.2)
B. Population conditional on attaining age 65				
US	93.9	62.9	5.24%	6.79%
UK	98.3	65.8	5.05%	6.74%
% (US-UK)/US	(4.43)	(4.40)	18.3	4.6

Source: Derived from Mitchell and McCarthy (2001).

Notes: Data on compulsory annuitants for the UK refer to those compelled to annuitize a certain fraction of pension benefits; in the US, data cover pensioners in retirement plans. Columns 1 and 3 use US male population period mortality as reference category, while columns 2 and 4 assume a 5% return for base annuity.

4 Valuing Annuity Flows

Having in hand the requisite data on annuitant mortality and annuity payouts, we are now in a position to provide information on the financial value – money's worth – of annuity products, along with their utility values for risk-averse consumers.

4.1 Empirical Results on Money's Worth

Table 7 summarizes findings on money's worth ratios from a range of US and UK studies on nominal annuity products that use *population* mortality tables. Panel A indicates that, in exchange for a $1 initial investment in a retail annuity product, the typical age-65 US male could anticipate receiving a pretax value of $0.835 from a life annuity, with the value slightly higher in a tax-qualified plan ($0.85). In other words, the "loading" implied by these figures amounts to about 15 cents per dollar premium. Women's longer life expectancies drive up their money's worth to around 89 cents on the dollar, so the loading implied is lower, at around 11 cents. These US figures are comparable to those for the UK given in Panels B and C, where loadings again stand at around 10-15 cents on the dollar.

Table 7. Money's Worth of Nominal Annuity Payouts, Men age 65, UK and US (Using population mortality tables)

	Money's Worth By Age:	Retail Annuity	Tax-qualified Annuity*
A. US (1998):	**Men:**		
	Age 65	0.835	0.850
	Women:		
	Age 65	0.888	0.875
B: UK (1998):	**Men:**		
	Age 65	0.861	0.901
	Women:		
	Age 65	0.851	0.902

Sources: Panel A: Poterba and Warshawsky (1999: T2). Panel B. Finklestein and Poterba (1999: 34-5).

Notes: All results use Treasury interest rates and population mortality tables. *Refers to compulsory annuity purchase from UK personal pension accounts.

These results would not be precisely relevant for a consumer considering whether to annuitize his retirement wealth voluntarily, since, as demonstrated above, people buying annuities tend to live longer than average. This difference, termed *adverse selection*, is captured in Tables 8 and 9, where we report money's worth values obtained with *annuitant* as well as *population* tables. In the US case shown in Table 8, the male annuity purchaser could expect a benefit flow some 7-13 percentage points higher than would someone in the population at large. In other words, the insurance load drops from 10-15 cents on the dollar to 7-9 cents on the dollar after correction for the differential survival rate of annuitants.

Results for men and women are quite similar. Evidence for the UK, Australia, Canada, and Switzerland appears in Table 9, where once again the loading implied with population mortality is much attenuated by using annuitant mortality tables. Indeed these annuity products appear quite valuable for annuitant purchasers, with estimated values close to and occasionally exceeding 1.0 in a few cases. In general, therefore, a financial assessment of these life-annuity products indicates that (a) adverse selection explains a substantial portion of the load perceived by potential purchasers in the general population, and (b) these products are rather valuable to annuitants, though in most cases their money's worth is still less than 1.0.

Table 8. Money's Worth of US Nominal Annuity Payouts Using Population and Annuitant Mortality Tables, After-Tax Calculations (US 1995)

	Mortality Table Employed	
	Population	Annuitant
A. Men:		
Age 55	0.852	0.934
Age 65	0.814	0.927
Age 75	0.783	0.913
B. Women:		
Age 55	0.880	0.937
Age 65	0.854	0.927
Age 75	0.846	0.919

Source: Mitchell, Poterba, Warshawsky and Brown (1999).

Note: Computations use the Treasury yield curve and 15% marginal tax bracket.

Table 9. Comparative Evidence on Money's Worth of Nominal Annuity Payouts: Single-Premium Nominal Life Annuities Offered to 65-Year-Olds

	UK		*Australia*		*Canada*		*Switzerland*	
Mort.Table	**Pop**	**Ann**	**Pop**	**Ann**	**Pop**	**Ann**	**Pop**	**Ann**
Men	0.897	0.966	0.914	0.986	0.925	1.014	0.965	NA
Women	0.910	0.957	0.914	0.970	0.937	1.015	1.115	NA

Source: James and Vittas (1999). Note: Computations use the Treasury-yield curve.

Thus far, the discussion has focused on nominal annuities, since they tend to be the dominant form of annuities in much of the developed world. Yet real annuity products are on offer in some countries, and it is useful to consider how to value them, since retirees presumably seek to preserve not only nominal but real consumption levels over the retirement period. Inflation is a concern in old age to the extent that even small inflation rates undermine real consumption paths over time. Thus a 3.2% annual rate (the US average from 1926-97) could halve the real value of a nominal annuity over 22 years. To some extent, annuity sellers can offset this concern by means of graduated products that grow over time at some fixed rate. On the other hand, inflation is a concern in retirement because of uncertainty in its realization. Hence it is important to allow for persistence in inflation over time, when assessing the money's worth of a real annuity product. Of course, for risk-averse consumers, the impact of inflation uncertainty would be expected to make real annuities even more attractive.

The results in Table 10 review results for real annuities on offer in the UK and the US, where we see that the money's worth of a real annuity is below that for a nominal annuity. In the US there is only a single firm offering a real annuity, while in the UK many more insurers offer real products. The fact that the UK discount associated with the real annuity is smaller, on the order of 5 percentage points, suggest that a more active market in the products narrows price differentials.

Table 10. Expected Discounted Values for Nominal and Real Annuities, UK and US

Money's Worth / By Age:		**Nominal Annuity**	**Real Annuity**
A. Compulsory UK Market (1998)			
Men:			
	60	0.921	0.867
	65	0.908	0.854
	70	0.889	0.836
Women:			
	60	0.928	0.876
	65	0.907	0.857
	70	0.886	0.836
B. US Retail Market (1998)			
Men:			
	65	0.864	0.702
	75	0.830	0.720
Women:			
	65	0.889	0.708
	75	0.887	0.762

Source: Brown, Mitchell and Poterba (forthcoming). Computations use population mortality table and pretax benefits. Data from UK represent 14 companies; data from US from one company.

The fact that there is any discount in the money's worth of real versus nominal annuities probably indicates that, at the margin, non-purchasers are unwilling to pay extra to ensure a flat real consumption stream during retirement.[7] This finding may challenge a widely-held view that pension participants should be required or encouraged to elect real life annuities in lieu of programmed withdrawals.

Another way to think about annuity values assesses money's worth ratios for particular population subgroups having mortality experiences that differ from the average. Thus in the US, for instance, blacks tend to have higher mortality probabilities than do whites, and hispanics are more likely to live longer than average. Brown (1999) takes into account differential ethnic mortality patterns, and his results are summarized in Table 11. He finds that US white and hispanic women tend to be subsidized on average when they purchase annuities, while US black males receive less. In a related study, Finkelstein and Poterba (2000) find that annuity purchasers who buy graduated products tend to live longer than those who buy flat nominal products.

Table 11. Money's Worth for Nominal and Real Annuities by Ethnicity (US 1998)

Money's Worth By Ethnicity:		Nominal Annuity	Real Annuity
Men:			
	White	0.947	0.930
	Black	0.873	0.847
	Hispanic	1.027	1.024
Women:			
	White	1.059	1.077
	Black	0.994	1.003
	Hispanic	1.105	1.136

Source: Brown (2000). Computations use age/sex/race population mortality tables derived by the author and the Treasury rate.

What these findings imply is that differential mortality expectations influence how people value these life-income products, and these different expectations, in turn, shape what types of annuity products people buy when given a choice. Despite this, or sometimes because of this, government policy often requires that a particular mortality table be used for purposes of pricing life payouts, to limit

7 Of course it might also indicate that non-purchasers are constrained in some way from entering the market; this is not directly testable, given our data.

perceived adverse selection and/or discrimination against members of particular groups. This is the case in the UK, for example, where *unisex* mortality tables (using a blend of male and female rates) must be used for pricing pension annuities,[8] and in the US, where annuity taxation requires that unisex tables be used to assess the taxable portion of the benefit (BMPW, 1999). As a result, government policy toward mortality tables in annuity pricing tends to generate a degree of cross-subsidization, if the product purchase is mandatory, and exacerbates adverse selection, if the product purchase is voluntary.

3.2 Equivalent Wealth Computations

Above we argued that risk-averse consumers may value annuities more highly than the simple financial money's worth approach implies, since people would be expected to derive utility from insurance protection *per se*. In this section we report results on the **annuity-equivalent-wealth** values, which represent how much wealth a consumer would need, *if he did not have access to an annuity market,* to achieve the same lifetime expected utility level that he could achieve if he used that wealth to purchase a nominal annuity.

In order to evaluate this measure, we must assume a particular form for the consumer's expected utility function and compare the utility value of full annuitization with no annuitization. We conduct these evaluations using a range of risk-aversion parameters, since there is no clear guidance from the literature as to just how risk-averse consumers may be. We also acknowledge that the value of a new life annuity depends on the existence of other types of life-income streams. This is relevant where an older consumer could anticipate receiving lifelong retirement benefits from a company and/or governmental pension scheme, so the marginal utility of a new life annuity would be expected to be less than for someone entirely lacking in access to any retirement annuity.[9]

---

8 Strictly speaking, unisex tables must be used in pricing only the "protected rights" component of the pension system; see Blake (1999).

9 This calculation is carried out by first finding the maximum utility V^* the consumer could attain when he can fully annuitize his wealth, and then re-computing the dynamic programming problem solving for utility V when annuities are not available. Then one can evaluate the additional wealth (i.e. ΔW) that would have to be given to the consumer to leave him at an equivalent level of utility: $V(W^* + \Delta W | \text{no annuities}) = V^*$, and the annuity-equivalent-wealth would then be $\alpha = ((W^* + \Delta W)/W^*)$. For more discussion see Brown (2001). The calculations in Table 12 therefore represent not the marginal value placed on the first dollar of an annuity, but the full utility value of annuitization (and hence would be expected to understate the marginal value of the first dollar of annuitization).

The results of our analysis appear in Table 12. What the table shows is that annuity-equivalent-wealth values prove to be quite substantial. Going down the first column, for instance, a consumer with a logarithmic utility function (CRRA = 1) and no other pre-existing annuity would assess the annuity-equivalent wealth of a real life annuity at 1.502. What this means is that this consumer would be indifferent between having $1 in a real lifetime annuity, and $1.50 in non-annuitized wealth. At higher levels of risk aversion, the annuity-equivalent-wealth value rises to 2, meaning that access to a real annuity is worth double the wealth that value invested in a nominal annuity. In other words, annuities are valued more highly than in strictly money's worth terms.

Table 12. Annuity-Equivalent-Wealth for Real and Nominal Annuities

Coefficient of Relative Risk Aversion (CRRA)	Consumer with No Pre-Existing Annuity Wealth			Consumer with Half of Initial Wealth in Pre-Existing Real Annuity		
	Real Annuity	Nominal Annuity: i.i.d. inflation	Nominal Annuity: Persistent Inflation	Real Annuity	Nominal Annuity: i.i.d. inflation	Nominal Annuity: Persistent inflation
1	1.502	1.451	1.424	1.330	1.304	1.286
2	1.650	1.553	1.501	1.441	1.403	1.366
5	1.855	1.616	1.487	1.623	1.515	1.450
10	2.004	1.592	1.346	1.815	1.577	1.451

Source: Derived from Brown, Mitchell and Poterba (forthcoming).

Note: The annuity-equivalent wealth calculation for the nominal annuity assumes inflation takes one of six possible values, roughly capturing the distribution of inflation outcomes over the 1926-1997 period. Inflation shocks are assumed independent across periods in the i.i.d. case and follow a stylized AR(1) process in the persistent inflation case.

Our results also indicate that a real annuity is worth more than a nominal annuity to the risk-averse consumer. Thus with CRRA=1, the consumer would value the real annuity at 5-8 cents more on the dollar as compared with a nominal annuity, whereas with CRRA=10, the gap is 40-60 cents. The effect differs with the inflation process, however, since the annuity-equivalent wealth for the consumer is 1.451 in the case of i.i.d. inflation, and 1.424 in the case of persistent inflation. Not surprisingly, the real annuity is more highly valued when the inflation process is more persistent.

It is interesting that the monotonic relationship between annuity-equivalent wealth and risk aversion does not hold for nominal annuities in the presence of uncertain inflation. This is because risk aversion works in opposing directions in the face of inflation uncertainty. On the one hand, higher risk aversion makes the consumer value an annuitized payout more highly, since he avoids the risk of outliving his resources, and this is all that matters in the case of the real annuity. On the other hand, a risk-averse consumer dislikes the uncertainty introduced into the nominal annuity stream by stochastic inflation. Increased variability in the real value of the nominal annuity payouts reduces utility, and this effect is larger for those with the highest degree of risk aversion. At low levels of risk aversion, the first effect dominates, and the annuity-equivalent wealth for a nominal annuity rises with risk aversion. For example, moving from CRRA=1 to CRRA=2, the annuity-equivalent wealth increases from 1.451 to 1.553 in the i.i.d. inflation case, and from 1.424 to 1.501 in the persistent inflation case. However, as risk aversion rises, the second effect becomes stronger, and the annuity-equivalent wealth begins to decrease with risk aversion (BMP, forthcoming).

Our results are also informative regarding how consumers might value annuities if some substantial portion of their retirement assets were already annuitized. In the extension of our model, we posit that one-half of retiree wealth is annuitized. This seems reasonable in view of the finding that the median US older household holds about this fraction of its retirement resources in the form of expected future Social Security benefits, and these are payable only as a real life annuity (Moore and Mitchell, 2000). In this case, a consumer who is already half annuitized still values access to the additional annuity, but the value is lower than before. For instance, a relatively non-risk-averse half-annuitized consumer (CRRA=1) is indifferent between \$1 in a real annuity and \$1.33 in non-annuitized wealth. In other words, he would require only a 33% increment to his wealth to be made as well-off as if he had a real annuity, compared to 50% in the case where he lacked an annuity altogether. Clearly having a pre-existing real annuity affords the consumer insurance against very low consumption values and drives down the value of additional privately-purchased annuities.

If the annuity on offer were *nominal* instead of real, and inflation followed an i.i.d. process, the results prove to be similar. More striking is the result with persistent inflationary process, since, at high levels of risk aversion, having a real pre-existing annuity provides insurance against very low consumption. Here the utility cost of having persistent high inflation eat into the nominal annuity is lower, so that the willingness to buy the nominal annuity would be greater.

Because many investors seem to believe that equity-linked variable annuities are superior to traditional nominal annuities, we have also compared annuity-equivalent-wealth measures for these products. In the US the variable annuity has been around since the 1950's, initiated by the TIAA-CREF and VALIC companies (Brown, 2000). In the UK, "unit-linked" or "with-profit" annuities appear to be more recent entrants on the insurance scene (Blake, 1999).

In terms of the retiree's perception, a higher expected mean return on a variable annuity product might make it more appealing, though the greater risk could provide an offset. To assess this trade-off, we must model the mean return associated with portfolios of risky securities and the variability of returns around this mean. Accordingly, we have developed an indexed portfolio of common stocks using two alternative assumptions on the mean real return on equities: 6% real (i.e., a 3 percent premium over the indexed bond return), and 9% real. (Both assume an equity premium smaller than the U.S. historical 1926-1997 average, to be on the conservative side.) We also assume the standard deviation of real returns on equities equals its historical average value of 20.9 percent per year.[10] We compare these with fixed real retirement annuities in Table 13.

Table 13. Annuity-Equivalent Wealth for Equity-Linked Variable Annuity Products

Coefficient of Relative Risk Aversion	No Pre-Existing Annuities		Pre-Existing Annuity Equal to Half of Initial Wealth	
	Real Stock Return 6 %	Real Stock Return 9 %	Real Stock Return 6 %	Real Stock Return 9 %
	Annuity-equivalent Wealth			
1	1.623	2.024	1.567	1.953
2	1.499	1.901	1.570	1.957
5	0.921	1.355	1.443	1.789
10	0.331	0.622	1.261	1.563

Source: Derived from Brown, Mitchell, & Poterba (forthcoming).
Notes: See Table 12.

We give results with and without pre-existing annuitization. When real returns on equities average 6%, the consumer having logarithmic utility would value equity-linked variable payout annuities more highly than real annuities. Without a pre-existing annuity, his annuity-equivalent wealth from accessing a variable annuity is $1.62, larger than that for the real annuity previously described ($1.50). On the other hand, more risk-averse people who might have placed all their wealth in a variable annuity would be worse off than not annuitizing at all. This can be seen by annuity-equivalent wealth values below unity.

These findings imply that one segment of the older population might favor a variable annuity, linking payouts to equity returns in retirement, rather than a real annuity. The higher anticipated returns from the variable annuity would

---

10 For further elaboration of the estimating procedure, see BMP (2000) and BMPW (2001). Real equity returns are assumed independent over time.

compensate these consumers for (some of) the inflation risk borne. But more risk-averse retirees would still be expected to prefer a fixed life annuity over a variable product (and a real product over a nominal one).

5 Can Annuity-Market Performance Be Enhanced?

Our analysis thus far indicates that annuities deserve an important role in many retiree portfolios. This is particularly evident if government-based social security benefits cannot be relied on to support a major portion of old-age consumption. Nevertheless, people might value annuities less than our models imply for several reasons.

One explanation for under-purchasing of annuities is that they may be seen as "too expensive." As we have shown, however, adverse selection is rather small in the insurance markets examined. It is true that insurers levy charges to cover administrative costs and insurer reserves, contributing to the perception that these products are costly. Yet our analysis indicates that, in the US and UK at least, these charges are relatively low and appear to be falling over time. As a result, the types of loads currently charged by insurers would not be expected to offset the substantial utility values associated with insurance against longevity risk. Yet as new markets for annuities develop in Europe and Asia, it will be important to pay close attention to loads and adverse selection to ensure that these do not discourage market growth unnecessarily.

Public policy can also play a key role in structuring the way in which these markets develop. In the UK and Sweden, for instance, *unisex* mortality tables must be used to compute pension payouts (for those benefits provided under the "protected rights" legislation). In this instance, expected mortality differences between men and women may not be used to price retiree payouts differentially. While this produces cross-subsidization between male and female retirees on average, it also restricts the amount of adverse selection that occurs at the annuitization stage. In the US, while unisex tables are required for employer-pension calculations, they are not imposed on individual annuitants purchasing their life payouts on the retail market. Furthermore, many pension-covered workers are permitted to take a lump-sum cashout from the employer pension at retirement. As a result, an informed and rational consumer anticipating a shorter life expectancy (e.g. a male retiree) has an economic incentive to cash out his employer pension rather than take it as a life-annuity stream, in order to avoid remaining in the (increasingly female) group mortality pool. In this sense pension and insurance regulations can powerfully affect the appeal of life annuities versus other retirement assets (Mitchell, 2000b).

Another factor shaping the demand for annuities is retiree interest in maintaining financial liquidity in old age. Having relatively large amounts of cash on hand is deemed especially important in countries such as the US, where nursing-home benefits are not provided by the government as a matter of course. With over one-quarter of the 65+ population needing long-term care at some point, and nursing-home fees exceeding $50,000 per year, it is not surprising that older Americans would want to keep a portion of their assets liquid instead of annuitizing all their wealth. In response, some insurers have begun to devise new financial products that will integrate a life annuity with nursing-home insurance for the elderly (Warshawsky, Spillman and Murtaugh, forthcoming). The demand for cash in old age is also generating some interest in *reverse annuity mortgages*, financial contracts that allow an older person to obtain an annuity by pledging the net equity in his home (Caplin, forthcoming). Of course, such products have been a "hard sell" since the owner-occupied home is often used as a tax-preferred vehicle for bequests.[11]

Above, we argued that sensible regulation and supervision can strengthen the development of annuity markets. Conversely, institutional and regulatory problems have sometimes impeded their growth (Bodie, Hammond and Mitchell, forthcoming). For example, in the developing-country arena, insurers are frequently poorly supervised and regulated, so retirees may not view them as reliable institutions for old-age assets (Mitchell, 2000a). In the developed world, regulation varies across countries (e.g. within Europe) and by state (in the US) regarding reserve and reporting requirements, investment restrictions, guaranty-fund-programmes, and tax policy. This makes it difficult for investors and annuity consumers to evaluate insurance competitiveness across national boundaries (cf Lemaire, 1997).

Another area where policymakers can fundamentally influence annuity-market functioning is in the intersection between public and private old-age programmes. For example, Doyle, Mitchell and Piggott (2001) devise annuity loadings for Australia and Singapore and find adverse selection far stronger in Australia. This difference is attributed, in part, to the existence of a means-tested and quite generous old-age safety net in Australia, while no minimum consumption is guaranteed in Singapore. In other words, the extent to which people seek to annuitize retirement accumulations from defined-contribution plans will be influenced by guarantees offered by the public sector.

---

11 Among economists, there is little agreement over how to model bequests (Gale and Slemrod, forthcoming), so it is not clear precisely how the bequest motive should be incorporated into a model of the demand for annuities. Brown (2001) argues that most specifications imply that full annuitization is no longer optimal. To the extent that some people may be over-annuitized (e.g. through Social Security), the purchase of life insurance may be beneficial. Further mandatory annuitization acts like a tax on such persons.

When enumerating possible obstacles to annuity markets, the discussion would be incomplete without recognizing that older people might be labouring under some misinformation regarding the relative advantage of annuity investments versus other financial products. For one thing, people might underestimate their longevity prospects, which would lead them to undervalue protection against longevity. In fact, however, survey evidence suggests that older people are quite well-informed about their survival probabilities, and their expected survival patterns track actuarial tables quite closely (Hurd and McGarry, 1995). Of course, there remains the problem of cohort mortality risk: that is, if an entire cohort lives longer than anticipated by those selling the products, insuring firms would experience excess costs. Some have argued that this problem can best be resolved by having governments issue "survivor bonds", inasmuch as private insurers have no financial mechanism adequate to diversify this type of risk (Blake, Burrows, and Orszag, forthcoming). A similar argument was made by financial experts and economists seeking to encourage the US government to issue inflation-linked bonds (Brynjolfosson and Fabozzi, 1999). Retiree demand for these products in the US has been modest but, of course, inflation has been very low of late. In other pension markets, including the UK and Israel, inflation-linked bonds have become a mandatory component of the retiree portfolio and demand for these assets has been more substantial.

A final factor driving the size and development of annuity markets is how people feel about alternative investments. Retirees may believe that they can manage their money better than do institutional investors, a perspective that may have contributed to what the British have called pension "misselling." Here participants in company-based defined-benefit plans switched their DB accruals into individual-account-defined-contribution plans at the behest of insurers and other financial advisers, and in some cases this switch proved financially deleterious (Ward, 1996). At issue and still not unsettled in this arena is how well those who switched plans understood what the switch meant, including their foregone defined-benefit plan accrual and differences in how longevity risk was handled in the two environments.

Due to adverse selection in annuity markets, as well as participant inability to make informed decisions on retirement decumulation, some policymakers have argued that all pension participants should be required to annuitize at least a portion of their assets at retirement.[12] For this reason it is of interest to examine the case of Chile, which more than two decades ago closed down its traditional national pay-as-you-go defined-benefit programme in favour of a mandatory defined-contribution individual-accounts system. As such, Chile is providing an

---

12 For example, several members of the 1994-6 Social Security Advisory Council in the US urged mandating annuities if a portion of the national Social Security system were to be privatized; see http://www.ssa.gov.

interesting test-case for those seeking to design better old-age retirement programmes. Of particular interest in the present context is the way in which the retirement asset *decumulation* process is working, now that the Chilean system has been in place long enough to have people reaching retirement age with a reasonably long working career under the new programme. Thus far, early retirees and disabled persons constitute the bulk of the retired population: only 38% of programme affiliates in payment status left work via regular retirement, while 45% were early retirees and 17% were on disability (Diaz and Edwards, 1999). Of most interest, however, are the payout patterns. With over a decade of data available, it appears that a minority - 40% - of Chilean beneficiaries have elected a life annuity, while 60% opted for the "programmed-withdrawal" or graduated-payment approach that does not guarantee benefits for life (see Figure 2). While the fraction of recipients electing life annuities has grown over time, these are mostly early rather than normal retirees. Whether, and how, annuity-pricing policy under the Chilean system and other national, mandatory, defined-contribution pension programmes influence these outcomes is a subject deserving of further study.

Figure 2. Distribution of Payouts by Type: Chilean Pension System, 1988-1997

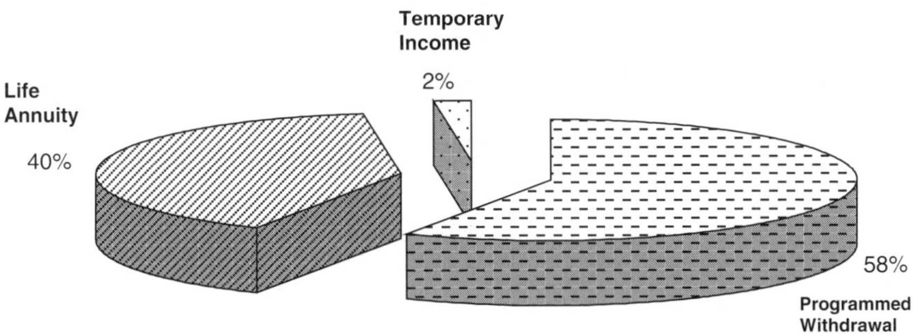

Source: Derived from Diaz and Edwards (1999).

6 Final Thoughts

This overview of annuity-market functioning suggests several lessons. One is that annuity products have an important role to play in helping people manage their assets into retirement. Risk-averse individuals will value annuities at well above their purchase price. We also find that that adverse selection is not large and insurance loadings have been falling over time, in countries where we have had adequate data to measure these. Last, there is a wide range of annuity products that can enhance retirement security. Here, we emphasize the role of inflation-indexed annuities and related products protecting real consumption streams into very old age.

Without a doubt, many developed and developing nations will face the need to restructure their retirement systems in the next two decades: Germany has already made some progress in this vein, Greece is confronting political pressures around this topic, and the US has announced a new effort to examine individual accounts. During this reform process, it will be critical to devote careful attention to the role of annuity markets in this new environment. Protecting economic security during a long retirement period requires more attention to the long-term consequences of regulatory, financial, and social-insurance programmes than has heretofore been noted by most reformers. Additional study would be helpful on the risks affecting retiree consumption and how people can make better-informed decisions about decumulating assets in retirement. New financial products to help retirees draw down their lifetime savings more sensibly would be helpful, but with them arise new challenges for financial system and insurance regulators.

References

Blake, D. 1999. Annuity Markets: Problems and Solutions. *The Geneva Papers on Risk and Insurance*. July: pp. 359-375.

Blake, D., W. Burrows and J. M. Orszag. Forthcoming. Survivor Bonds and Compulsory Annuitization: Helping Reduce the Cost of Pension Provision. In Z. Bodie, B. Hammond and O., S. Mitchell, eds. *Innovations in Retirement Financing*. Pension Research Council. Philadelphia, PA: University of Pennsylvania Press.

Bodie, Z., B. Hammond and O. S. Mitchell, Ed.. Forthcoming. *Innovations in Retirement Financing*. Pension Research Council. Philadelphia, PA: University of Pennsylvania Press.

Brown, J. 2000. *How Should We Insure Longevity Risk in Pensions and Social Security?* Issue in Brief, Center for Retirement Research, Boston College, Boston, MA.

Brown, J. 2001. *Redistribution and Insurance: Mandatory Annuitization with Mortality Heterogeneity*, JFK School, Harvard, Cambridge, MA.

Brown, J., O. S. Mitchell and J. Poterba. 2000 [BMP2000]. The Role of Real Annuities and Indexed Bonds. In an Individual Accounts Retirement Program. In *Risk Aspects of Investment-Based Social Security Reform*. Ed. J. Y. Campbell and M. Feldstein. NBER: University of Chicago Press: pp. 321-360.

Brown, J., O. S. Mitchell and J. Poterba. Forthcoming [BMP forthcoming]. Mortality Risk, Inflation Risk, and Annuity Products. In Z. Bodie, B. Hammond and O. S. Mitchell, eds. *Innovations in Retirement Financing*. Pension Research Council. Philadelphia, PA: University of Pennsylvania Press.

Brown, J., O. S. Mitchell, J. Poterba and M. Warshawsky. 2001 [BMPW 2001]. *The Role of Annuity Markets in Financing Retirement*. MIT Press.

Brown, J., O. S. Mitchell, J. Poterba and M. Warshawsky. 1999. Taxing Retirement Retirement Income: Nonqualified Annuities and Distributions from Qualified Accounts. *National Tax Journal*. LII,3, September: pp. 563-592.

Brown, J. R. and J. M. Poterba. 1999. *Joint Life Annuities and Annuity Demand by Married Couples*. NBER Working Paper No. 7199.

Brown, J. R. and M. J. Warshawsky. 2000. *Longevity-Insured Retirement Distributions from Pension Plans: Market and Regulatory Issues*. Working Paper prepared for the Brookings/SIEPR/TIAA-CREF Institute Conference on Public Policies and Private Pensions. Washington, D.C: September, pp. 21-22.

Brynjolfsson, J. B. and F. J. Fabozzi, eds. 1999. *The Handbook of Inflation-Indexed Bonds*. F.J. Fabozzi Associates. New Hope, PA.

Caplin, A. Forthcoming. Turning Assets into Cash Problems and Prospects in the Reverse Mortgage Market. In Z. Bodie, B. Hammond and O. S. Mitchell, eds. *Innovations in Retirement Financing*. Pension Research Council. Philadelphia, PA: University of Pennsylvania Press.

Diaz, A. and S. Edwards. 1999. *Benefits of the Social Security System: Critical Examination and Proposals*. Paper presented at the Pensions Conference, Santiago, Chile, August.

Disney, R. 2000. Crises in Public Pension Programmes in OECD: What are the Reform Options? *Economic Journal* 110, 461. pp. F1-F41.

Disney, R. and P. Johnson. 2000. *Pension Systems and Retirement Incomes Across OECD Countries*. Institute for Fiscal Studies Working Paper. London.

Doyle, S., O. S. Mitchell and J. Piggott. 2001. *Annuity Values in Defined-contribution Retirement Systems: The Case of Singapore and Australia*. NBER Working Paper No. W8091.

European Commission. 1997. *Supplementary Pensions in the Single Market: A Green Paper*. http://europa.eu.int/comm/internal_market/eu/finances/pensions/gppensen.pdf

Feldstein, M. and E. Ranguelova. 2000. *Accumulated Pension Collars: A Market Approach to Reducing the Risk of Investment-Based Social Security Reform*. NBER Working Paper No. W7861.

Finkelstein, A. and J. Poterba. 2000. *Adverse Selection in Insurance Markets: Policyholder Evidence from the U.K. Annuity Market.* NBER Working Paper No. W8045.

Gale, W. G. and J. B. Slemrod. Forthcoming. Rethinking the Estate and Gift Tax: Overview. In *Rethinking Estate and Gift Taxation*. Eds. W. Gale, J. Hines and J. Slemrod. Washington, D.C: Brookings.

Hurd, M. D. and K. McGarry. 1995. Evaluation of Subjective Probability Distributions in the HRS. *Journal of Human Resources*. 30(0), pp. 268-92.

James, E. and D. Vittas. 1999. *Annuities Markets in Comparative Perspective.* Presented at the World Bank Conference on New Ideas About Old Age Security, Washington, D.C.

Jennings, R. M. and A. P. Trout. 1982. *The Tontine: From the Reign of Louis XIV to the French Revolutionary Era.* Huebner Foundation Monograph Number 12.Wharton School, University of Pennsylvania. Philadelphia, PA.

Lemaire, J. with the assistance of K. Subramanian. 1997. *Insurance Regulation in Europe and the United States*. Huebner Foundation Monograph Number 16. Wharton School, University of Pennsylvania. Philadelphia, PA.

Mitchell, O. S. 2000a. Building an Environment for Pension Reform in Developing Countries. In *Foundations of Pension Finance*. Eds. Z. Bodie and E P. Davis. London: E. Elgar. pp. 480-503.

Mitchell, O. S. 2000b. Developments in Pensions. *Handbook of Insurance*. Ed. G. Dionne. Kluwer Academic Publishers, Boston. pp. 873-899.

Mitchell, O. S. 1999. *New Trends in Pension Benefit and Retirement Provisions.* NBER Working Paper No. W7381.

Mitchell, O. S. and F. Barreto. 1997. After Chile, What? Second-Round Social Security Reforms in Latin America. *Revista de Analisis Economico,* 12(2) November 1997, pp. 3-36.

Mitchell, O. S., B. Hammond and A. Rappaport. 2000. *Forecasting Retirement Needs and Retirement Wealth.* Pension Research Council. Philadelphia, PA: University of Pennsylvania Press.

Mitchell, O. S. and D. McCarthy. 2001. *Estimating International Adverse Selection in Annuities.* Pension Research Council Working Paper. Wharton School. Philadelphia, PA: University of Pennsylvania.

Mitchell, O. S., J. Poterba, M. Warshawsky, J. Brown [MPWB]. 1999. New Evidence on the Money's Worth of Individual Annuities. *American Economic Review*. December, pp. 1299-1318.

Moore, J. and O. S. Mitchell. 2000. Projected Retirement Wealth and Saving Adequacy. In *Forecasting Retirement Needs and Retirement Wealth.* Eds. O.S. Mitchell, B. Hammond and A. Rappaport. Pension Research Council. Philadelphia, PA: University of Pennsylvania Press. pp. 68-94.

Murthi, M., M. Orszag and P. Orszag. *The Value for Money of Annuities in the UK.* Economics Department Working Paper, Birkbeck College. August 1999.

Poterba, J. M. and M. J. Warshawsky. 1999. *The Costs of Annuitizing Retirement Payouts from Individual Accounts.* NBER Working Paper No. W6918.

Ward, P. 1996. *The Great British Pensions Robbery*. Waterfall Books, 2nd ed. Lancashire, UK.

Warshawsky, M., J. B. Spillman and C. Murtaugh. Forthcoming. Integrating the Life Annuity and Long-term Care Insurance: Theory, Evidence, Practice, and Policy In: Z. Bodie, B. Hammond and O. S. Mitchell, eds. *Innovations in Retirement Financing*. Pension Research Council. Philadelphia, PA: University of Pennsylvania Press.

Discussion

David Blake

1 Introduction

Olivia's paper provides a timely reminder of the importance of the decumulation phase of retirement provision. More and more countries are adopting the defined-contribution (DC) model of funded pension provision. But virtually the entire public debate on DC-pensions has been confined to the accumulation phase.

With the exception of the UK, where a vigorous debate has taken place since the mid-1990s, any discussion of the decumulation phase has been almost entirely lacking. The reason for this is very clear to me: the UK along with Holland and Ireland are the only countries in the world with DC plans that are actually **pension** plans. Every other country has **asset-accumulation** plans, but not **pension** plans. By this I mean that such plans do not provide, as an essential feature of their design, retirement-income security for the remaining life of the plan member. This is because there is no mandatory requirement for members of these plans to purchase a life annuity at some point during their retirement, either at the point of retirement or by a certain age. Retirees can take the entire accumulated fund as a cash lump sum or they can draw down an income from the fund without any restrictions on minimum or maximum drawdown. In the UK, DC-pension-plan members are obliged to convert the accumulated fund into a life annuity by the age of 75 at the latest.

The life annuity is the **only** financial contract in existence that provides retirement-income security for however long the plan member lives. When I went to Olivia's annual conference at the Pension Research Council a year ago, I discovered that the UK has the world's largest pension-annuity market. The size of the UK pension-annuity market is £6bn ($8.5bn) p.a. In contrast, the size of the US pension-annuity market is just $2bn p.a., less than one-quarter the size of the UK market and the US is the world's largest market for DC 'pension' plans: more than half of US workers and nearly half of US pension assets are in DC plans. But these plans are in reality tax-sheltered asset-accumulation plans, not pension plans, because there is no requirement to purchase a life annuity. As long as workers maintain the plans until a certain age (59.5 years), they can take out the entire lump sum in cash.

This obviously leads to a moral hazard problem, since there is nothing to prevent members spending this money 'too quickly' and then falling back on the state (this is known as 'double dipping'). On the other hand, there is nothing to prevent, wealthier plan members spending the money 'too slowly' (since they can live comfortably on other resources) and leaving the bulk of their 'pension pot' as a tax-subsidised bequest to their children. Yet neither of these actions is compatible with the original purpose of a pension plan, namely the provision of retirement-income security.

The US is not the only country in this position. The last time I saw Olivia was a month ago in Tokyo where we were discussing the new Japanese pension laws with economists from Australia. In Japan and Australia, as well as in the US, there is no requirement to purchase an annuity with the accumulated fund. Today we are meeting in Germany, and again there is no requirement to purchase an annuity when a plan member retires.

I think the reason for this is clear. Such countries have relatively high social security payouts, and so there is an effective safety net in place if plan members do spend their money too quickly. On the other hand, the number of wealthy individuals leaving bequests from tax-subsidised assets is currently sufficiently small that this has not generated any real political outcry. However, I cannot see this lasting. The whole purpose of introducing private funded pension plans is to deal with the demographic ageing problem, and once the baby boomers start to retire after 2010, we could begin to see huge sums of pension assets being subject to inter-generational transfers, rather than being used for their primary purpose, namely intra-generational transfers from those who die early after retirement to those who live a long time after retirement.

2 The UK Experience

The debate concerning the role of annuities in pension provision has been going on in the UK since the mid-1990s, prompted by the fall in nominal interest rates during the early 1990s. Annuities in a low-nominal-interest-rate environment were perceived to be poor value for money. DC-plan members retiring at the end of the 1990s complained about having to convert their lump sums into annuities that were providing less than half of what annuitants retiring in 1990 had received. This is because the income from the annuity is related to interest rates on the retirement date. Plan members find themselves suddenly switching from an equity exposure to an interest-rate exposure.

Notwithstanding the money-illusion aspect of this argument, the UK government made a major concession in 1995, when it permitted DC-pension-plan members to delay the purchase of an annuity from retirement age to a maximum age of 75 (1995 Finance Act). Instead, plan members could implement an 'income drawdown' strategy, whereby they could keep the fund fully invested in higher-returning assets (mainly equities) and draw an income from the fund of between 35% and 100% of the corresponding annuity up to a maximum age of 75, at which time a life annuity has to be purchased with the residual assets.

The introduction of drawdown prompted a search for new investment-linked retirement-income programmes (ILRIPs) to replace annuities. The search was started by the Retirement-Income Working Party in the UK (see Blake and Hudson, 2000). The motivation for the search rested on the following facts: bond yields were now very low, retirees were facing the risk of low interest rates at retirement, equities historically generated higher returns, and retirees were living longer. The question that needed to be answered was: would it be better to maintain some equity exposure during retirement?

3 An Analysis of Investment-linked Retirement-Income Programmes

A detailed analysis of key ILRIPS was conducted by Blake, Cairns and Dowd (2000). This study compared various ILRIPs with an annuity by using stochastic simulations of discounted utilities. We then compared the simulated utility distributions, while allowing for two possibilities at the time of death: leave a bequest on death or surrender the bequest to the insurer in exchange for an annual mortality bonus payable while the plan member remains alive; the mortality bonus is a cross-subsidy to those who live a long time from those who die early in retirement. In each case individuals were assumed to retire at 65.

The various programmes considered were:

- Annuity (the benchmark programme)
- Fixed-income drawdown, with annuity at 75
- Flexible-income drawdown, with annuity at 75
- Flexible-income drawdown, with deferred annuity payable from 75
- Unit-linked drawdown, with annuity at 75
- Collared-income drawdown, with annuity at 75
- Floored-income drawdown, with annuity at 75
- All except the first programme come in bequest or mortality-bonus variants.

The annuity programme:

- Converts the fund into an annuity on retirement
- Is the only programme that protects the policy-holder against outliving his resources
- Has risks for the policy-holder:
  - Low interest rates on retirement
  - Inflation risk (if a level annuity is chosen)
- Has risks for the insurer:
  - Reinvestment risk (if the insurer is unable to invest the lump sum in suitable long-maturing, income-generating assets on the retirement date)
  - Mortality-improvement risk (arising because annuitants live longer than anticipated).

The fixed-income programme (FXD):

- Converts the retirement fund to a managed fund (100% equity weighting)
- Provides the same income as under the annuity, where funds are available
- Converts the residual fund to an annuity at 75
- Has the danger that poor investment returns will exhaust the fund before 75, so that the policy-holder outlives his resources (this happened in 10% of the simulations).

The flexible-income programme (FLX):

- Converts the retirement fund to a managed fund (with four different choices of equity weighting: 25%, 50%, 75% and 100%; the remainder is in bonds)
- Provides an income that depends on investment performance, so that poor performance implies a lower income
- Involves no danger that poor investment returns will exhaust the fund before 75, although there is a danger of low interim income. This is because the income each year is determined by dividing the fund size available by the life expectancy of the annuitant for that year. The life expectancy adjusts in the light of mortality experience and is not fixed at the retirement age, as is the case with a standard annuity. For example, someone with a life expectancy one year of 10 years might have a life expectancy the following year of 9.3 years because of improved mortality.
- Converts the residual fund to an annuity at 75.

The flexible-income programme with a deferred annuity (DEF):

- Purchases an annuity at retirement to start at 75 which pays the same income as a traditional annuity from 75 on
- Converts the residual fund at retirement to a managed fund
- Receives a flexible income from the managed fund
- Involves no danger of outliving resources, although there is a danger of low interim income.

The unit-linked programme (UNI):

- Uses the retirement fund to purchase units in a managed fund, with the number of units depending on mortality forecasts
- Sells some units each year, with the income depending on the price of units
- Involves no danger of outliving resources, although there is a danger of low interim income.
- Uses the residual fund to purchase an annuity at 75.

The collared-income programme (COL):

- Is similar to the flexible-income programme but involves a zero-cost collar to smooth returns
- The collar is a combination of long put and short call options: the puts produce a floor on the income received, while the calls produce a ceiling
- Invests in a managed fund (100% equities) subject to the floor and ceiling
- Uses the residual fund to purchase an annuity at 75.

The floored-income programme (FLR):

- Is similar to the flexible-income programme but involves a put option to provide downside protection
- Sacrifices some upside potential to guarantee a minimum return
- Uses the residual fund to purchase an annuity at 75.

We used the following modelling assumptions:

- Equity returns in the managed fund follow geometric Brownian motion
- Returns on equities are parameterised according to UK experience
- 1% annual charge on all distribution programmes
- Analysis in real (RPI-adjusted) terms.

The policy-holder's utility function is the discounted present value of future per-period utility levels, until death, where the per-period utility function is the negative exponential:

$$U(.) = 1 - \exp\,[-\gamma\ \text{(ILRIP income} - \text{annuity income)}]$$

where a low value for γ reflects a low level of risk aversion. The incorporation of the level of annuity income in the utility function reflects the use of the annuity as a benchmark.

The objective of the exercise is to compare the utilities from alternative programmes on the basis of the following simulation exercise:

- 2,000 simulations of equity returns
- 45 simulations of age of death: 65, 66, ..., 110
- Mortality independent of equity returns
- 90,000 mortality/investment scenarios

The results when the programmes involve bequests is shown in the figures below.

The best programme has three key features:

- It never involves bequests, thereby highlighting the value of the mortality bonus (the principal feature of a life annuity)
- It involves a simple mix of equities and bonds, and does not involve derivatives or deferred annuities
- *It depends on the retiree's degree of risk aversion:*
  - Low risk aversion allows a high equity exposure
  - Greater risk aversion lowers the equity exposure
  - Extreme risk aversion leads to the standard annuity (a 100% bond-based investment) as the optimal programme.

So we have provided additional quantitative evidence backing up Olivia's hypothesis concerning the importance of annuities in retirement. However, I would like to reiterate the statement that life annuities (whether equity-enhanced or not) are an essential component of a DC-pension plan, and are not simply one of three possible choices alongside cash lump sums or drawdown.

Expected discounted utility:
programmes with mortality bonuses

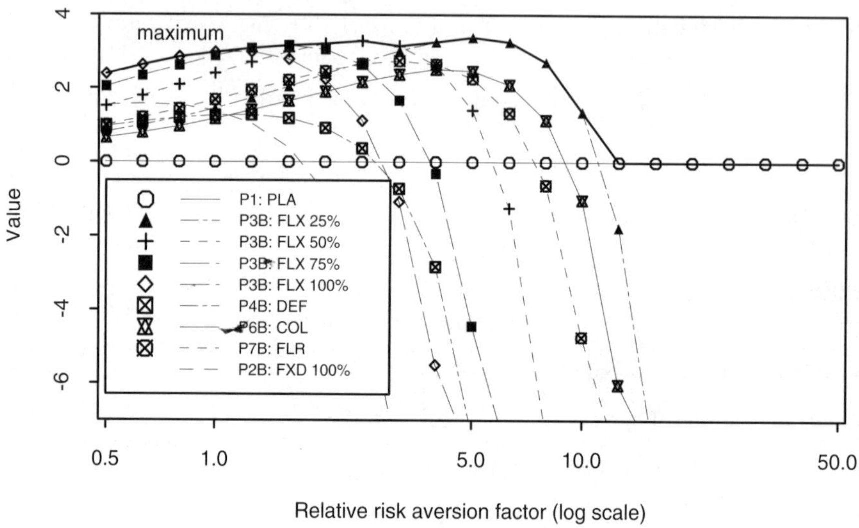

Expected discounted utility:
programmes paying a bequest on death

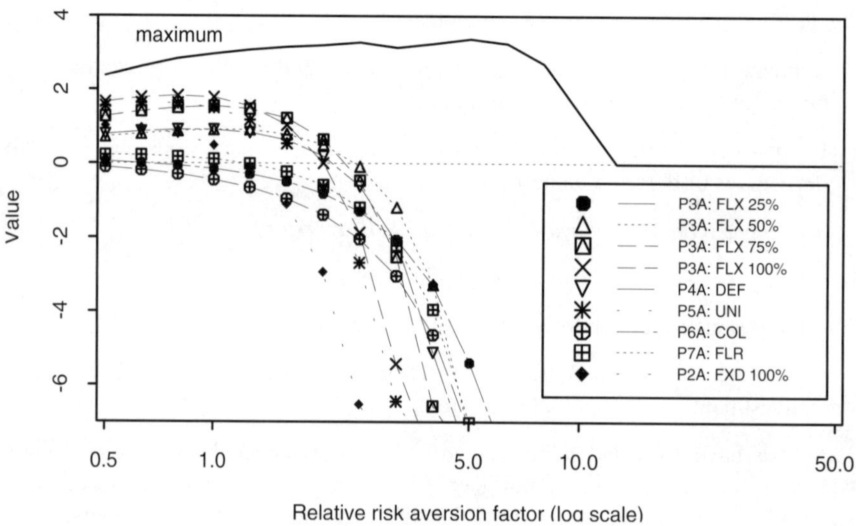

References

Blake, D. and R. Hudson. 2000. Retirement-Income Working Party: Full Technical Report, Pensions Institute, London (*www.pensions-institute.org*).

Blake, D., A. Cairns and K. Dowd. 2000. Stochastic Pension-Plan Design during the Distribution Phase, Pensions Institute, London (*www.pensions-institute.org*).

Discussion

Friedrich Breyer

Olivia Mitchell has given us an excellent survey of the state of the art in the economics of annuities. My remarks will be organized in two parts. In Section 1 I will comment on a small number of points in her paper where I think additional clarification might be useful. In Section 2 I shall touch on a point which is related to her discussion of asymmetric information in annuity markets, although it is concerned more with public than with private pension systems - a point which to my knowledge has been overlooked so far in the literature.*

1 Comments on Mitchell's paper

1. The comparison of the monetary values of real versus nominal annuities in Section 3 demonstrates that it is costly to produce the inflation protection inherent in real annuities. In this context it has to be remarked that this is not true of unfunded pension systems, where the retirement benefit is by definition a real figure. In the same vein, it is stated that a further annuity is less valuable if the purchaser is already partly annuitized. This case applies to the present situation in many countries of the developed world: the political debate there centres around a partial replacement of unfunded by funded pensions. Quite apart from the obvious transition problem, the results of this paper imply that rate-of-return comparisons between unfunded and funded pensions are misleading if it is forgotten that the monetary equivalent of a nominal annuity is

   a) lower than that of an equal real annuity,

   b) reduced if the purchaser is already annuitized.

2. I have missed a discussion of the implications of a positive bankruptcy probability of the insurance firm, which here appears to be assumed as zero. In the real world it is always positive, even if small. This might affect the results. Any government regulation designed to keep this probability low must necessarily reduce the expected rate of return. Moreover, if the government provides a survival guarantee for the insurance companies, then the expected costs of this guarantee must be the borne by the next generation's taxpayers which reduces the degree of funding.

---

* An Exception is Philipson and Becker (1998)

2 Additional Considerations: Asymmetric Information in Compulsory Annuity Systems

2.1 The Problem

Mitchell explicitly discusses one particular form of asymmetric information in annuity markets, namely adverse selection. While it is true that this may become a problem in the market for private annuities, it is not a problem for public annuities because these are typically compulsory. However, in this case there arise other economic problems, including the second prominent type of asymmetric information, moral hazard.

This has to do with what Mitchell calls the "non-discriminating" nature of the benefit formula: typically, in calculating the monthly retirement benefit in a public pension system, the lifetime income stream is taken into account,[1] but - unlike private annuities – the purchaser of a contract is not assessed with respect to his remaining life expectancy. Even obvious criteria such as gender are not used to discriminate the benefit rate. This poses a number of economic problems if longevity - in contrast to the traditional view, on which e.g. the contribution by Bohn to this volume is based – is not considered a purely exogenous "risk" :

1. *Moral Hazard*: While gender is not subject to change, other determinants of longevity are, and this has consequences: somebody who lives a healthy lifestyle, and thereby increases his life expectancy, is not treated differently with respect to the monthly benefit. Thus, he increases the expected value of his annuity and places a financial burden on others. When more and more people go jogging, it is mainly because they want to live as long as possible – preferably at the expense of future generations of tax-payers.

2. *Equity of Income Distribution*: It has been shown in many empirical studies that longevity is positively correlated with income. (for Germany: Klein, 1995; Reil-Held, 2000; for Sweden: Gerdtham and Johannesson, 2000; for the U.S.: Menchik 1993). Thus, the more wealthy get a higher rate of return on their pension contributions than the less well-off, which makes the social security tax highly regressive.

   In the same vein, women get a better return than men because their remaining life expectancy at age 65 is about 4 years higher that of men. So instead of giving a compensation for the unequal survival chances provided by nature, the pension system distributes material resources in a systematically correlated fashion.

---

[1] The close „tax-benefit linkage" is regarded by many as one of the strengths of the German pension system.

3. *Intergenerational Burden Shifting*: In the past few decades, remaining life expectancy at age 65 has increased dramatically. Since retirement benefits of the persons who are already retired do not react to these changes, the financial burden of this development always has to be borne by the next generation. This effect may actually get stronger in the future if the biomedical "revolution" continues at the speed shown recently.

2.2 A Possible Remedy

A solution to the problem described above could be found in the following reform of the mandatory public pension systems: of the total retirement claims of each person, only a part is paid out as an annuity. The level of this annuity must be sufficient to rule out that the pensioners claims additional social assistance, i.e. it must at least cover the subsistence level. The remainder is paid out as a lump sum.

Retirees who want to maintain an equal consumption stream until death can convert part or all of the lump-sum payment into an annuity, but they have to do so in the market for private annuities. In this market, health checks will be performed, so that the more healthy will receive a smaller annuity for a given sum than the less healthy.

But it is by no means clear that everybody will want to annuitize his or her entire wealth. First, the part of wealth which the owner wants to leave as a bequest should not be converted in an annuity. Secondly, some big expenses (e.g. for a trip aroud the world) can be enjoyed only in good health. Thus it makes sense to spend part of the lump-sum payment already in the early part of the retirement period.[2] If this is not possible owing to complete annuitization, the consumption stream is suboptimal, and unplanned bequests are likely.

There may also be an effect on the labour supply of older workers: As has been shown in many studies, the present retirement benefit formula in most countries provides strong incentives to retire early (see, e.g., Börsch-Supan and Schnabel 1998, Gruber and Wise 1998). The lump-sum part of the retirement benefit would now have an income effect on older workers: those in good health would experience an income loss, which - if leisure is a normal good – would raise their labour supply. Those in bad health would experience an income gain, and they would wish to reduce their labour supply. Now it is likely that the former effect is much larger than the latter because most people in bad health already retire very early.

---

[2] Empirical evidence (Börsch-Supan and Stahl 1991) shows that the wealth of the very old is typically increasing, which can be accounted for by age-related physical consumption constraints.

A possible objection to the proposal made here is that the lump-sum payment would mean a reduction in the "longevity insurance", which insures already a new-born against the risk to being born with "healthy genes". There are two defences against this objection:

- It is questionable if the unborn would want this type of insurance, which leads to net gains in the "good" state of nature, which is associated with good health and therefore probably high productive capacity, and to net losses in the "bad" state which is associated with bad health.
- Moreover, this insurance is characterized by severe moral hazard and thus has an unfavourable relation between premium costs and expected benefits.

2.3 Concluding Remarks

The possible reform discussed here is in two respects a mixture of the components typically found in pension systems:

- a mix of "defined benefit" (the annuitized part) and "defined contributions" (the lump-sum part),
- a mix of public (again the annuitized part) and private pensions (if the lump-sum part is also converted into an annuity).

I have tried to show that it would make the unfunded compulsory pension systems more equitable and more efficient, and would, moreover, provide an incentive for healthy older workers to increase their labor supply. So it is one of the rare instances of a free lunch.

References

Börsch-Supan, A. and R. Schnabel. 1998. Social Security and Declining Labor-Force Participation in Germany. *American Economic Review,* 88 (2), pp. 173-178.

Börsch-Supan, A. and K. Stahl. 1991. Life-Cycle Savings and Consumption Constraints. *Journal of Population Economics,* 4, pp. 233-255.

Gerdtham, U.-G. and M. Johannesson. 2000. Income-related Inequality in Life-Years and Quality-Adjusted Life Years. *Journal of Health Economics,* 19, pp. 1007-1026.

Gruber, J. and D. Wise. 1998. Social Security and Retirement: An International Comparison. *American Economic Review,* 88 (2), pp. 158-163.

Klein, T. 1995. Mortalität in Deutschland - Aktuelle Entwicklungen und soziale Unterschiede. In W. Zapf, J. Schupp and R. Habich, ed. *Lebenslagen im Wandel: Sozialberichterstattung im Längsschnitt,* Frankfurt/New York: Campus.

Menchik, P. L. 1993. Economic Status as a Determinant of Mortality among Black and White Older Men: Does Poverty Kill?, *Population Studies,* 47, pp. 427-436.

Philipson, T. J. and G. S. Becker. 1998. Old-Age Longevity and Mortality-Contigent Claims, *Journal of Political Economy 106,* pp. 551–573.

Reil-Held, A. 2000. *Einkommen und Sterblichkeit in Deutschland: Leben Reiche länger?,* Beiträge zur angewandten Wirtschaftsforschung No. 580-00, Institut für Volkswirtschaftslehre und Statistik, Universität Mannheim, May.

Retirement Savings in an Ageing Society: A Case for Innovative Government Debt Management

Henning Bohn

1 Introduction

Most retirees, especially in Europe, receive the bulk of their retirement income from the government. As life expectancy keeps rising and birth rates are declining, publicly-financed retirement becomes increasingly costly, both in absolute terms and as a share of national income. (See OECD, 1998; Raffelhüschen, 1998; Disney, 2000, for recent surveys.) While it is debatable how society should respond to these cost pressures, it seems clear that individual savings will be part of the solution. Most retirees will have to rely more on private savings to finance their increasingly long retirement. This paper is about the optimal composition of retirement savings and about the role of fiscal policy in determining the aggregate allocation of risk.[1]

The question of how retirement savings should be structured has one quick answer, the laissez-faire solution: let individuals choose their own portfolios of stocks, bonds, and other privately issued securities that best reflect their individual preferences over risks and returns. While appealing on an individual level, the laissez-faire answer is seriously incomplete on a macroeconomic level. On aggregate, savers must hold the capital stock and the government debt. Hence, debt policy plays a key role in determining savers' equilibrium portfolios.

The market portfolio available to savers implicitly determines how society allocates aggregate risks over time and over generations. The main sources of aggregate risk are the uncertainty about future productivity growth, uncertainty about population growth, uncertainty about longevity, and the uncertain valuation of capital assets. Productivity uncertainty is naturally shared across cohorts through the correlation of labour income and capital income, though not necessarily in the most efficient way (Bohn, 1998). The other risks are not naturally shared. Population growth raises the marginal product of capital while it

---

1 Questions about allocating risk arise at virtually any level of government taxes and transfers. To avoid distracting discussions about the future scale of transfers, I take the income share of intergenerational transfers as given.

reduces the marginal product of labour. Longevity and asset-valuation uncertainty mainly affect retirees, not workers. Government policy can reallocate these risks across cohorts, either by changing the market portfolio (through debt management) or through taxes and transfers. The paper will examine the interaction of these policies.

The public policy debate about population ageing and pension reform in Europe and in the U.S. has focused on two main options, reforms intended to limit payroll contributions ("privatization") and reforms directed towards partial prefunding.[2] A cap on payroll taxes necessarily requires benefit cuts, explicitly or in disguise, and is often accompanied by tax incentives to encourage private savings. Prefunding involves building up a trust fund that finances part of the contributors' own retirement benefits. If comparably calibrated, both options yield identical expected intergenerational transfers. Apart from questions of corporate control, the key differences involve risk-sharing: with prefunding, the publicly-managed trust fund can serve as risk-sharing device, e.g., to share stock-market uncertainty between cohorts (Bohn, 1999a). With privatization, government influence is more indirect, working through the supply of bonds and potentially through taxation. In either case, the government's problem is to manage the overall portfolio of marketable liabilities (bonds), implicit liabilities (pension promises), and financial assets (trust funds, if any).

The management of the government's portfolio raises three main issues. First, as households rely more on private savings, the structure of net government debt becomes increasingly important. Traditionally, government bonds are considered the typical safe asset. Safe debt does not make the world safer, however, nor does it contribute to risk-sharing. On the contrary, a fixed burden of safe government debt in a world with uncertain future incomes makes the after-tax income of future generations more volatile. Safe debt provides safety to retirees only by exposing future generations to increased productivity and population uncertainty. Safe debt also fails to protect retirees against longevity risk. A wage-indexed pension system, in contrast, shares all these risk across generations. If the pension-replacement rate is reduced due to ageing, a natural remedy is to introduce wage-indexed and longevity-indexed bonds that mimic the risk-sharing features of public pensions. Moreover, once such bonds are introduced, their supply can be optimized to share risk even better.

Second, the structure of public pensions remains relevant and deserves attention. Should public pensions move to a defined-contributions model or maintain defined benefits? And in the latter case, what should be the benefit formula? Defined benefits imply that unexpected shocks must be absorbed by future

---

2 I will not consider full funding nor complete privatization plans, because these more radical options seem unrealistic in developed countries with large unfunded systems. See Disney (2000) and McHale (2001) for recent more detailed reviews of pension reform.

generations, either directly through variable payroll taxes, or indirectly through debt finance or other taxes. I will argue that a switch to fixed contribution rates would be welfare-reducing.[3] Public pensions with annuitized, wage-indexed, defined benefits have important risk-sharing benefits even at a reduced replacement rate. They provide for the three most important aggregate risks-uncertainty about growth, uncertainty about population growth, and uncertainty about longevity.

Third, the role of taxes, especially capital income taxes, needs rethinking. The government can in principle implement a perfect sharing of aggregate risks through appropriately designed tax-transfers systems-provided they are efficiently and credibly administered. The European welfare state can be interpreted sympathetically as such a system. Capital income taxes are particularly important in this context, because they can reallocate risk in an ex ante non-distortionary way (Zhu, 1992; Chari et al., 1994; Bohn, 1994). Unfortunately, capital income taxes also create troubling time-consistency problems (Fischer, 1980) that discourage savings and limit their usefulness as risk-sharing tools. Public debt with explicit contractual contingencies is therefore preferable to state-contingent taxes. Such debt provides a perfect substitute for risk-sharing purposes and it is less subject to capital-levy-problems.

Capital income taxes may have a supporting role if they can be administered in a way that does not discourage savings, i.e., in a non-discretionary manner and accompanied by offsetting savings incentives (Gordon, 1985; Smetters, 2000). Undistorted savings incentives can be achieved in two ways: by making interest income and capital gains tax-exempt or by allowing workers to make pre-tax contributions to special accounts that are taxed at withdrawal. In the latter case, return risk is shared with future generations through the tax system. This helps to share the underlying productivity, demographic, and asset-valuation risks. In addition, the tax deferral implies that a larger stock of debt is available for optimal debt management.

In a setting with pension-trust fund, the same risk-sharing principles apply to the government's net position. If the trust fund holds government bonds, these bonds cancel out. If the trust fund holds private assets, the government holds a leveraged position. Most interesting is the case of debt-financed stock-market investments within a defined-benefits pension plan. Then stock-market risks and returns are effectively transferred to future generations, i.e., asset-pricing uncertainty is shared across generations (Bohn, 1997; 1999a).

---

[3] If one interprets the baby bust as a demographic shock, the defined-benefits principle would call for higher payroll taxes. Reforms that limit contributions and cut benefits to retirees (e.g., recent German reforms) can be interpreted as shifts towards defined contributions. Hence, the issue is relevant even in countries without explicit debate about defined benefits versus defined contributions.

The paper is organized as follows. Section 2 presents a simple representative agent overlapping-generations (OG) model and explains the main effects of public debt, public pensions, and taxation on the allocation of risk. Section 3 examines population ageing and discusses alternative policy responses: business-as-usual (letting public retirement plans grow), reforms that limit payroll taxes, and prefunding. Section 4 examines risk-sharing in a heterogeneous agents OG model, a stylized "welfare state," and its limitations as the population ages. Section 5 concludes.

2 An Overlapping-Generations Perspective on Retirement Finance

Overlapping-generations (OG) models provide a natural framework for examining retirement issues. This section presents a stochastic version of Diamond's (1965) two-period-life OG model, building on Bohn (1998; 1999a; 2001). The model provides a coherent equilibrium framework and is sufficient to discuss life-cycle savings and retirement issues without distracting technical complications. I will start with a laissez-faire economy and then introduce government activities one-by-one.

2.1 The Basic Overlapping-Generations Economy

In the basic OG model, individuals enter the economy as adults and live for two periods. In their first period, individuals work, consume, pay taxes, and save for retirement. In the second period, individuals are retired, consume, and receive transfers from the government. Generation t consists of N_t identical individuals entering the workforce in period t. The labour supply per worker is normalized to 1, so that N_t is also the aggregate labour supply. In period t+1, generation t is retired. To capture a variable life-expectancy, I model retirement as a fractional period of length λ_t, following Auerbach and Hassett (1998) and Bohn (2001). Individuals maximize a utility function

$$(1) \qquad U_t = u(c_t^1) + \rho \cdot \lambda_{t+1} \cdot u(c_{t+1}^2)$$

where c_t^1 and c_{t+1}^2 are consumption flows (per unit time). Period utility is assumed increasing, concave, and homothetic; $\rho>0$ is a time-preference factor, assumed low enough to guarantee dynamic efficiency.

The aggregate output Y_t is produced with capital and labour,

$$(2) \qquad Y_t = K_t^{\alpha} \cdot (A_t \cdot N_t)^{1-\alpha}$$

where K_t is the economy's capital stock, A_t is productivity, and $\alpha \in (0,1)$ is the capital share. The assumption of Cobb-Douglas production is standard and motivated by balanced growth. Capital in the next generational period, K_{t+1}, is a function of old capital K_t and new investment spending I_t,

$$(3) \qquad K_{t+1} = I_t + v_t^k \cdot K_t ,$$

where v_t^k is the residual value of old capital.[4]

To examine the allocation of risk, I assume the economy faces four fundamental types of risk: uncertain productivity (A), an uncertain population (N), uncertain longevity (λ), and an uncertain value of capital (v^k). To avoid dealing with idiosyncratic risk, I assume that individuals within a cohort can share risks perfectly.[5]

The aggregate risks at issue here are quantitatively huge. A one per cent change in annual productivity growth, for example, will raise or lower the next generation's income by more than 30% (30 years ahead). Breaks in productivity growth of more than one per cent per year have been observed, e.g., during the productivity slowdown of the 1970s and during the recent internet revolution. Variations in birthrates and jumps in longevity have financial effects of a similar magnitude. (After all, this is why we discuss ageing.) On a generational scale, these risks are so large that stock-market crashes and booms (v^k_t shocks) might look small in comparison.[6] Nonetheless, valuation risk is explicitly modelled because it raises some special issues and because stock-market uncertainty is such a well-known source of risk.

---

4 Fluctuations in v^k_t can be interpreted as stochastic depreciation or in terms of Tobin's q. For the latter, K should be interpreted as the value of capital. If K' is physical capital and $K_{t+1}=K'_{t+1}\cdot q_t$ is its market value, then (3) is equivalent to $K'_{t+1} = I_t/q_t + v^k_t \cdot q_{t-1}/q_t \cdot K'_t = I_t/q_t + v'_t \cdot K'_t$, where v'_t is one minus the physical depreciation rate; and (2) can be replaced by $Y_t = K'^{\alpha}_t \cdot N_t^{1-\alpha} \cdot A'^{1-\alpha}_t$, where A'_t is the underlying physical productivity. If one defines $A_t=A'_t/q_{t-1}^{\alpha/(1-\alpha)}$ as measured productivity, one recovers equation (2). Thus, the parameter $v^k_t = q_t/q_{t-1} \cdot v'_t$ combines physical depreciation and changes in the market value of capital.

5 If uncertain longevity includes an idiosyncratic element (survival risk), this implicitly assumes the existence of annuities markets; see Bohn (2001). An economy with heterogeneous agents would raise the same aggregate issues; see Section 4.

6 In the U.S., for example, the recent NASDAQ crash (about 60% decline from March 2000 to March 2001) cost investors about \$3 trillion. If one interprets the post-1970 productivity slowdown as a 1% growth reduction from 1970-1995 (with normal growth resuming after 1995), GDP is $1.01^{25}-1 = 28\%$ or about \$2.8 trillion below the previous growth path. Over a generation (30 years), the income loss is more than \$85 trillion. In this sense, a \$3 trillion stock-market decline is a small number.

For this section, I will abstract from all government interventions. This is in the interest of clarity, to lay out the equilibrium allocation of risk in a laissez-faire economy.[7] Then the budget constraints and equilibrium conditions are as follows: workers earn a wage w_t equal to the marginal product of labour, $w_t = (1-\alpha)\cdot K_t^{\alpha}\cdot A_t^{1-\alpha}\cdot N_t^{-\alpha}$. Workers divide their labour income into consumption c_t^1 and savings s_t, $w_t = c_t^1 + s_t$. On aggregate, savings are invested in capital, $K_{t+1} = N_t \cdot s_t$. The gross return on savings is the marginal product of capital,

$$(4) \qquad R_{t+1}^k = \alpha \cdot K_{t+1}^{\alpha-1}\cdot (A_{t+1}\cdot N_{t+1})^{1-\alpha} + v_{t+1}^k .$$

The retirement income $R^k_{t+1}\cdot s_t$ is allocated to the fractional period λ_{t+1}, yielding a consumption flow of

$$(5) \qquad c_{t+1}^2 = R_{t+1}^k / \lambda_{t+1} \cdot s_t .$$

The dynamics of the OG model are driven by the effective capital-labour ratio $K_t/(A_t \cdot N_t)$. Let $k_{t-1} = K_t/(A_{t-1}\cdot N_{t-1})$ denote its predetermined component and let $a_t = \ln(A_t/A_{t-1})$ and $n_t = \ln(N_t/N_{t-1})$ denote the stochastic growth rates of productivity and population. Then the wage can be written as

$$(6) \qquad w_t = (1-\alpha)\cdot k_{t-1}^{\alpha}\cdot \exp\{(1-\alpha)\cdot a_t - \alpha\cdot n_t\},$$

showing that wages depend positively on productivity shocks and negatively on the size of the workforce. Similarly,

$$(7) \qquad R_t^k = \alpha\cdot k_{t-1}^{\alpha-1}\cdot \exp\{(1-\alpha)\cdot a_t + (1-\alpha)\cdot n_t\} + v_t^k .$$

shows that the return on capital responds positively to productivity shocks, population shocks, and shocks to the value of capital. Retiree consumption depends positively on all these shocks, and negatively on longevity.

Overall, productivity shocks are naturally shared between workers and retirees. Population shocks, in contrast, affect retirees and workers in opposite directions. Current longevity and valuation shocks affects only retirees, not workers, while

[7] Some complicating factors and extensions are discussed in Bohn (1998, 2001). Throughout, I abstract from international interactions because they would considerably complicate the analysis (see Baxter and King, 2001 for an analysis) and because ageing is a world-wide phenomenon that cannot be avoided by going abroad. In the model, foreign assets can be interpreted as investment opportunities that matter to the extent that they alter the aggregate return on capital, eq. (4) below; they would enter like additional v^k_t shocks. Country risk and informational frictions suggest that international investments have a quite limited risk-sharing role, especially if the model is applied to Europe as a whole; see Kraay et al (2000) on country risk and Gordon and Bovenberg (1996) on informational issues.

news about future disturbances affects only workers, who may have to adjust their savings. Thus, there is substantial scope for improved risk-sharing.

The laissez-faire allocation is examined in more detail in Bohn (1998). Since this paper is more concerned with policy reforms in economies with substantial government participation, I will add a government sector before providing a quantitative analysis.

2.2 Public Pensions and Public Debt

Pensions and debt are the two classic intergenerational transfer schemes, and therefore central to intergenerational risk-sharing. This section adds public pensions, public debt, and a rudimentary tax system to the laissez-faire economy.

Public pensions are commonly operated as defined-benefit (DB) pay-as-you-go (PAYG) systems, financing formula-fixed pension benefits with current payroll taxes. In a growing economy, taxes and benefits are conveniently stated in terms of the payroll tax rate θ_t and the replacement rate β_t. At time t, real benefits $\beta_t \cdot w_t$ are paid to N_{t-1} retirees during their lifetime, the fraction λ_t of the period, and financed by taxes $\theta_t \cdot w_t$ on N_t workers. This implies the basic PAYG constraint

$$(8) \qquad \theta_t = \beta_t \cdot \lambda_t / \exp\{n_t\},$$

saying that the tax rate required to finance a given replacement rate is proportional to the retiree-worker ratio $\lambda_t/\exp\{n_t\}$.

The general government budget (taxes, spending, and debt) raises a multitude of new issues. Importantly, taxes can be used a risk-sharing devices in ways that can interact with the income and incentive effects. For clarity, this section assumes a very simple tax system, a wage tax designed to maintain a stable debt-GDP ratio. The goal is to present a stylized model of an economy with “limited government”, to be contrasted with a stylized “welfare state” in Section 4. The welfare state will be endowed with a much more sophisticated tax system. To anticipate, the limited government setting will provides key insights about the allocation of risk in a welfare state after pension reforms that force the middle class to save privately.

The simple tax system is motivated by two results from tax theory. First, the income effects of taxation depend only on a cohort’s generational account (the present value of taxes; see Auerbach et al., 1999), not on the timing of tax payments over the life cycle. Second, taxes that distort savings would be inefficient. Given the revenues that the government wants to extract from some individual, this revenue is most conveniently extracted in the form of a wage tax (τ^{w}_{t}).[8]

Let end-of-period government debt be D_{t+1}, the gross return on old debt R^{d}_{t}, and abstract from government spending.[9] The basic government budget equation is then

$$(9) \qquad D_{t+1} = R_t^d \cdot D_t - \tau_t^w \cdot w_t \cdot N_t .$$

Workers' budget equation is now $c_t^1 + s_t = w_t \cdot (1 - \theta_t - \tau_t^w)$. On aggregate, savers must hold all capital plus government bonds, $N_t \cdot s_t = K_{t+1} + D_{t+1}$. Their retirement consumption is augmented by public pensions and by income from bonds,

$$(10) \qquad c_{t+1}^2 = R_{t+1}^k / \lambda_{t+1} \cdot K_{t+1}/N_t + R_{t+1}^d / \lambda_{t+1} \cdot D_{t+1}/N_t + \beta_{t+1} \cdot w_{t+1} .$$

Since taxes are bounded by wage income, a sustainable debt policy requires a bounded debt-income (or debt-wage) ratio. Without much loss of generality, one may assume a debt path consistent with balanced growth, i.e., a stationary time series for the ratio $d_t = D_{t+1} / (N_t \cdot A_t)$. Combined with the budget constraint, this implies

$$(11) \qquad c_t^1 + K_{t+1}/N_t = w_t \cdot (1 - \theta_t) - R_t^d \cdot d_{t-1} \cdot A_{t-1} / \exp\{n_t\} = w_t \cdot (1 - \theta_t^*) ,$$

where

$$(12) \qquad \theta_t^* = \theta_t + \frac{R_t^d \cdot A_{t-1}}{w_t \cdot \exp\{n_t\}} \cdot d_{t-1}$$

summarizes the overall level of intergenerational transfers.[10] This defines the model with basic government sector.

---

8 In this OG model, the inefficiency of taxes on savings is due to homothetic utility; for other arguments, see Chamley (1986), Lucas (1990), and Chari et al. (1994). Given the inelastic labour supply, the wage tax is lump sum; this is convenient, but inessential.

9 Government spending could be added easily but it would merely complicate the analysis. Shocks to real government spending are not separately modelled because they can be interpreted as negative productivity shocks (see Bohn 1998).

10 Aggregate transfers to retirees are $N_t \cdot w_t \cdot \theta^*_t$, which is (-1) times the retirees' generational account. The government budget constraint implies that generational accounts for all generations add up to zero. Hence, $N_t \cdot w_t \cdot \theta^*_t$ is also the sum of the generational accounts of generations t+i, summed over i≥0. Either way, θ^*_t has a generational accounting interpretation.

For workers, taxation can increase or decrease the volatility of after-tax income. For retirees, public pensions and the income from selling government debt weaken the role of capital income. Pensions and debt may, however, introduce new sources uncertainty. How risk is shared depends on the form of government debt and on the pension system.

Government debt is commonly viewed as a safe claim, a safe income source for retirees (if one abstracts from inflation risk). For retirees, safe debt clearly reduces the volatility of consumption. However, to keep debt over output stationary, debt growth must be held down whenever output growth is low, which requires higher taxes τ^{w}_{t} on workers. High output growth, in contrast, allows more debt growth and lower taxes. Algebraically, if $d_{t-1}>0$ and R^{d}_{t} is predetermined, eq. (12) shows that the fraction θ^{*}_{t} of wages paid to the government depends inversely on wages and population growth. Hence, safe debt magnifies the impact of productivity and population shocks on workers and makes their consumption more volatile. Safe government debt shifts productivity and population risk from retirees to workers.

Innovative debt management is promising in this context because appropriately-indexed government bonds would yield a different allocation of risk. If the government issued GDP-indexed debt, for example, the return R^{d}_{t} would be proportional to $w_t \cdot \exp\{n_t\}$, cancelling out the effects of productivity and population growth in (12). This would make retiree income more positively correlated with productivity and population growth, however. The wisdom of issuing these or other types of bonds is a question to be examined; it is the central question for optimal debt management in an OG setting.[11]

Public pension plans typically promise defined benefits that are either wage-indexed (e.g. in Germany and France) or indexed to a combination of prices and wages (e.g. in the UK and US).[12] Wage-indexed benefits offer a fixed replacement rate $\beta_t=\beta$ and they provide retirement income proportional to the current wage. With wage-indexed pensions, the payroll tax depends on demographic shocks, but not on macroeconomic disturbances (see eq. (8)). Importantly, the payroll tax must rise whenever longevity increase and/or workforce growth declines. This demographic link is a key feature of defined-benefits pensions; it yields an automatic sharing of longevity and population shocks.

Price-indexed public pensions share longevity risk just like wage-indexed pensions, but risks correlated with wages are shared differently. If real benefits

11 In a representative agent setting, optimal debt management usually focuses on tax distortions; see Barro (1979), Lucas-Stokey (1983), Bohn (1990), Missale (1999). The presentation here is complementary, focusing on broad periods and abstracting from tax distortions.

12 In the UK and US, benefits are price-indexed after retirement and wage-indexed up to the date of retirement; in terms of broad generational periods, this amounts to a mixture of price and wage indexation.

$\beta_{t+1} \cdot w_{t+1}$ are predetermined at time t, the replacement rate β_{t+1} must respond inversely to all shocks affecting w_{t+1}, e.g., to productivity and population shocks. Due to the PAYG constraint, the tax rate θ_{t+1} must also vary inversely with w_{t+1}, which makes workers after-tax income more volatile than the pre-tax wage. Thus, safe pensions-like safe debt-provide safety to retirees by shifting aggregate risks to future generations of workers.

2.3 The Quantitative Allocation of Risk: Measurement and Welfare

How efficient is the allocation of aggregate risks with and without government intervention? This section will provide a quantitative perspective on risk-sharing.

For a formal analysis, explicit assumptions about policy and about the stochastic shocks are needed. This section focuses on risk-sharing in a stochastic steady state, as baseline. The stochastic shocks are assumed to follow a vector Markov process and the policy rules are assumed to be stationary. Then the model variables have a simple Markov structure. Unless noted, I assume that productivity growth, population, and valuation shocks are i.i.d. on a generational time scale; that debt is safe and grows at the same average rate as GDP (d_t is either constant or a function of k_t); and that pensions offer either a constant replacement rate ($\beta_t=\beta$) or a replacement rate that depends only on current shocks. Longevity shocks (λ) are treated as essentially permanent (AR=0.9), because medical innovations that affect retirees longevity also tend to increase workers' life expectancy. (Some results for i.i.d. longevity shocks are shown for comparison.)

In terms of efficiency units, the allocation at time t is then a mapping from the vector of disturbances (a_t, n_t, λ_t, $v^k{}_t$) and the lagged effective capital-labour ratio k_{t-1} to the model's endogenous variables. As usual in growth models, per-worker variables are obtained as efficiency-unit values multiplied by the productivity trend (A_t); aggregate variables are efficiency units scaled by $N_t \cdot A_t$. The mapping from state variables to endogenous variables is generally non-linear and does not have a closed form solution. One can, however, derive the elasticities of the endogenous variables with respect to the Markov state variables. The tables will show elasticity values that are evaluated at a log-linear approximation around the model's deterministic steady state.

For any variable ξ, the elasticity with respect to a state variable s is denoted by $\pi_{\xi,s}$. Steady-state values are denoted by symbols without time subscripts. If applied to consumption ($\xi=c^1,c^2$), the elasticities $\pi_{c1,s}$ and $\pi_{c2,s}$ measure the percentage exposure of the cohort's consumption to shocks of type s. The risk exposure of factor incomes and after-tax incomes can be measured similarly by inspecting $\pi_{w,s}$, $\pi_{Rk,s}$, etc. For retirees, after-tax income equals consumption. For workers, income and consumption typically move in the same direction, but not always, since savings are endogenous.

As is well known, risk-sharing in OG economies is generally not ex ante Pareto efficient (Peled, 1982; Gale, 1990). Though ex ante inefficiency does not necessarily justify government intervention, ex ante efficiency provides a natural criterion for comparing *alternative* policies if one takes it as given that governments do intervene (Bohn, 1998). For standard time-separable homothetic preferences, ex ante efficiency has strong implications: the consumption of workers' and retirees' cohorts should be equally exposed to all types of shocks. That is, the elasticity coefficients $\pi_{c1,s}$ and $\pi_{c2,s}$ should satisfy $\pi_{c1,s} = \pi_{c2,s}$ for all shocks *s*. Bohn (1998) has shown that these consumption elasticities should have the same sign even if the assumption of time-separable homothetic preferences is dropped (e.g., in the case of non-expected utility or age-dependent risk aversion). Hence, coefficients pairs $\{\pi_{c1,s}, \pi_{c2,s}\}$ with opposite signs and pairs with zero and non-zero values provide clear indications of inefficiency.

Table 1 displays elasticity coefficients for several numerical versions of the OG economy. The benchmark is an economy with a traditional wage-linked defined-benefit pension system and safe debt. The pension is assumed PAYG with a 10% payroll tax that supports a 40.4% replacement rate. The government debt is calibrated to generate an intergenerational transfer of 5% of wages on average, which corresponds to an annual debt/GDP ratio of about 50% (see Bohn, 1999a). Thus, steady-state intergenerational transfers are θ^*=15% of wages. The demographic parameters are an average retirement period of λ=1/3 and an average annual population growth of 1%. The macro assumptions are a capital share of α=30%, homothetic utility $u(c)=c^{1-1/\varepsilon}/(1-1/\varepsilon)$ with elasticity of intertemporal substitution ε=0.75, average annual productivity growth of 1.5%, an average real annual return on capital of 5%, and an average share of v^k/R^k=10% of old capital in the return on capital.[13] These "round" numbers are used to emphasize the conceptual nature of the results. The objective is to understand the principal role of the main fiscal policy instruments and their interaction. This requires strong assumptions, not only about parameter values, but, perhaps more importantly,

---

13 The ε=0.75 elasticity of substitution and the v^k/R^k=10% share of old capital are perhaps worth a comment. The 0.75 elasticity is at the upper end of empirical estimates (Ogaki and Reinhart, 1998) while the v^k/R^k=10% is rather low, less than the share of raw land in wealth. Lower ε and/or higher v^k values would raise the relative exposure of workers to productivity risk under laissez-faire and therefore make innovative debt management look better (see Bohn, 1998). The high-ε and low-v^k parametrization is therefore conservative, to avoid exaggeration. The low v^k is also intended to correct for the omission of bequests; namely, if one took it as given that certain assets are bequeathed without market transactions (say, illiquid real estate and durable goods), such assets are best excluded for risk-sharing purposes and for calibrating v^k.

Table 1. **The Allocation of Risk in a basic OG Economy with Public Debt and Pensions**

Policies: **Risk/Exposure**	**Laisse-faire** (No pensions, no debt)	**Economy with DB Pensions** (No debt)	**Benchmark** (DB Pension and Safe Debt)	**Price-indexed Pensions**
Column:	(1)	(2)	(3)	(4)
Productivity (a_t)				
Workers $\pi_{c1,a}$	0.717	0.717	**0.763**	0.838
Retirees $\pi_{c2,a}$	0.599	0.634	**0.591**	0.479
Risk-sharing:	Gap = 0.118	Gap = 0.083	**Gap = 0.172**	Gap = 0.359 Increased/Worse
Asset Value (v^k_t)				
Workers $\pi_{c1,v}$	0.000	0.000	**0.000**	0.000
Retirees $\pi_{c2,v}$	0.144	0.094	**0.076**	0.076
Risk-sharing:	All Risk on Retirees	All Risk on Retirees	**All Risk on Retirees**	= Benchmark
Workforce (n_t)				
Workers $\pi_{c1,n}$	-0.283	-0.178	**-0.129**	-0.162
Retirees $\pi_{c2,n}$	0.599	0.463	**0.431**	0.479
Risk-sharing:	Different signs: Gap = 0.882	Different signs: Gap = 0.641	**Different signs: Gap = 0.560**	Gap = 0.641 Increased/Worse
Longevity (λ_t)				
Workers $\pi_{c1,\lambda}$	-0.389	-0.384	**-0.388**	-0.388
Retirees $\pi_{c2,\lambda}$	-1.000	-0.829	**-0.840**	-0.840
Risk-sharing:	Gap = 0.611	Gap = 0.445	**Gap = 0.452**	= Benchmark
Memo: λ_t = i.i.d.				
Workers $\pi_{c1,\lambda}$	0.000	-0.105	**-0.108**	-0.108
Retirees $\pi_{c2,\lambda}$	-1.000	-0.829	**-0.840**	-0.840
Risk-sharing:	All Risk on Retirees	Gap = 0.724	**Gap = 0.732**	= Benchmark

Legend: Table entries show elasticities of worker and retiree consumption with respect to macroeconomic disturbances.

about aspects of reality that are excluded from the model. Most results below are robust to reasonable parametric variations.

In Table 1, Col.1 shows the laissez-faire allocation without debt and pensions. Col.2 adds a wage-linked defined-benefit PAYG pension system and Col.3 (the benchmark) adds safe government debt. Col.4 replaces the wage-linked pension system by a price-indexed pension system.

In the laissez-faire allocation, productivity shocks raise both wages and the return on capital and therefore raise retiree and worker consumption, though not by the same amount. The numerical elasticity values depend on a number of modeling choices discussed in more detail in Bohn (1998). I found that under fairly general conditions, $\pi_{c1,a}$ is well above $\pi_{c2,a}$: Under laissez-faire, workers are more exposed to productivity risk than retirees. The parameters here are chosen to be consistent with this finding.

Unexpected growth in the work force reduces the capital-labour ratio and therefore reduces wages while raising the return on capital. The consumption elasticities with respect to n_t-shocks therefore have opposite signs, $\pi_{c1,n}<0$ versus $\pi_{c2,n}>0$. Conversely, a baby bust is a plus for workers and a negative for retirees.

Similar linkages arise in larger, multi-period-lives OG models (e.g. Auerbach-Kotlikoff, 1987; De Nardi et al., 1999) and in representative agent models (e.g. Cutler et al., 1990). Knabb (2001) has carefully examined factor-price effects in an OG model with education and endogenous human capital and shown that human capital accumulation would magnify the wage effects. The linkage between population and wages is also consistent with findings in labour economics and in the empirical growth literature (see Bohn, 2001 for references). Overall, this literature emphasizes one key point: *Demographic changes by themselves have significant macroeconomic effects, primarily through factor prices; fiscal policy merely adds to these effects and therefore cannot reasonably be considered in isolation.*

Unexpected longevity shocks have a strong negative impact on retirees, with elasticity $\pi_{c2,\lambda} = -1$. Retiree consumption declines one-for-one if they must divide their savings over a longer life-span. Workers are affected much less and only because of the assumed autocorrelation of longevity shocks, which signals to workers that they need to save for a longer retirement period. (Workers would be unaffected by i.i.d. longevity shocks, as noted at the bottom of the table.) Finally, retirees bear the entire risk of asset-valuation shocks; workers are unaffected.[14]

[14] The list of shocks could be extended and their interaction be generalized in many ways. For example, different shocks might be correlated (e.g., if asset values depend on productivity growth, or if fertility depends endogenously on, say, income growth); all the shocks may have temporary and more or less permanent components; and advance information may be available about some shocks (say, current births signalling the future number of workers). One could also expand the model, e.g., add endogenous labour supply, add more periods of life, or add international considerations. I have addressed some of these extensions in other work; e.g., see Bohn (1998) on endogenous labour supply, Bohn (1999a) on the correlation of productivity and valuation risk, Bohn (2001) on advance information about birth-rate and longevity shocks; others are left for future work, because they would excessively complicate this paper.

In the allocation with public pensions (Col.2), the gaps between worker and retiree exposures for all shocks are less than under laissez-faire. The wage-indexation of public pension raises retirees' exposure to productivity shocks, moving $\pi_{c2,a}$ closer to $\pi_{c1,a}$. The annuitized nature of public pensions reduces retirees' exposure to longevity shocks. The defined benefit principle triggers automatically triggers payroll tax adjustments that help share the impact of population shocks (Smith, 1982). This is because population growth reduces pre-tax wages and reduces payroll taxes, and the latter increases the after-tax wage. Increased longevity triggers higher payroll taxes and gives workers a negative exposure to such shocks. Asset-valuation risk is still carried entirely by retirees, but their exposure is reduced because capital income is crowded out by pension income.[15] Overall, *wage-indexed public pensions have strikingly beneficial risk-sharing effects that no discussion of retirement reform should ignore.*

When safe public debt is added (Col.3), productivity risk is shifted from retirees to workers, making the gap in elasticities even wider than in the laissez-faire economy. The increased intergenerational transfers further reduce the role of valuation shocks. Since population growth reduces the per capita burden of debt, debt reduces the negative exposure of workers to n_t-shocks; but it does not help share longevity shocks. While these effects might appear mixed, one should remember that debt and pensions are substitutable tools of intergenerational redistribution. If the safe debt were replaced by pensions, the productivity and demographic risks would be more evenly shared.

The risk-sharing effects of public pensions depend importantly on the wage indexation. In countries where public pensions are inflation-indexed, i.e., fixed in real terms, the allocation of risk would differ. This is shown in Col.4. Productivity and population shocks are clearly shared less evenly than in Col.3, suggesting that price-indexed pensions are inferior to wage-indexed pensions.

2.4 Innovative Debt Management

The role of alternative types of public debt is demonstrated in Table 2. It shows the effects of replacing safe debt by GDP-indexed debt (Col.2); wage-indexed debt (Col.3); and longevity-contingent debt (Col.4); for comparison, Col.1 reproduces the benchmark setting with safe debt.

Wage- and GDP-indexed debt both improve the sharing of productivity shocks. They differ with respect to demographic shocks. GDP-indexed debt magnifies the effect of population shocks on workers because such debt rises in value when the

---

15 Here, the quantitative effect depends on the elasticity of substitution. A higher elasticity would imply slightly less crowding out. The qualitative insights are robust, however.

Table 2. Policy Alternatives

Policies: **Risk/Exposure**	**Safe Debt** (For comparison =Table 1, Col.3)	**GDP- indexed Debt**	**Wage-indexed Debt**	**Longevity-indexed Debt**
Column:	(1)	(2)	(3)	(4)
Productivity (a_t)				
Workers $\pi_{c1,a}$	0.763	**0.725**	**0.725**	0.763
Retirees $\pi_{c2,a}$	0.591	**0.647**	**0.647**	0.591
Risk-sharing:	Gap = 0.172	**Gap = 0.078** **Reduced/Better**	**Gap = 0.078** **Reduced/Better**	= Benchmark
Asset Value (v^k_t)				
Workers $\pi_{c1,v}$	0.000	0.000	0.000	0.000
Retirees $\pi_{c2,v}$	0.076	0.076	0.076	0.076
Risk-sharing:	All Risk on Retirees	= Benchmark	= Benchmark	= Benchmark
Workforce (n_t)				
Workers $\pi_{c1,n}$	-0.129	**-0.167**	**-0.113**	-0.129
Retirees $\pi_{c2,n}$	0.431	**0.487**	**0.407**	0.431
Risk-sharing:	Different signs: Gap = 0.560	**Gap = 0.654** **Increased/Worse**	**Gap = 0.521** **Reduced/Better**	= Benchmark
Longevity (λ_t)				
Workers $\pi_{c1,\lambda}$	-0.388	-0.388	-0.388	**-0.415**
Retirees $\pi_{c2,\lambda}$	-0.840	-0.840	-0.840	**-0.760**
Risk-sharing:	Gap = 0.452	= Benchmark	= Benchmark	**Gap = 0.345** **Reduced/Better**
Overall Assessment:	=Benchmark	Mixed Effects	More Efficient	More Efficient

Legend: The table entries show elasticities of worker and retiree consumption with respect to macroeconomic disturbances for different policies. Except for the specified policy alternative, the assumptions of Section 2.3 apply. The comparisons in Cols.2-4 are relative to the benchmark setting from Table 1, Col.3, which is reproduced in Col.1. Entries that differ from the benchmark are shown in **bold**, to highlight the changes.

workforce increases and wages fall. Wage-indexed debt, in contrast, is devalued when the workforce increases and therefore helps share population risk. It seems to dominate GDP-indexed debt.[16]

---

16 GDP-indexed debt was first proposed by Shiller (1993). Shiller considered agents with infinite horizon-planning and debt indexed to the present value of GDP. Since agents here have a finite planning horizon, they are not interested in claims maturing after their death, suggesting finite-maturity claims that do not expose retirees to new valuation risks.

Longevity-indexed debt would help protect retirees against shocks to life expectancy and thereby improve risk-sharing (see Col.4). Longevity-indexed debt would essentially replicate the annuitization of pensions, but through tradable securities and with the potential to share risk in a more flexible way.

To implement longevity-indexed debt, one might consider a variety of security designs, depending on practical considerations and microeconomic objectives. One option would be for the government to sell annuities directly to consumers. This would likely trigger adverse selection, which might be undesirable. Adverse selection may, on the other hand, be socially desirable if the government wants to make annuities available to individuals who would be subject to statistical discrimination in the private market. Alternatively, the government could issue bonds with payoffs indexed to aggregate survival statistics. Such bonds might be attractive to insurance companies that sell individual annuities. By providing reinsurance, the government would encourage the development of private annuities markets and leave it to the private sector to cope with adverse selection. Either way, an increased aggregate supply of annuities would improve aggregate risk-sharing.

2.5 The Risk-Sharing Role of Capital Income Taxes

The analysis would be incomplete without a comment on the risk-sharing role of the tax system (Gordon, 1985; Smetters, 2000).

Since the tax on workers (τ^w_t) was already assumed to be proportional to wage income, the main extension is to include taxes on capital income (broadly defined, including interest, dividends, capital gains). Taxes on the capital income of retirees can share risk across generations if fluctuations in revenues are used (on the margin) to reduce wage taxes. A limiting case, perfect risk-sharing through discretionary taxes and transfers, will be discussed in Section 4 as an interpretation of the welfare state. Here I will merely introduce a fixed-rate capital income tax to illustrate the principal effects and to improve the calibration.

The risk-sharing effects of capital income taxes are illustrated in Table 3. Col.1 reproduces the original benchmark from Section 2.3. Col.2 shows the allocation of risk in an economy with the same government debt and the same net intergenerational transfers (θ^*=15%), but with a 25% marginal tax rate on capital income (instead of zero), a 16% payroll tax (instead of 10%), and a 65% replacement rate (instead of 40%). These policy parameters are motivated by

Table 3. The Role of Capital Income Taxes

Setting: **Risk/Exposure**	**Initial Benchmark** (For comparison =Table 1, Col.3)	**Modified: With capital Income taxes**
Column:	(1)	(2)
Productivity (a_t)		
Workers $\pi_{c1,a}$	0.763	0.770
Retirees $\pi_{c2,a}$	0.591	0.579
Risk-sharing:	Gap = 0.172	Gap = 0.191
Asset Value (v^k_t)		
Workers $\pi_{c1,v}$	0.000	0.008
Retirees $\pi_{c2,v}$	0.076	0.064
Risk-sharing:	All Risk on Retirees	Risk is shared: Gap = 0.055
Workforce (n_t)		
Workers $\pi_{c1,n}$	-0.129	-0.057
Retirees $\pi_{c2,n}$	0.431	0.324
Risk-sharing:	Different signs: Gap = 0.560	Different signs: Gap = 0.381
Longevity (λ_t)		
Workers $\pi_{c1,\lambda}$	-0.388	-0.421
Retirees $\pi_{c2,\lambda}$	-0.840	-0.744
Risk-sharing:	Gap = 0.452	Gap = 0.324

Legend: Table entries show elasticities of worker and retiree consumption with respect to macroeconomic disturbances. Col.1 reproduces the original benchmark allocation from Section 2.3 (Table 1, Col.3). Col.2 shows elasticity values for the modified economy with 25% capital income taxes.

German payroll taxes and capital income taxes.[17] Capital-income-tax receipts are allocated to generations to keep total intergenerational transfers unchanged at 15%. I will refer to this as the European calibration (whereas the Section 2.3 case with 10% payroll tax may be called a U.S. calibration). One important caveat applies: to focus on risk-sharing, I assume that capital investment is sufficiently tax favored that savings incentives remain undistorted despite the income tax; the incentive aspects are discussed in Section 4 in more detail.

---

[17] Mendoza et al., (1994, Table 3) show effective capital-income-tax rates of about 25% for Germany. German payroll taxes were about 11% of GDP in 1995 (Disney, 2000), i.e., about 16% of wages. The capital income tax is assumed to be offset by investment incentives sufficient to avoid distorted savings. See Section 4 for a formal exposition of the model with capital income taxes. Mendoza et al. do not clearly distinguish marginal from average taxes (cf. Gordon, 1985), so the 25% value should be interpreted as illustrative.

The main point of Table 3 is that capital income taxes help to share asset-valuation risk, a source of risk that none of the other policy tools was able to address. In addition, the model with larger pension system provides a somewhat better sharing of demographic risk. The similarity of Col.1 and Col.2 highlights the general applicability of the model. Without much loss of generality, the Col.2 calibration is used as a starting point for the analysis below.

This summarizes the principal risk-sharing effects of government policy in the OG economy. The analysis was placed in a balanced growth setting for convenience. The next section considers demographic change from a different perspective that includes "unbalanced" growth.

3 Retirement Savings in an Ageing Society

From a welfare perspective, there are two ways to interpret population ageing. Up to now, I have treated longevity and birth-rate changes as stochastic shocks that call for an appropriate risk-sharing response. This interpretation is natural when a demographic change first occurs and is probably sufficient for temporary shocks. In the case of permanent shocks, however, a risk-sharing response may be difficult to implement in practice. Taken seriously, optimal insurance against permanent shocks would require extremely long-term fiscal transfers, "insurance payments" from/to a current cohort to/from an infinite sequence of future generations. Instead, society may interpret some aspects of demographic change as a known, uninsurable trend. This alternative interpretation of the ageing process is pursued in this section.

3.1 Population Ageing

Longevity and birth rates, the two source of population ageing, have very different time-series properties (see Lee and Tuljapurkar, 1998). Increased longevity is a gradual, monotone process largely driven by improvements in medical technology. It increases individuals' need to save for retirement and it puts increasing pressure on retirement systems that guarantee benefits for life. Uncertainty about longevity is mainly about the speed, not the direction. Declining birth rates also contribute to an ageing population, but the decline is far from smooth or monotone in most countries. Sometimes birth rates decline rapidly (during baby busts), sometimes the declining trend is interrupted by increases (baby booms). Baby boom and bust phenomena are therefore reasonably interpreted as insurable events that should be covered by intergenerational risk-sharing. The longevity trend, in contrast, is more difficult to insure against and arguably poses the more serious demographic problem.

The longevity trend, as a non-stationary phenomenon, also creates new complications for economic analysis because it is inherently inconsistent with balanced growth. Working-age families not only have to save more, but to save a larger fraction of their income. To maintain defined-benefits pensions, governments would have to raise increasing tax revenues, not just in line with growth, but rising as a percentage of GDP.

Two potentially offsetting trends should be discussed, though they will not be at the centre of analysis. First, if longevity is driven by medical technology, one may wonder if the same improvements in medical technology also reduce the incidence of disability and enable people to defer retirement. This would raise workers' lifetime supply of labour and reduce the retiree-worker ratio. This endogeneity of lifetime work effort is another reason why the longevity trend is probably best treated as a not fully insurable phenomenon. In practice, however, the age of retirement has declined throughout Europe and in the U.S. Even if this is due in part to removable policy distortions (Gruber and Wise, 1999), a longer working life is unlikely to stop the rise in retiree-worker ratios. Hence, I will continue to treat the relative length of the retirement period (λ) as exogenous.

Secondly, lower birth rates reduce the youth-dependency rate. Sinn (2000) has argued that it is only fair to ask households that spend less on children to increase their retirement savings. In Bohn (2001), I have added a childhood period to the two-period OG model and examined demographic risk in a setting that includes youth-dependency effects. I find that youth-dependency effects modify the effects of birth-rate shocks, but do not overturn the basic results. Hence, this paper omits youth dependency. [18]

For the analysis, I use the same model as in Section 2, but with trends in the demographic variables. Formally, let $\lambda_t = \exp(\overline{\lambda}_t + \hat{\lambda}_t)$ be the sum of a time trend $\overline{\lambda}_t$ and an autocorrelated stochastic component $\hat{\lambda}_t$. Similarly, let $n_t = \overline{n}_t + \hat{n}_t$ be the sum of a time trend $\overline{n}_t$, and a stochastic component $\hat{n}_t$ which is again assumed i.i.d. The demographic trends are inherently non-stationary and the resulting macroeconomic dynamics are inconsistent with balanced growth. The OG setting is fortunately simple enough to accommodate time trends and unbalanced growth. Assuming Markov-type shocks and policy rules that are functions of the

---

18 Spending on children raises the broader issues of altruism and endogenous fertility. Forms of altruism that yield Ricardian neutrality would make fiscal transfers uninteresting and are therefore excluded (see Bohn 1998 and 2001 for more discussion). The relevance of endogenous fertility is unclear. Endogenous fertility per se is inessential, provided the endogenous changes are deterministic or driven by exogenous variables (say, changes in tax policy to encourage births). For endogenous fertility to matter, the endogenous components would have to be stochastic and correlated with other stochastic elements of the model.

underlying trends and shocks, the model still has an autoregressive structure, not time-independent, but still tractable.

In terms of efficiency units, the equilibrium allocation at time t is a mapping from the capital-labour ratio k_{t-1}, the vector of disturbances (a_t, v_t, $\hat{n}_t$, $\hat{\lambda}_t$) and the time-t and time-(t+1) demographic trend values ($\bar{n}_t$, $\bar{\lambda}_t$, $\bar{n}_{t+1}$, $\bar{\lambda}_{t+1}$), to the endogenous variables. Ageing in this context means a rise in mean longevity ($\bar{\lambda}_{t+1}$) and/or a decline in mean workforce growth ($\bar{n}_{t+1}$). This raises two issues: what are reasonable policy responses to the shifting means? What is different about policy in a high-$\bar{\lambda}$ /low-$\bar{n}$ environment, compared with the setting of Section 2?

3.2 Business as usual

One policy option is to continue the existing government programmes *as is*, i.e., to maintain PAYG retirement benefits and to keep the regular budget on a sustainable path. Most public retirement programmes promise defined benefits for life. No change is therefore best interpreted as a constant replacement rate (β_t=β). With PAYG financing, this implies a payroll-tax rate $\theta_t = \beta \cdot \lambda_t / \exp\{n_t\}$ that increases when λ_t rises and when n_t declines. For the regular budget, a sustainable path is usually interpreted as a stable debt-GDP ratio, which I will proxy by d_t=d. Overall, business as usual implies a rising total tax rate θ^*_t, i.e., growing intergenerational transfers.

In the public policy discussion, especially among academics, it has become fashionable to dismiss the business-as-usual option as infeasible because it would imply an unbalanced growth path. This sense of unsustainability is perhaps reinforced by the upward trend in retiree medical cost. Increasing taxes may indeed be politically undesirable, as I will assume below, and burdensome because of the increasingly adverse effects on labour-supply incentives. One should be careful about the reasoning, however. From an economic perspective, unbalanced growth is not necessarily unsustainable. From a political perspective, public pensions remain hugely popular. Bohn (1999b) has argued that this is not accidental. Starting at about age 40 (for U.S. data), the present value of public pension and medical benefits starts exceeding the present value of future payroll taxes. As the population ages and payroll taxes rise, the critical age where future benefits and taxes balance out will increase. But the age of the median voter rises, too, and fast enough that the fraction of voters with positive expected net benefits will remain well above 50 percent for a variety of projections (in the U.S., for the Social Security Administration's medium, high, and low-cost-scenarios). The same logic should apply to European PAYG systems. Altruism and redistributional arguments provide further support for public pensions (see Bohn,

1999b). Thus, as a cautionary note, a contraction or demise of public pensions is far from inevitable.

Table 4, Col. 1, illustrates the implications of keeping the replacement rate constant as the population ages. The demographic scenario is a drop in mean population growth to $\bar{n}_t = \bar{n}_{t+1} = 0$, combined with an average longevity of $\bar{\lambda}_t = \bar{\lambda}_{t+1} = 0.5$, starting from the policy setting in Section 2.5 (Table 3, Col.2). An unchanged 65% replacement rate now requires an average payroll-tax rate of more than 32%. From a risk-sharing perspective, this policy looks remarkably good. With the expanded pension system, new demographic shocks are shared almost perfectly. Workers actually bear slightly more demographic risk than retirees, suggesting that policy is shifting too much risk from retirees to workers. The sharing of productivity and valuation shocks remains roughly similar, as in Table 3, Col.2.

The good risk-sharing properties of the business-as-usual policy should not, however, obscure the key problem. Business-as-usual means sharply higher payroll taxes. In the calibrated example, intergenerational transfers would more than double to $\theta^*=0.313$. In practice, some of the model's "consumption" would be tax-financed public goods, and one would have to add redistributional transfers. Thus, business-as-usual imposes a heavy tax burden on workers.

For examining alternative policy options, I will rule out increased intergenerational transfers. For the sake of economic analysis, I assume instead that a political decision is made to hold expected intergenerational transfers constant as the population ages (at $\theta^*=15\%$ in the numerical examples). How should these limited transfers be structured to provide optimal risk-sharing? The next two section will examine two classes of policy alternatives: limiting payroll taxes and partial prefunding. Both can be packaged in ways that turn out to have very different risk-sharing implications.

Table 4. Basic Responses to an Ageing Population

Given:	Rising Taxes	Holding Intergenerational Transfers Constant		
	($\theta^* = 31.3\%$)	($\theta^* = 15\%$)		
Policies: **Risk/Exposure**	"Business as usual"	Wage-indexed DB Pension	Price-indexed DB pension	Capped taxes (DC Pension)
Column:	(1)	(2)	(3)	(4)
Productivity (a_t)				
Workers $\pi_{c1,a}$	0.787	0.788	0.905	0.788
Retirees $\pi_{c2,a}$	0.591	0.525	0.358	0.525
Risk-sharing:	Gap = 0.197	Gap = 0.263	Gap = 0.546 Worse than Col.2	= Col.2
Asset Value ($v^k{}_t$)				
Workers $\pi_{c1,v}$	0.014	0.019	0.019	0.019
Retirees $\pi_{c2,v}$	0.063	0.112	0.112	0.112
Risk-sharing:	Gap = 0.049	Gap = 0.093	= Col.2	= Col.2
Workforce (n_t)				
Workers $\pi_{c1,n}$	0.214	-0.045	-0.095	-0.212
Retirees $\pi_{c2,n}$	0.188	0.287	0.358	0.525
Risk-sharing:	Gap = 0.026	Gap = 0.332	Gap = 0.454 Worse than Col.2	Gap = 0.737 Worse than Col.2
Longevity (λ_t)				
Workers $\pi_{c1,\lambda}$	-0.658	-0.523	-0.523	-0.466
Retirees $\pi_{c2,\lambda}$	-0.597	-0.762	-0.762	-1.000
Risk-sharing:	Gap = 0.061	Gap = 0.239	= Col.2	Gap = 0.534 Worse than Col.2
Overall assessment:	Reduced gaps with high θ^*	Less efficient than Col.1	Less Efficient than Col.2	Less Efficient than Col.2

Legend: The table entries show elasticities of worker and retiree consumption with respect to macroeconomic disturbances under different policy assumptions. All columns describe the "aged" economy with zero population growth and 50% retiree-worker ratio. Col.1 shows the "business as usual" scenario; Columns 2-4 show the alternatives described below.

3.3 Limiting Payroll Taxes

Cost-cutting has been a key objective of most recent pension reforms, e.g., in Germany, France, and Italy in the 1990s. Since reduced pensions tend to encourage private savings, these reforms can be interpreted as partial "privatization" scenarios. For this paper, the main questions are about the implied changes in the allocation of risk. I will suggest two interpretations, a literal and a functional one.

Taken literally, many recent reforms have included movements away from wage-indexation, e.g., shifts from wage- to price-indexation in Italy and France and from gross wage to net wage indexation in Germany (see McHale, 2001). In each case, the apparent intent was to shift to a slower-growing index and thereby to reduce the average replacement rate over time. As noted in Section 2, such changes in indexation have profound effects on the allocation of risk. They remove a mechanism that shares future productivity and population risks.

In Table 4, Cols.2-3 demonstrate the quantitative effects of alternative cost-cutting reforms. For the stylized interpretation, I assume that average payroll taxes are held constant to prevent an increase in intergenerational transfers. For θ=16%, $\bar{\lambda}$ =0.5 and $\bar{n}$ =0, the PAYG constraint implies an average replacement rate of β=32%, i.e., a much lower value than in the "young" economy of Section 2. Col.2 shows the risk-sharing features of a wage-indexed pension system with fixed 32% replacement rate and a variable payroll tax with an average rate of 16%. Col.3 illustrates a price-indexed system of the same size. The price-indexed system clearly displays a more uneven generational exposure to productivity and other shocks, confirming the negative results in Section 2.

An alternative, functional interpretation of recent reforms is to focus on their intent, namely to cap payroll taxes. The hallmark of a defined-benefit system is that unexpected shocks are not allowed to affect the benefits of current retirees. Benefit changes must be phased in over time. In a country where benefits are adjusted through repeated "reforms" whenever unexpected cost increases occur, the pension system is best interpreted economically as a system with fixed, "defined" contributions (DC), regardless of the statutory benefit formulas.

The risk-sharing implications of a DC pension system are shown in Col.4 of Table 4, assuming a fixed payroll-tax rate and variable benefits. The DC system clearly fails to share demographic risk. It is inferior in this respect to both DB systems (Col.2-3). The fixed tax rate combined with PAYG implies that benefits are effectively proportional to wages. Hence, the DC system shares productivity risk as effectively as the wage-indexed DB system, but only because DC makes indexation clauses meaningless. A price-indexed system with DC-type adjustments to demographic shocks would be even worse than a pure DC system.

To conclude, all interpretations of piecemeal, cost-cutting pension reform imply that movements away from wage-indexed defined benefits have adverse effects on risk-sharing. If retirement benefits must be reduced, an outright, transparent benefit reduction that leaves a wage-indexed defined-benefit system in place seems more efficient than cost-cutting through shifts in indexation.

3.4 Partial Prefunding

Now consider the second main reform trend, the movement towards prefunding through a publicly-owned trust-fund (e.g. in Sweden, U.S, Canada). The two key questions for a trust-fund system are how the fund is invested and who the true beneficiaries are. In accounting terms, a trust fund buildup means that worker pay more in payroll taxes than current retirees receive in transfers. The surplus is invested and helps to pay pension benefits in future periods, i.e., helps to keep future payroll taxes lower.

The question about true beneficiaries is again about defined benefits versus defined contributions. Within a defined-contributions system, trust-fund returns would accrue to retirees, as if there was no trust fund and as if the retirees were holding the portfolio themselves (Bohn, 1997). The trust fund would be irrelevant for risk-sharing. Within a defined-benefits system, in contrast, a public trust fund has significant risk-sharing because the investments returns effectively accrue to future generations (Bohn, 1997; 1999a).

The economic analysis is most straightforward if the trust fund holds government bonds (or close substitutes, like high-quality corporate bonds). In this case, the fund is economically equivalent to a reduction in net government debt. For given intergenerational transfers θ^*, this allows the replacement rate β to be higher than in the privatization scenarios. Table 5, Col.2, shows the risk-sharing implications of a trust fund that holds the entire gross public debt, reduces net debt to zero, and keeps the replacement rate at 42%, above the pure PAYG value of 32%. For comparison, the PAYG allocation is reproduced in Col.1. Since trust funds with different sizes would have proportional effects, the 100%-of-debt case is sufficient to explain the impact of prefunding. Col. 2 shows that productivity, population, and longevity risks are better shared with partial prefunding than in the pure PAYG case. For the intuition, recall that safe debt shifts productivity risk and population risk from retirees to workers. The trust fund does the reverse, and thereby equalizes retiree and worker exposure to a_t- and n_t-shocks. In addition, the increased replacement rate provides more annuitized benefits and helps share longevity risk.

One alternative is to invest public trust-fund balances in the stock market, i.e., in claims on capital. Public equity investments are promising in this context because they allow a flexible sharing of valuation risk (Bohn, 1999a), which augments risk-sharing through capital income taxes (Smetters, 2000). This is demonstrated in Table 5, Col.3. Pensions, gross public debt, and the 100% of-debt size of the trust fund are the same as in Col.2, but the trust fund holds claims on capital. This leaves gross government debt in the hand of the public, i.e., involves a leveraged equity investment.

Table 5. The Role of Pension Trust Fund Investments

Policy:	**Pure PAYG Pension** (For comparison =Table 4, Col.2	**Trust Fund in Gov. Bonds** **Debt = Safe**	**Trust Fund in Equities** **Debt = Safe**	**Trust Fund in Equities** **Debt = Wage-indexed**
Risk/Exposure Column:	(1)	(2)	(3)	(4)
Productivity (a_t)				
Workers $\pi_{c1,a}$	0.788	0.741	0.773	0.735
Retirees $\pi_{c2,a}$	0.525	0.577	0.534	0.586
Risk-sharing:	Gap = 0.263	Gap = 0.165	Gap = 0.238	Gap = 0.148
		Better than PAYG	Worse than Col.2	Smallest Gap/Best
Asset Value (v^k_t)				
Workers $\pi_{c1,v}$	0.019	0.020	0.030	0.030
Retirees $\pi_{c2,v}$	0.112	0.112	0.099	0.099
Risk-sharing:	Gap = 0.093	Gap = 0.092	Gap = 0.069	Gap = 0.069
		Like PAYG	Better than Col.2	Like Col.3/Best
Workforce (n_t)				
Workers $\pi_{c1,n}$	-0.045	0.008	0.039	0.055
Retirees $\pi_{c2,n}$	0.287	0.213	0.170	0.148
Risk-sharing:	Gap = 0.332	Gap = 0.205	Gap = 0.131	Gap = 0.093
		Better than PAYG	Better than Col.2	Smallest Gap/Best
Longevity (λ_t)				
Workers $\pi_{c1,\lambda}$	-0.523	-0.535	-0.535	-0.535
Retirees $\pi_{c2,\lambda}$	-0.762	-0.688	-0.688	-0.688
Risk-sharing:	Gap = 0.239	Gap = 0.153	Gap = 0.153	Gap = 0.153
		Better than PAYG	Same as Col.2	Like Col.3/Best
Overall assessment:		Better than PAYG	Better than PAYG	Combines best of Col.2 and Col.3

Legend: The table entries show elasticities of worker and retiree consumption with respect to macroeconomic disturbances under different policy assumptions. All columns describe the "aged" economy with zero population growth and 50% retiree-worker ratio. Col. 1 shows the basic PAYG/defined-benefits policy from Table 3, Col.2, for comparison. Col. 2-4 all assume a wage-indexed defined-benefit pension system with partial prefunding, but with different investment policies.

Since future generations effectively hold part of the capital stock through the trust fund, they have increased exposure to valuation risk and increased exposure to population risk. However, equity investments also shift productivity risk to future generations, which increases the gap between worker and retiree exposures.[19] Innovative debt management provides an even better solution that avoids the trade-off between valuation and productivity risk. As illustrated in Table 5, Col.4, the adverse shift of productivity risk can be avoided if government debt is wage-indexed instead of safe; population risk would also be better shared.

Overall, Table 5 yields two general and arguably robust lessons about debt management. First, a combination of "risky" trust-fund investments (here, equity) combined with "risky" debt (here, wage-indexed) can produce better risk-sharing results than simpler alternatives. Second, compared to safe debt, innovative debt management can improve welfare in the direction of taking a "short" position in productivity risk (exemplified by wage-indexed debt) and by taking a "long" position in valuation risk (exemplified by equities).

For comparison, Bohn (1990) used tax-smoothing arguments in a Ricardian economy to argue for a government short position in either GDP-indexed bonds or in stocks. Intergenerational risk-sharing suggests a refinement. Wage-indexed debt, wage-indexed public pensions, GDP-indexed debt and a short position in equities all give the government a short position in productivity (a_t). This improves risk-sharing *and* is desirable on tax-smoothing grounds. With regard to demographic shocks, there is a tradeoff between tax-smoothing and risk-sharing. Risk-sharing tends to favour defined-benefit pensions (best) or wage-indexed debt (second best, because they lack annuitization), both because they yield a long position in "population" (n_t). Tax smoothing, in contrast, suggests a short position in n_t-shocks, both in Bohn (1990) and in OG model here. (To see the former, note that θ^*_t in (12) would be "smoothed" by a positive correlation between R^d_t and n_t.) Valuation shocks were implicitly treated as noise in Bohn (1990), which reduced the government's optimal equity position without overturning the negative sign. Here, valuation shocks are treated as a separate risk factor; the optimal policy is then a long position in equity (in v^k_t) combined with productivity-contingent debt that ensures a net short position in productivity.

In the U.S., proposals to invest government trust funds in the stock market have encountered strong opposition because of corporate control concerns. While European countries may not be as averse to government influence over corporations, the amount of trust-fund-equity holdings is probably limited in practice. The rest of the paper will therefore focus on privatization scenarios.

The representative agent analysis used so far implicitly assumes that all individuals of a cohort have equal access to financial markets. This is clearly

---

19 For calibrated U.S. data, I found that leveraged equity investments yield a net welfare gain despite the adverse shift in productivity risk (Bohn, 1999a) This is a somewhat fragile result, however, because it relies on an estimated covariance of shocks based on only about 4 generational observations (120 years of data).

unrealistic for most European countries, and an abstraction even for the U.S. The next section will turn to a heterogeneous agent setting.

4 Risk-sharing with Heterogeneous Agents

Retirement funding in Europe differs significantly from the laissez-faire model. Government pensions are sufficiently generous that much of the population has no need to save privately for retirement. Redistributional policies and the presence of non-savers have potentially significant effects on the allocation of risk. This section develops a stylized "welfare-state" OG model to capture these issues.[20] The main points are that a welfare state has potentially excellent risk-sharing properties; that risk-sharing through welfare-state taxation is difficult to sustain with an ageing population; and that population ageing is therefore likely to trigger a transition to a setting where most citizens make their own savings decisions and where the risk-sharing principles of Sections 2-3 apply.

4.1 A Welfare State OG Model

Consider the same OG model as before except that each generation consists of a continuum of individuals with heterogeneous labour productivity.

Individuals are identified by their relative productivity level x. The wage income of a type-x individual is $x \cdot w_t$. Let x have a distribution F(x) with unit mean so that w_t is the average wage. For comparison with Sections 2-3, note that heterogeneity per se is inessential if individuals have standard time-additive homothetic preferences. A laissez-faire economy with heterogeneity then has the same per-capita allocation as the representative-agent OG economy except that individuals with productivity $x \neq 1$ save and consume x times as much as individuals of type x=1. The same applies to an economy with proportional taxes and transfers. Hence, non-proportional redistributive taxes and transfers are critical to obtain qualitatively new results.

A second critical assumption is that individuals cannot borrow against public pensions or other transfers. At some level of public retirement benefits, individuals will stop saving and drop out of financial markets. Their consumption path and their exposure to aggregate risks are determined by the government. In a prototypical welfare state, this applies to bulk of the population. People expect the government to protect them from risk and to provide for their retirement. Such a

[20] Welfare-state is used as a label and not in a judgmental way. The general model also applies to the U.S. (which also has a sizeable fraction of non-savers), but the interpretation focuses on a European-style welfare state.

welfare state can have a very different allocation of risk from an economy with universal private savings.[21] To be clear about the label: by a welfare state, I mean a system of public pensions and transfers generous enough so that most people don't save.

It is beyond the scope of this paper to formalize the limits of income redistribution, to motivate redistribution, or to examine the possible allocations in great detail. The objective is merely to provide a framework for this about aggregate risk-sharing in a heterogeneous agent environment. The desired level of cross-sectional redistribution is therefore taken as given. The focus is on the aggregate, intertemporal risk-sharing implications.

Welfare-state governments tend to impose a variety of taxes, e.g., on payrolls, on general labour income, on capital income, and on consumption (or value added). For the worker cohort, let $\theta_t(x)$ be the payroll-tax rate, let $\tau_t^I(x, s_t(x))$ be the savings incentives (if any) as a fraction of savings $s_t(x)$, and let $\tau_t^1(x)$ be the total of other taxes and transfers as a fraction of the wage (negative in case of net transfers). All taxes and transfers may depend on productivity x, either directly - without explicit modelling - through observable proxies like income. Because savings are the model's key endogenous variable, savings-related taxes and incentives are distinguished from others and allowed to vary with the level of savings. A worker's budget equation is then

$$c_t^1(x) = x \cdot w_t \cdot [1 - \theta_t(x) - \tau_t^1(x)] - s_t(x) \cdot [1 - \tau_t^I(x, s_t(x))] \tag{13}$$

For retirees, let $\beta_{t+1}(x)$ denote the pension replacement rate, let $\tau_{t+1}^2(x)$ represent all other taxes on retirees without savings (e.g., consumption taxes), and let $\tau_{t+1}^R(x, s_t(x))$ be the tax rate on capital income. Then

$$\begin{aligned} c_{t+1}^2(x) = {} & \beta_{t+1}(x) \cdot x \cdot w_{t+1} - \frac{\tau_{t+1}^2(x)}{\lambda_{t+1}} \cdot x \cdot w_{t+1} \\ & + \frac{R_{t+1}^s(x)}{\lambda_{t+1}} \cdot s_t(x) \cdot [1 - \tau_{t+1}^R(x, s_t(x))] \end{aligned} \tag{14}$$

is the budget equation of retirees, where $R_{t+1}^s(x)$ is the gross return on savings.

21 To emphasize, the key issue is market participation, not heterogeneity or redistribution per se. Regardless of government intervention, if all individuals of a cohort can trade state-contingent claims, they will equate their marginal rates of substitutions, i.e., share risk within a cohort. The allocation of risk between cohorts would then resemble the representative agent OG allocation. This section considers the opposite case, an economy where only a minority engages in private retirement savings.

For tractability and to focus on financial market participation, I will make assumptions designed to confront individuals with a discrete choice, to be *non-savers* who accept the age-consumption profile implied by the government's tax-transfer system, or to be *savers* who have full access to financial markets. Financial markets are assumed to be complete so that savers can and will perfectly pool their risks. (This implicitly determines how $R^s_{t+1}(x)$ varies with x.) All savers are assumed to face the same effective marginal tax rate on all financial investments, to avoiding market segmentation. To produce a discrete choice, my main assumption is that a certain minimum level of savings is required to access to financial markets, namely savings of at least $\sigma \cdot w_t$, where $\sigma>0$ is exogenous.[22] One may interpret σ as the minimum collateral needed to operate on contingent claims markets, or as the minimum equity needed to establish a business. In many countries, only business-owners can take advantage of savings incentives such as investment-tax credits and accelerated depreciation. Such incentives are especially important if capital income is taxed at high statutory rates, as is common in welfare states, and the government relies on special incentives ($\tau^I>0$) to avoid gross distortions in capital accumulation. With this motivation, I assume that individuals with savings $s_t(x) \geq \sigma \cdot w_t$ face a common tax rate $\tau^R_{t+1}(x, s_t(x)) = \tau^R_{t+1}$ and incentives $\tau^I_t(x, s_t(x)) = \tau^I_t$. This yields a common effective tax rate on savings, $\tau^s_{t+1} = (\tau^R_{t+1} - \tau^I_t)/(1 - \tau^I_t) \geq 0$. For savings below $\sigma \cdot w_t$, assume the effective tax rate is high enough to discourage savings.

The population then divides into two groups, non-savers and savers with $s_t(x) \geq \sigma \cdot w_t$. Non-savers have a consumption stream that is entirely determined by government policy.[23] Their consumption-growth rate is

$$(15) \qquad \frac{c^2_{t+1}(x)}{c^1_t(x)} = \frac{w_{t+1}}{w_t} \cdot \left(\frac{\beta_{t+1}(x) - \tau^2_{t+1}(x)/\lambda_{t+1}}{1 - \theta_t(x) - \tau^1_t(x)} \right).$$

---

22 The minimum is assumed to be proportional to w_t to maintain balanced growth; $\sigma>0$ is a technical necessity because existence problems would arise if individuals could gain access to complete markets with infinitesimal savings. One may assume that σ is small so that with uniform taxation only a small fraction of individuals would save less than $\sigma \cdot w_t$.

23 A discussion of private pension plans is beyond the scope of this paper, but they should be mentioned here. Traditional corporate pension plans with defined benefits would also provide an essentially exogenous retirement income and further reduce the incentives to save individually. Defined-contribution plans (as increasingly common in the U.S.), in contrast, should be considered a form of individual savings. In a welfare state, pension plans tend to have defined benefits and are usually subject to tight regulations. Hence, they are reasonably treated as government-controlled for the purposes of this analysis.

Typical public pensions offer replacement rates that decline with income. Hence, it is reasonable to assume that the ratio on the r.h.s. of (15) is decreasing in x (at least weakly). For any $\sigma>0$, the ratio of minimum savings over type-x's wage, σ/x, is strictly decreasing in x, which makes individual incentives to save strictly increasing in x. Hence, there are cutoff values x^0_t and $x^+_t \geq x^0_t$ such that individuals with productivity $x \leq x^0_t$ are non-savers, individuals with $x \in (x^0_t, x^+_t]$ save the minimal amount, and individuals with $x > x^+_t$ save more than $\sigma \cdot w_t$. If σ is small and the cross-state pattern of pension benefits is reasonably efficient, x^+_t will be close to x^0_t, which makes the corner solution $s_t(x) = \sigma \cdot w_t$ unimportant. The main point is that individuals sort themselves into savers and non-savers by productivity (or, equivalently, by income).

For any value of x, savings incentives are a decreasing function of net retirement transfers $\beta_{t+1}(x) - \tau^2_{t+1}(x)/\lambda_{t+1}$ and of taxes $\theta_t(x) + \tau^1_t(x)$. In a welfare state, pensions and transfers are sufficiently generous and taxes sufficiently high that the cutoff x^0_t lies in the upper end of the income distribution, i.e., that most people don't save.

On aggregate, the heterogeneous agent OG economy works as follows: since savers have access to complete markets, their consumption across states of nature is perfectly correlated. The marginal rate of substitution of the unconstrained savers defines a well-defined pricing kernel for valuing all financial assets. Denoting aggregate savings by $S_t = (\int_{x \geq x^0_t} s_t(x) dF) \cdot N_t$, capital-market equilibrium requires $K_{t+1} = S_t - D_{t+1}$. This condition combined with savers' first-order optimality condition for capital investment determines the equilibrium capital stock. The PAYG constraint for pensions is now $(\int \beta_t(x) \cdot x dF) \cdot \lambda_t \cdot N_t) = (\int \theta_t(x) \cdot x dF) \cdot N_t$. The general government-budget constraint is $(\int \tau^1_t(x) \cdot x dF) \cdot w_t \cdot N_t - \tau^I_t \cdot S_t + (\int \tau^2_t(x) \cdot x dF) \cdot w_t \cdot N_{t-1} + \tau^R_t \cdot (\int R^s_t(x) \cdot s_{t-1}(x) dF) \cdot N_{t-1} + D_{t+1} = R^d_t \cdot D_t$. Assuming policy depends only on current x-values, on current aggregate shocks, and on the initial capital-labour ratio, the economy has again a Markov structure with the same aggregate state variables as the representative agent OG model.

For the interpretation, a simplified distributional scenario is perhaps instructive: suppose there are three groups, some "poor" low-productivity types $x=x^L$, a large middle class $x=x^M$, and a minority of "rich" high-productivity types $x=x^H$, where $x^L<x^M<x^H$. As starting point, assume pensions are generous enough that neither the poor nor the middle class have sufficient incentives to save ($x^M<x^0_t$ for all t). The government controls their exposure to risk through taxes and transfers. Only the rich save on their own ($x^H>x^+_t$). Their risk exposure is influenced indirectly, as in Sections 2-3: they hold all savings, savings must add up to capital plus government debt, and the government controls taxes and the supply of debt securities. The next section will examine optimal risk-sharing in this setting.

4.2 Efficient Policies in the Welfare State

This section considers an "ideal" welfare state, a setting where the government allocates aggregate risks perfectly, sets taxes efficiently, and attains redistributional goals at its discretion.

Efficiency (ex ante Pareto optimality) imposes strong restrictions on the timing of taxes and transfers and on capital income taxes. First, since distortions in the capital stock are inefficient, the ideal welfare state should set the marginal tax rate on savings at zero, $\tau^s=0$ (i.e., $\tau^R=\tau^I\geq 0$). Intuitively, whatever revenue the government wants to extract from high-income individuals is more efficiently raised through wage taxes or constant consumption taxes rather than capital income taxes.

Second, efficiency calls for equal marginal rates of substitution across income groups. Otherwise, utility could be increased by shifts in tax timing across groups. Unconstrained savers already have a common marginal rate of substitution. The marginal rate of substitution of non-savers is determined by the consumption-growth rate, which is government-controlled as shown in equation (15). An efficient tax-transfer system therefore requires that the ratio

$$(16) \qquad \frac{\beta_{t+1}(x)-\tau_{t+1}^{2}(x)/\lambda_{t+1}}{1-\theta_t(x)-\tau_t^{1}(x)}$$

is the same for all non-savers and that the resulting marginal rate of substitution matches the marginal rate of substitution of savers. If σ is near zero, this means that non-savers are just on the verge of starting to save (at least if savings are not taxed).[24] For each cohort and productivity level, this determines the time profile of taxes and transfers. Since efficiency requires equal consumption growth for all groups and all states of nature, this growth rate equals the growth rate of aggregate consumption:

$$\frac{c_{t+1}^{2}(x)}{c_t^{1}(x)} = \left(\int c_{t+1}^{2}(z)\, dF(z)\right) / \left(\int c_t^{1}(z)\, dF(z)\right) \text{ holds for all x.}$$

This aggregation condition has an important implication: *The welfare state with efficient redistribution has exactly the same macroeconomic dynamics as a representative agent OG economy.* Hence, the main insights from Sections 2-3 continue to apply.

---

24 With efficient risk-sharing though transfers, individuals have no incentive to save merely to gain access to financial markets. This makes the assumption $\sigma>0$ unnecessary. For non-zero σ, efficiency requires a discontinuous downward jump in retirement benefits at x^0_t so that the minimum savings constraint is non-binding, i.e., all savers are unconstrained ($x^0_t=x^+_t$).

Given efficiency, the only new issue is redistribution within a cohort. Intra- and intergenerational distribution can be distinguished by their impact on generational accounts. Policies that change the retirees generational account are inter-generational transfers; all others are intra-generational. For comparison with (12), intergenerational transfers can be expressed in terms net payments to the government as a share of the wage, θ^*_t; this is again the negative of retirees' generational account divided by $N_t \cdot w_t$. Using the government-budget equation, θ^*_t can be written as

$$(17)\, \theta_t^* = \left(\int x \cdot \theta_t(x) dF\right) + \left(\int x \cdot \tau_t^1(x) dF\right) - \tau_t^I \cdot \left(\int s_t(x) dF\right) + \frac{D_{t+1}}{N_t \cdot w_t}$$

$$= \left(\int [\beta_t(x)\lambda_t - \tau_t^2(x)] \cdot x dF\right) \cdot e^{-n_t} - \tau_t^R \cdot \left(\int R_t^s(x) \cdot s_{t-1}(x) dF\right) \cdot e^{-n_t} + \frac{R_t^d \cdot D_t}{N_t \cdot w_t}$$

This captures the distributional equivalence of payroll taxes, other taxes, and new debt for workers and of pension benefits, taxes, and payments on government debt for retirees. This equivalence combined with the multiplicity of policy instruments in (16) implies that the government has many different ways to implement efficient risk-sharing at a given level of inter- and intra-generational redistribution. The redundant instruments can be set arbitrarily.

Table 6, Col.1 shows the efficient aggregate allocation of risk in the calibrated economy of Section 2 (with 1% annual population growth and $\overline{\lambda}$ =1/3).

Given the multiplicity of policy instruments, optimal taxes and transfers for type-x agents are best specified in terms of total transfers $\theta^*_t(x)$, the agents' contribution to the intergenerational transfer in (17). Non-saving workers contribute $\theta^*_t(x)=\theta_t(x)+\tau^1_t(x)$ to the generational account. The optimal elasticities of $1-\theta^*_t(x) = c^1_t(x)/(x \cdot w_t)$ with respect to the shocks are uniquely determined and shown in Col.2.[25] The same elasticities apply to the net replacement rate of non-saving retirees, to $\beta_t(x)-\tau^2_t(x)/\lambda_t$. Table 6 suggests that efficient taxes on non-savers should increase with productivity, decrease when asset values are high (when tax receipts from capital owners are high), decrease with population growth, and increase with longevity. Note in particular that the efficient risk-sharing response to population ageing (low n_t, high λ_t) calls for higher taxes, confirming the findings in Section 2.

---

25 The elasticities of $\theta^*_t(x)$ with respect to shock are equal to $-(1-\theta^*_t(x))/\theta^*_t(x)$ times the elasticities of $1-\theta^*_t(x)$. I show the elasticities of "1-tax rate" because they do not depend on x. Since $(1-\theta^*_t(x))/\theta^*_t(x)$ varies with x, the elasticities of $\theta^*_t(x)$ vary with x and are sensitive to the level of taxes. Analogous calculations for savers are omitted because they would mix taxes and debt purchases, and therefore be difficult to interpret.

Table 6. Efficient Risk-sharing and Taxes in the Welfare State

Elasticities for: **Risk/Exposure**	Efficient consumption	Tax factors $1-\theta^*_t(x)$ of workers
Column:	(1)	(2)
Productivity (a_t)		
Workers $\pi_{c1,a}$	0.691	-0.009
Retirees $\pi_{c2,a}$	0.691	
Risk-sharing:	Efficient	
Asset Value (v^k_t)		
Workers $\pi_{c1,v}$	0.031	+0.031
Retirees $\pi_{c2,v}$	0.031	
Risk-sharing:	Efficient	
Workforce (n_t)		
Workers $\pi_{c1,n}$	0.097	+0.397
Retirees $\pi_{c2,n}$	0.097	
Risk-sharing:	Efficient	
Longevity (λ_t)		
Workers $\pi_{c1,\lambda}$	-0.503	-0.503
Retirees $\pi_{c2,\lambda}$	-0.503	
Risk-sharing:	Efficient	

Legend: Col.1 shows the consumption elasticities with respect to shocks when taxes and transfers are used efficiently to share all risks perfectly. Col.2 shows the elasticities of $1-\theta^*_t(x)$ with respect to shocks that are required to implement perfect risk-sharing. The elasticities of $\theta^*_t(x)$ with respect to the shocks would have the opposite signs.

4.3 Ageing, Credibility, and Middle-Class Savings

While a welfare state can implement an efficient allocation with perfect risk-sharing, it is doubtful that the efficiency benchmark is attained in practice, and it is unlikely that the ideal welfare state can survive population ageing.

One problem is that the abundance of discretionary policy interventions raises credibility and commitment issues. In practice, welfare states tend to impose high taxes on capital income. While statutory taxes may overstate the effective tax rates, e.g., if investment is subsidized through tax credits or generous depreciation allowances, any positive taxation will depress capital accumulation and result in first-order welfare losses.[26] Lack of commitment provides a plausible explanation for the observed high capital income taxes. The cost of commitment problems has arguably increased in recent decades due to increased international capital mobility and tax competition.

Ageing is a serious problem for the welfare state because the government must finance the *entire* consumption of retirees to maintain control over their exposure to risk. This becomes increasingly costly as the population ages. The PAYG constraint (8) shows that population ageing reduces the ratio of replacement rate to payroll-tax rate. Unless the adjustment takes place primarily through higher taxes, not reduced benefits, this will reduce the consumption-growth rate of non-savers (see (15)-(16)) and therefore motivate the middle class to start saving and perhaps, to lobby for tax reforms that facilitate retirement savings.

Thus, maintaining the welfare state as the population ages requires either a huge increase in redistribution from the rich to the middle class and poor (holding (16) and θ^* constant) or a massive increase in intergenerational redistribution (holding (16) constant by raising θ^*). If higher taxes on the rich are impractical (taxes that can easily be imposed on the rich are probably in effect already, and an increased temptation to tax the rich would reinforce the capital-levy problem) and if increased intergenerational redistribution is politically undesirable, this leaves only one solution: the middle class must start to save for its own retirement.[27]

If this solution is accepted, it triggers a quantum leap in the allocation of risk - not just a marginal shift, as in the representative agent OG model. Several policy recommendations follow. First, private savings require a credible commitment not to tax the returns. As more people are supposed to save, avoiding capital-levy problems becomes increasingly important. This suggests caution about discretionary taxation, especially of retiree incomes. A better alternative is to share risk through government bonds with contractually well-defined, non-discretionary indexation clauses. Second, if the poor are a sufficiently small group, reforms that encourage middle-class savings will release resources that can be used to maintain government support for the poor. To avoid the Samaritan's dilemma, the government should probably maintain considerable control over the

---

26 Lucas (1990) has argued that distortionary capital income taxes that reduce the capital stock induce quantitatively large welfare losses. The welfare state faces many other problems, of course, e.g., rigidities in making policy adjustments. Rigidities are of second-order importance, however, if policy instruments are set optimally on average and the welfare function is flat near the optimum. Hence I emphasize systematic biases that yield first-order welfare losses, such as capital income taxes.

27 Another alternative is massive government savings, e.g., in a pension-trust fund. To the extent that large, equity-holding trust funds or direct government ownership of capital are politically acceptable, public savings is obviously a substitute for private savings. The same considerations as in Section 3 would apply; hence this option is ignored here.

saving of potential welfare recipients. Thus, public pensions will continue to have a role as forced saving mechanism, but only at the bottom of the income distribution.

The need to avoid discretionary retiree taxation has several implications of its own. Discretion must be constrained for all components of retirees' generational account, otherwise the government could impose indirect capital levies. Credible plans are therefore needed not only for taxes, but also for pension benefits and debt returns.

With regard to pensions, credibility requires clear pension rules, including commitments about their taxation. This might also be useful to minimize labour-supply distortions by convincing workers that payroll taxes are in part contributions to a "savings plan" that offers a well-defined return. The results in Sections 3 suggest that a wage-indexed defined-benefits plan remains the best choice, especially if the choice is among simple, transparent systems.

With regard to government debt, credibility suggests a move away from nominal bonds that are subject to discretionary monetization and towards securities that explicitly are contractually indexed to relevant macroeconomic variables. The recent trend towards price-indexed debt (e.g., in the UK, US and Sweden) documents the feasibility of indexation. Unfortunately, price-indexed debt makes no contribution to risk-sharing and mainly eliminates the informal shock-absorber role of nominal debt. Since the heterogeneous agent economy faces the same aggregate risk-sharing problems as the representative agent economy, the same forms of indexation are welfare-improving, notably wage-indexation and linkages to longevity. An indexation of government debt to these variables would be preferable to price-indexation.

As a caveat, note that optimal debt may consist entirely of nominal bonds if the monetary authority is viewed as credible with regard to maintaining low average inflation *and* willing to pursue a state-contingent inflation policy that induces optimal variations in the real value of nominal bonds. However, my interpretation of current trends in monetary policy is that central banks are trying to build credibility by targeting a fixed low inflation rate, and would be unwilling to vary inflation in the interests of debt policy. This makes nominal debt equivalent to price-indexed debt and hence inferior to wage- and longevity-indexed debt.

With regard to capital income taxes, policy-makers around the world seem to take one of two approaches. One is to exempt capital income from taxation ($\tau^I=\tau^R=0$), at least up to a generous limit. The other is to set up special retirement-savings plans, exempt the contributions from wage taxes, and subject the withdrawals to normal income taxes ($\tau^I\approx\tau^R>0$, as exemplified by the U.S 401k and IRA plans). If credible, both approaches allow for approximately undistorted savings choices. I would conjecture that the savings-plan approach is more credible, however, because it grants a tax break up front and "back-loads" the taxes. The two approaches are also different with regard to risk-sharing. The savings plan approach provides an automatic sharing of rate-of-return risk. In addition (though related in equilibrium) the tax deferral implies a higher government debt that can be used for debt management.[28] A movement towards tax-exempt capital income, in contrast, would eliminate the risk-sharing features inherent in the existing income-tax system.

From a political perspective, the increased number of savers might help to make savings incentives more credible. Savers will presumably vote against taxes on their savings. But this does not resolve the temptation to tax the savings of others. In part, credibility is therefore about transparency and uniformity, so that savers are indeed politically protected by the presence of many other voters/savers facing the same rules. My interpretation of welfare-state saving incentives is that high statutory capital-income-tax rates are, at least in part, offset by a multitude of sometimes opaque investment incentives-ranging from outright investment-tax credits to tax breaks for locating in certain communities or regulatory permits that effectively convey valuable market power. Such incentives may actually work if they involve large corporate investors powerful enough to negotiate directly with public officials. However, as we move to a setting with more small savers, such opaque incentives must be replaced by a more open and transparent system.

---

28 A substitution of labour-income taxes by a uniform consumption tax would imply a similar back-loading of taxes and a sharing of return risk. In Europe, the value-added tax can be interpreted in this way, but only if the tax rate is held constant to avoid a tax on savings.

4.4 Examples of efficient and nearly-efficient risk-sharing

Table 7 provides several scenarios of efficient risk-sharing that combine debt-management, pension, and tax considerations.

Table 7. Optimal Debt Management with an Aged Population: Perfect Risk-sharing

Policy Setting: **Risk/Exposure**	Optimal Debt Combined with DB pension & standard taxes	Optimal Debt combined with tax-deductible retirement savings	Optimal Debt Combined with DC pension & untaxed savings
Column:	(1)	(2)	(3)
Productivity (a_t)			
Workers $\pi_{c1,a}$	0.681		
Retirees $\pi_{c2,a}$	0.681		
Risk-sharing:	Efficient	= Col.1	= Col.1
Asset Value ($v^k{}_t$)			
Workers $\pi_{c1,v}$	0.057		
Retirees $\pi_{c2,v}$	0.057		
Risk-sharing:	Efficient	= Col.1	= Col.1
Workforce (n_t)			
Workers $\pi_{c1,n}$	0.092		
Retirees $\pi_{c2,n}$	0.092		
Risk-sharing:	Efficient	= Col.1	= Col.1
Longevity (λ_t)			
Workers $\pi_{c1,\lambda}$	-0.569		
Retirees $\pi_{c2,\lambda}$	-0.569		
Risk-sharing:	Efficient	= Col.1	= Col.1
Debt Management:			
Factor loadings:			
Productivity $\pi_{Rd,a}$	2.095	1.129	1.755
Valuation $\pi_{Rd,v}$	-0.737	-0.397	-1.107
Population $\pi_{Rd,n}$	-2.626	-1.415	-6.167
Longevity $\pi_{Rd,\lambda}$	2.601	1.401	5.801

Legend: The table entries show the elasticities of worker and retiree consumption with respect to macroeconomic disturbances with optimal debt policy, and the associated debt-policy parameters.

Table 7 shows the efficient allocation of risk in the "aged" economy with $\overline{n}$ =0 and $\overline{\lambda}$ =0.5, and how it can be implemented via optimal debt policy at the original level of intergenerational redistribution (θ^*=0.15%). The efficient allocation is similar to the efficient allocation of the "young" economy shown in Table 6, but it would be very costly to implement without middle-class saving.[29] Specifically, the bottom rows show the factor loadings on the debt return R^d_t that are required to attain efficiency when combined with various tax and pension policies.

In Col.1, I assume debt policy is combined with the same wage-indexed PAYG pension (β=32%) and 25% capital income tax as in Section 3 (Table 4, Col.2). Some of the factor loadings are clearly high and might create concerns about practical feasibility. This issue can be addressed because the required factor loadings depend on the pension and tax system. Col.2 displays the debt-policy parameters when capital accumulation takes place through a retirement-savings plan that allows a 25% deduction of contributions and taxes the returns at the same rate. Optimal factor loadings are then much lower. Factor loadings would have to be much higher, on the other hand, if debt policy were accompanied by less-effective pension and tax policies. This is illustrated in Col.3, which assumes a DC pension (capped payroll taxes) and zero capital income taxes.

Another way to avoid policies that rely on high factor loadings is simply to constrain the factor loadings. Two examples are presented in Table 8. In both, factor loadings are limited to ±1.0 for productivity, population, and longevity, and to zero for valuation risk (i.e., ruling out claims on capital).

In Table 8, Col.1 again considers debt policy in a setting with wage-indexed PAYG pension and 25% capital income tax (as in Table 4, Col.2 and Table 7, Col.1). Col.2 displays a non-nested alternative, an optimal debt policy with the same constraints on factor loadings, but in a setting with an equity-holding trust fund (as in Table 5) instead of a retirement-saving plan. This alternative yields better sharing of valuation risk but a less efficient sharing of productivity risk.

Even with the limited factor loadings, both policy settings in Table 8 yield better risk-sharing than any of the "simple" policies discussed in Section 3. To highlight the comparison, Col.3 shows the "capped taxes" scenario from Table 4, Col.4, which is my preferred interpretation of piecemeal pension reform. For productivity, workforce, and longevity risk, the gaps in the risk exposures between

---

29 In the calibrated model, payroll tax would have to rise from θ=16% to θ=28% to keep the ratio $\beta/(1-\theta)$ constant (which would be required according to (16)), i.e., θ^* would have to rise from 15% to 27%, almost as much as in the business-as-usual scenario of Section 3.

Table 8. Optimal Debt Management with Constrained Factor Loadings

Policy Setting: **Risk/Exposure**	Constrained Optimal Debt & Tax-deductible retirement savings	Constrained Optimal Debt & Trust fund holding capital	"Capped taxes" & Safe Debt (For comparison =Table 4, Col.4)
Column:	(1)	(2)	(3)
Productivity (a_t)			
Workers $\pi_{c1,a}$	0.693	0.731	0.788
Retirees $\pi_{c2,a}$	0.663	0.609	0.525
Risk-sharing:	Gap = 0.03 Almost efficient	Gap = 0.122	Gap = 0.263
Asset Value (v^k_t)			
Workers $\pi_{c1,v}$	0.019	0.029	0.019
Retirees $\pi_{c2,v}$	0.112	0.099	0.112
Risk-sharing:	Gap = 0.093	Gap = 0.07 Better than Col.1	Gap = 0.093
Workforce (n_t)			
Workers $\pi_{c1,n}$	0.052	0.089	-0.212
Retirees $\pi_{c2,n}$	0.149	0.096	0.525
Risk-sharing:	Gap = 0.097 Almost efficient	Gap = 0.007 Almost efficient	Gap = 0.737
Longevity (λ_t)			
Workers $\pi_{c1,\lambda}$	-0.556	-0.559	-0.466
Retirees $\pi_{c2,\lambda}$	-0.624	-0.613	-1.000
Risk-sharing:	Gap = 0.068 Almost efficient	Gap = 0.055 Almost efficient	Gap = 0.534
Debt Management:			
Factor loadings:	Constrained to [-1,1]	Constrained to [-1,1]	Zero (Safe Debt)

Legend: Cols.1-2 show the allocation of risk with constrained optimal debt policy. The optimal constrained factor loadings in both Col.1 and Col.2 are +1 on a_t, constrained zero on v_t, –1 on n_t, and +1 on λ_t. Col.3 shows the "capped taxes" scenario of Table 4, Col.4 for comparison.

retirees and workers are reduced by almost an order of magnitude (in Cols. 1-2 versus Col.3).[30]

30 In contrast, gaps in valuation risk are not much reduced. This is in part by assumption - the assumed zero factor loading on v_t -_ but not entirely. A full sharing of valuation risk would require either very high capital-income taxes or substantial public ownership of capital assets. Fortunately, the inefficiency is not as severe as the gap values might suggest because valuation risk is quantitatively much smaller than the

All the debt policies discussed in this section can be implemented in a heterogeneous-income environment. Assuming the poor do not save under either scenario, the government will still have to set their taxes and benefits optimally, as discussed in the previous section. For the middle class and the rich, the government has to do no more than to issue the appropriately-indexed government debt. Equilibrium asset prices will then direct savers to the efficient allocation. Savers' portfolios do not have to be identical for this to work. For example, if pensions are a larger share of middle-class than of rich retirees' incomes, middle-class retirees will naturally hold assets that provide hedges against the state-contingent features (if any) of the pension formula. Thus, efficient risk-sharing can be implemented in a market setting that does not impose unreasonable informational requirements on the government.

5 Concluding Comments

Population ageing in a welfare state is likely to trigger a transition to a policy-setting where most households will participate in financial markets. The government supply of savings instruments will then play a critical role in allocating risk. To gain an overall perspective on risk-sharing and on the role of innovative debt management, it is instructive to compare the allocation of aggregate risk with piecemeal reform/capped contributions (Table 8, Col. 3) to perfect risk-sharing and with the final privatization and prefunding scenarios in Table 8, Cols. 1-2.

To move from capped contributions to perfect risk-sharing, three elements interact: public debt management, public pensions and capital income taxes. Debt management alone would require an aggressive policy with high (perhaps implausibly high) factor loadings on public debt. Pension policy can help by maintaining a wage-indexed defined-benefit system that shifts demographic and productivity risks in the right direction. (In other words, if ageing makes the original replacement rates too expensive, cost-cutting should be done in a way that maintains wage-indexed defined benefits.) Capital income taxes can also help to share risk, though their role is limited because of time-consistency problems. Particularly promising are retirement-saving plans that provide tax breaks up front (say, tax-deductible savings) and subject the withdrawals to taxation. Pension policy provides additional degrees of freedom if pensions are partially prefunded and the trust fund is invested in capital assets.

---

other risks (see fn.6). The management of capital assets also raises moral-hazard issues that might justify stronger linkages between ownership and risk-bearing than this model.

Regardless of the specific pension and capital-income-tax setting, government bonds indexed to wages and demographic variables offer a flexible set of instruments for sharing risk. Table 8 provides two stylized examples that combine defined-benefits pensions and optimally-indexed (but leverage-constrained) public debt with either a retirement-savings plan (Col.1) or a partially prefunded pension system holding equities (Col.2). Both settings eliminate many of the inefficiencies of a safe-debt/capped-payroll-taxes scenario. Both rely on government debt indexed to productivity growth (positively), population growth (negatively), and longevity (positively). Optimal public debt management should supply to savers securities that reflect these risk exposures. In practice, wage-indexed and/or longevity- indexed debt would go in the right direction.

References

Auerbach, A. and K. Hassett. 1998. Uncertainty and the Design of Long-Run Fiscal Policy, University of California Berkeley, Working Paper.

Auerbach, A. and L. Kotlikoff. 1987. *Dynamic Fiscal Policy,* Cambridge, UK: Cambridge University Press.

Auerbach, A. and L. Kotlikoff and W. Leibfritz. 1999. *Generational Accounting around the World.* Chicago, University of Chicago Press.

Barro, R. 1979. On the Determination of Public Debt, *Journal of Political Economy*, 87, pp. 940-971.

Baxter, M. and R. King. 2001. The Role of International Investment in a Privatized Social Security System. in: J. Campbell and M. Feldstein, eds., *Risk Aspects of Investment Based Social Security Reform.* Chicago: University of Chicago Press, pp. 371-437.

Bohn, H. 1990. Tax-Smoothing with Financial Instruments, *American Economic Review*, 80, pp. 1217-1230.

Bohn, H. 1994. Optimal State-Contingent Capital Taxation: When is there an Indeterminacy? *Journal of Monetary Economics*, 34, pp. 125-137.

Bohn, H. 1997. Social Security Reform and Financial Markets. in: S. Sass and R. Triest, eds. *Social Security Reform: Links to Savings, Investment, and Growth.* Boston: Federal Reserve Bank of Boston, pp. 193-227.

Bohn, H. 1998. Risk-sharing in a Stochastic Overlapping Generations Economy, University of California Santa Barbara, Working Paper.

Bohn, H. 1999a. Should the Social Security Trust Fund hold Equities? An Intergenerational Welfare Analysis, *Review of Economic Dynamics*, 2 (3), pp. 666-697.

Bohn, H. 1999b. Will Social Security and Medicare Remain Viable as the U.S. Population is Ageing? *Carnegie-Rochester Conference Series on Public Policy*, 50, pp. 1-53.

Bohn, H. 2001. Social Security and Demographic Uncertainty: The Risk-Sharing Properties of Alternative Policies. in: John Campbell and Martin Feldstein, eds., *Risk Aspects of Investment Based Social Security Reform.* Chicago: University of Chicago Press, pp. 203-241.

Chamley, C. 1986. Optimal Taxation of Capital Income in General Equilibrium with Infinite Lives, *Econometrica*, 54, pp. 607-622.

Chari, V.V., L. Christiano and P. Kehoe. 1994. Optimal Fiscal Policy in a Business Cycle Model, *Journal of Political Economy*, 102, pp. 617-652.

Cutler, D., J. Poterba, L. Sheiner and L. Summers. 1990. An Ageing Society: Opportunity or Challenge? *Brookings Papers on Economic Activity*, pp. 1-56.

Diamond, P. 1965. National Debt in a Neoclassical Growth Model, *American Economic Review*, 55, pp. 1126-1150.

Disney, R. 2000. Crises in Public Pension Proframs in OECD: What are the Reform Options? *The Economic Journal*, 110, pp. F1-F23.

De Nardi, M., S. Imrohoroglu and T. Sargent. 1999. Projected U.S. Demographics and Social Security, *Review of Economic Dynamics*, 2 (3), pp. 575-615.

Fischer, S. 1980. Dynamic Inconsistency, Cooperation, and the Benevolent Dissembling Government, *Journal of Economic Dynamics and Control*, 2, pp. 93-107.

Gale, D. 1990. The Efficient Design of Public Debt. In: R. Dornbusch and M. Draghi, eds., *Public debt management: theory and history*. Cambridge,U.K.: Cambridge University Press, pp. 14-47.

Gordon, R. 1985. Taxation of Corporate Capital Income: Tax Revenues versus Tax Distortions, *Quarterly Journal of Economics*, 100, pp. 1-27.

Gordon, R. and L. Bovenberg. 1996. Why is Capital So Immobile Internationally? Possible Explanations and Implications for Capital Income Taxation, *American Economic Review*, 86, pp. 1057-1075.

Gruber, J. and D. A. Wise. 1999. eds. *Social security and retirement around the world.* Chicago: University of Chicago Press.

Lee, R. and S.Tuljapurkar. 1998. Population Forecasting for Fiscal Planning: Issues and Innovations, University of California Berkeley, Working Paper.

Lucas, R. 1990. Supply-side economics: An Analytical Review, *Oxford Economic Papers*, 42, pp. 293-316.

Lucas, R. and N. Stokey. 1983. Optimal Fiscal and Monetary Policy in an Economy without Capital, *Journal of Monetary Economics*, 12, pp. 55-93.

Knabb, S. 2001. Can Human Capital and Factor Price Movements Save Social Security? University of California Santa Barbara, Working Paper.

Kraay, A., N. Loayza, L. Serven and J. Ventura. 2000. Country Portfolios, MIT, Working Paper 00-16.

McHale, J. 2001. The Risk of Social Security Benefit-Rule Changes: Some International Evidence. in: J. Campbell and M. Feldstein, eds., *Risk Aspects of Investment Based Social Security Reform*. Chicago: University of Chicago Press, pp. 247-282.

Mendoza, E., A., Razin and L. Tesar. 1994. Effective tax rates in macroeconomics: Cross-country estimates of tax rates on factor incomes and consumption, *Journal of Monetary Economics*, 34, pp. 297-333.

Missale, A. 1999. *Public Debt Management*. Oxford, U.K.: Oxford University Press.

OECD. 1998. *Maintaining Prosperity in an Ageing Society*. Paris, France: Organization for Economic Cooperation and Development.

Ogaki, M. and C. Reinhart. 1998. Measuring inter-temporal substitution: The role of durable goods, *Journal of Political Economy*, 106 (5), pp. 1078-1098.

Peled, D. 1982. Informational Diversity over Time and the Optimality of Monetary Equilibria, *Journal of Economic Theory*, 28, pp. 255-274.

Raffelhüschen, B. 1998. Ageing, Fiscal Policy and Social Insurance: A European Perspective, Universität Freiburg, Working Paper.

Sinn, H.-W. 2000. Why a Funded Pension System is Useful and Why it is Not Useful, *International Tax and Public Finance*, 7, pp. 389-410.

Shiller, R. 1993. *Macro Markets*. Oxford, U.K.: Clarendon Press.

Smetters, K. 2000. The Equivalence of the Social Security's Trust Fund Portfolio Allocation and Capital Income Tax Policy, University of Pennsylvania, Working Paper.

Smith, A. 1982. Intergenerational Transfers as Social Insurance, *Journal of Public Economics*, 19, pp. 97-106.

Zhu, X. 1992. Optimal Fiscal Policy in a Stochastic Growth Model, *Journal of Economic Theory*, 58, pp. 250-289.

Discussion

Alan J. Auerbach

1 Introduction

There are many elements of the problem facing governments that must decide how to structure public pensions, in particular, and fiscal policy, more generally. In this thought-provoking paper, Henning Bohn focuses on one issue, in particular, how well do alternative policies succeed in spreading different risks across generations. Many of his conclusions are straightforward and unsurprising. Others require more thought to understand, but, in the end, are just as convincing. Still others, though, do depend somewhat on particular modelling assumptions. In my comments, I will review what I see as the paper's important insights, and comment on where I think results might not be robust.

2 Basic Approach

For all of his analysis, Bohn uses a two-period overlapping-generations model, in which each household works in the first period of life and consumes in both periods. The household's only decision is how much to save for old age; labor supply is fixed. Capital and labor are combined in a single production sector with Cobb-Douglas technology. The economy has four sources of uncertainty, associated with productivity (a), capital depreciation (v^k), population growth (n), and life expectancy (λ). These shocks affect different generations in different ways. Whereas a productivity shock affects both labor (the young) and capital (the old), a depreciation shock affects only the old. Whereas both population growth and longevity shocks affect the ratio of young to old, the growth shock affects the capital-labor ratio and hence factor prices, while the longevity shock does not.

Using this basic set-up, Bohn considers three versions of the model. First, he examines the risk-sharing properties of different policies in a stationary steady state. Second, he considers the impact of ageing on these results in a model parameterized to reflect an older population. Finally, he considers a model of heterogeneous households and limited participation in capital markets, an important addition given the large proportion of households entering retirement without significant portfolios of liquid assets.

A central aspect of the approach taken throughout the paper is that the effects of government activities are considered as a whole, rather than separately. That is, it is irrelevant from Bohn's perspective whether risk-sharing is provided through government debt policy, the use of capital-income taxes, or the design of public pensions. I think this is the right approach to take, but it can be contrasted with an approach that looks only at an economy's public pension scheme, or perhaps the combination of public pensions and private saving. Government alters risk-bearing in a multitude of ways, including many not included in this paper, so it makes little sense to look at only one in trying to evaluate the performance of pension policy.

In performing his evaluations, Bohn measures success or failure by the degree to which the risks borne by the contemporaneous young and old are pooled. That is, do they bear risks equally, as measured by the elasticity of their consumption with respect to a change in the risky variable? If not, then there are potential efficiency gains from altering fiscal policy by spreading the bearing of risk more evenly. This follows from the assumption of equal attitudes towards risk among the young and old. While one might take issue with this assumption, the analysis could easily be adapted to accommodate an alternative assumption regarding risk-aversion.

But there is another question about the efficiency of risk-bearing that I am unsure the paper addresses. In particular, with temporary shocks, one would think that the efficient allocation of risk would involve the spreading of such shocks over as many generations as possible. At the limit, the share of any shock borne by a contemporaneous generation would be zero. Thus, not only would the young and old bear the shock equally, but each generation would also bear only a tiny share of it. It would seem that the evaluation of risk-spreading undertaken here is important, but not complete.

3 Key Findings

3.1 Initial results

Within the framework of his basic model, Bohn considers the outcome for an economy in which the government maintains a constant debt-GDP ratio, using taxes on labor income, which are non-distortionary in this context. He then evaluates the impact of different pension rules and different types of debt. His first key result is that indexing public pensions to prices is worse than indexing them to wages. The intuition is that wage indexation spreads part of the risk from workers to retirees. This helps not only to dampen the wage effects of cohort-size fluctuations, but also in the case of productivity shocks; even though the old are already exposed to productivity shocks, they are more insulated than the young, because part of their consumption is financed by the sale of assets. While this result makes sense, I wonder how much it would be undercut if asset values also fluctuated with productivity. In particular, if the stock market boomed during periods of high productivity, wouldn't this leave the old with greater relative exposure to productivity shocks than the model suggests?

Bohn's second major point is that the innovative design of government debt can contribute to improved risk-sharing. In particular, wage-indexed debt is an improvement over regular inflation-indexed debt, for the same reason, just given, as for wage-indexed pensions. But one can do better still by indexing debt to longevity as well, paying higher returns when people live longer. This, indeed, is simply an argument in favour of government annuities.

Finally, Bohn finds that capital-income taxes can improve the allocation of all but productivity risk. As in the classic contribution of Domar and Musgrave, and as elaborated by many papers since, capital-income taxation transfers risk from the investor to the government. We customarily think of the risk-sharing as occurring within a generation, as government pools risks more efficiently than the private sector. But the argument here goes through even with efficient risk-pooling *within* any given generation; the point is to spread risks among generations. This argument is compelling, and often overlooked. On the other hand, there is an important piece missing here, in that Bohn's style of capital-income taxation is non-distortionary - only the excess return is being taxed, so that the investor is being offered fair portfolio insurance. This is not a realistic characterization of capital-income taxes as they are actually imposed, and it is hard to conclude that higher capital-income taxes would be welfare-improving simply on the basis of this model's results.

3.2 The impact of an ageing population

A similar issue arises when the paper considers how things change with an older population, on average. In particular, a comparison of the first two columns of Table 4 suggests that it is more efficient to let payroll taxes grow, so as to maintain an existing, pay-as-you-go public pension system, than to hold intergenerational transfers constant by cutting back pension benefits. The reason is that, by taxing wages more heavily, the government is spreading more of the risk associated with wages. But imposing very high taxes on workers would introduce significant distortions that, like realistic capital-income taxes, would detract from any risk-sharing benefits. Taken to the limit, we could use this paper's model to show that having 100 percent taxes on labor and capital income would be most efficient.

Finally, when considering the structure of trust-fund investments, Bohn finds that investing the trust fund in equities and offering wage-indexed debt offers an attractive combination of sharing different risks. While investing in equities puts too much productivity risk on workers, this is more than offset by shifting some of the risk back on to the elderly via wage-indexed bonds, which we saw earlier were an attractive vehicle for shifting cohort-size risk as well. I find this result to be among the paper's most interesting, because this combination policy, while certainly feasible, is not one that one sees in the public policy debate. However, one should also note that this particular arrangement is not unique. As Bohn shows in the paper's next section, one can use more complicated public debt contracts alone to achieve optimal risk-spreading allocations.

3.3 Limited market participation

In his last section of analysis, Bohn considers the impact of partial participation in capital markets, using a model of three income classes in which the rich always save privately, the poor never do, and those in the middle may or may not, depending on government pension policy. His results here are of two types, some based formally on the model and others of a more intuitive and speculative nature. As to the former, he extends his previous analysis to estimate the optimal dependence of government debt on different shocks. In doing so, he clarifies that there are many ways of achieving efficient risk-sharing between generations, and that debt alone will suffice if debt contracts are sufficiently complicated.

In a less technical manner, he points out that, with population ageing, the welfare state can be preserved only with substantially increased transfers from the rich (who do not participate) or the next generation. If neither of these outcomes is feasible, economically or politically, the welfare state must be scaled back, and the middle class must leave it for the land of self-sufficiency, saving for its own retirement. When that happens, argues Bohn, governments will face a more serious problem of time consistency, for a larger share of the population will have assets ripe for the picking. Without a formal model, this conclusion is difficult to evaluate, but it strikes me that the argument goes the other way - once the middle class buys into the concept of private provision, it should be much more difficult, rather than easier, for governments to impose capital levies on existing assets. Thus, I am much less concerned than Bohn that we need to worry about the capital-levy problem; I am much more concerned about the transition to this benign state. One thing this paper does not describe is economic transitions. One must wonder whether the road from the welfare state to the land of self-sufficient savers is blocked by tractors and tear gas.

4 Conclusions

This paper makes an important contribution to our understanding of how public pensions, tax policy, and debt management can achieve improved risk allocation in an ageing economy. Its contribution lies not only in identifying desirable policy directions, but also in helping to shape the way we look at the problem. We should view fiscal policy's effects comprehensively, not piecemeal, and understand that there are alternative ways of achieving similar outcomes. We should remember that, with the burden of taxation, comes the sharing of risk. We should understand that polices that reduce or eliminate risk for certain generations, such as the elderly, may simultaneously shift this risk to those even less well-positioned to bear it. Finally, we must realize that the rules of the game change as the population ages.

Discussion

Philipp C. Rother

The paper addresses an issue that is of obvious concern for the current discussions on pension reform in many countries, not least in Europe. The implications of government-debt management for the sharing of risk between generations will affect the utility levels of participants in the pension system and, thus, need to be included in the analysis of the various options for financing pensions. However, while these issues are important for the evaluation of pension reforms, it should be kept in mind, that some of the parameters set exogenously in this paper will also be affected by the design of the pension reforms. In particular, different ways of financing pensions have an impact on the demand for and supply of, labour and capital and, potentially, on productivity. The resulting growth effects - not considered here - would need to form part of a comprehensive assessment of pension-reform options.

In the following, I will discuss the paper's treatment of two types of risk, namely population risk and valuation risk, and highlight some issues regarding agents' behaviour towards risk and the indexation of pensions.

The issue of including asset-valuation risk in the analysis represents my first and foremost area of concern. The paper assumes that workers acquire all real assets through saving during their working lives. As pensioners, they live off the return on capital but also need to sell part of the assets, exposing them to valuation risk. In this setting, the paper argues that valuation risk should be shared between generations through some form of government intervention. In a more realistic setting, however, the risk to pensioners from valuation changes appears low, because they need to sell only a small fraction of the capital stock, and this can occur in a gradual manner, insulating them from temporary shocks. International diversification provides additional insurance against fluctuations in asset values.

With realistic assumptions regarding the productive duration of economic assets relative to human life expectancy, the share of assets that might need to be sold by pensioners appears to be rather small. In particular, as almost all economic assets have a productive duration shorter than pensioners' life expectancy at the time of retirement, the share of assets that pensioners could need to sell to younger generations seems quite limited. To be more specific, imagine there is only one type of asset that depreciates over 20 years. Workers invest their savings in these assets when young and live off their returns when old. In this setting, even the last unit of the asset a worker buys immediately prior to retirement will have completely depreciated before the worker's demise. There is no need to sell the assets and no valuation risk.

A notable exception to this line of reasoning applies to land and (to a lesser extent) buildings. As those assets generally may outlast the holding generation, they need to be sold to younger generations if the owners are to enjoy the full return on the asset, i.e., with no bequest motive. However, the stock of such assets relative to the entire stock of capital in the economy appears too small to induce the substantial risks discussed in the paper.

The second caveat concerns the time structure of asset sales, independently of their volume, namely the potential to smooth out the impact of asset-price fluctuations, when assets are sold gradually over a longer period of time.[1] With only two generations, pensioners need to sell the entire capital stock to the existing working-age population at the time of retirement. Consequently, the old have to accept the level of asset prices prevailing at one particular point in time, exposing their entire consumption during retirement to short-term valuation risk. In a richer generational set up, this apparent problem fades away. With more than one generation in retirement, each cohort of pensioners sells only a fraction of their assets each period. For example, with life expectancy certain, pensioners would only sell a fraction of one over the duration of retirement each period. Short-term asset price fluctuations would, thus, only affect that fraction of asset holdings that is sold during one period and, consequently, only that part of pensioners' consumption and utility.

It should be noted that the asset price fluctuations presented in the model are likely to be of a short-term nature. Asset prices are not linked to productivity, so that productivity changes leave them unaffected.[2] A potential explanation for this type of temporary fluctuation would be asset-price bubbles, which in practice generally last only relatively short periods compared to the time-spans considered for pension provision, namely an investment period of some 35 years and a potential sell-off period of some 15 to 20 years.

A third qualification of the paper's implications concerning asset-valuation risk relates to international asset diversification. The general rejection of international diversification to alleviate asset-price risks overlooks the fact that, even if ageing were a global phenomenon, ageing-induced international capital flows are beneficial as long as the speed of ageing differs across countries, for which there is good evidence (see also the contribution by Axel Boersch-Supan in this

---

1 For example, the need to sell a larger part of the capital stock could arise if, owing to incomplete financial markets, investors can only acquire indefinite stakes (shares) in ongoing enterprises, rather than shares in finite real assets.

2 However, one could think of some link expressed through a correlation matrix between asset valuation and productivity shocks with imperfect correlation between the two, leaving a part of asset-valuation fluctuations as original risk.

volume). Moreover, many countries with emerging capital markets face a shortage of capital relative to labour, making them attractive for investments from rapidly ageing economies.

My second comment concerns population risk. The paper treats population risk, i.e., the possibility that young generations have fewer children, as exogenous and suggests that the burden should be shared between the young and the old. The immediate impact of population risk on the macroeconomy stems from the highly stylized two-period structure of the agents' lives, namely, that "new-born" humans enter the workforce immediately. In a more general setting with three or more generations, such effects would be alleviated, as the number of workers entering the labour force would become predictable and the capital stock could be adjusted in time, so that fluctuations in wages and asset prices would be reduced. In addition, it seems that in a model of this kind, the number of children should be a decision variable of the young generation. Therefore, the consequences of fluctuations in the number of children should also be borne by the young generation, and not affect the old.

As third area of comment, the modelling of risk behaviour deserves further attention. One issue that comes to mind is that risk aversion might increase with age. While the empirical literature remains undecided whether, and to what extent older people are more risk-averse, such behaviour could affect some of the qualitative results in the paper. Similarly, a flexible labour supply would give agents more degrees of freedom to adjust to shocks, and might make the young generation more willing to accept a greater share of overall risks.

Finally, the paper's suggestion of using wage-indexed bonds for spreading productivity risk across generations appears problematic. The paper takes as given that pay-as-you-go pensions need to be wage-indexed, but provides no reason for this. In fact, there seems to be no clear theoretical justification for preferring either wage or price indexation of pensions, and evidence from pension systems around the world points to a wide range of possible indexation arrangements including one or more variables. Consequently, the the validity of the paper's proposition to use wage-indexed bonds for efficient risk-sharing also remains restricted to the case of wage-indexed pensions and cannot be applied in general. In particular, wage-indexed bonds let the bearer participate in the trend behaviour of productivity, as well as in unexpected deviations from the trend, i.e., the first and the second moments of the productivity process. However, this result is not supported by the paper's analysis, which is restricted to risk aspects, i.e., second moments.

Ageing and Financial Stability

E. Philip Davis*

1 Introduction

This article reviews some of the potential pitfalls for systemic financial stability that may arise from the process of population ageing in coming decades. Although the issues are of global relevance, they will most directly affect OECD countries, such as those in the EU. We draw on the extant theoretical and empirical literature on ageing and financial instability, including projections of the macroeconomic effects of ageing and experience of recent financial crises. Our focus is largely on the forms of widespread financial instability that may affect the macroeconomy, in line with the following definition of systemic risk "a sequence of events entailing heightened risk of a financial crisis, where a financial crisis is seen in turn as 'a major and contagious collapse of the financial system, entailing inability to provide payment services or to allocate funds for investment"[1].

Beyond its systemic consequences, such financial turbulence could certainly also entail risks to retirement-income security for individuals. However, such individual risks are not central to our discussion, not least because risks to retirement-income security go much wider than situations characterized by financial instability in this sense. Notably, individual risks in pay-as-you-go schemes are broadly political (that the "contract between generations" will not be honoured), while retirement incomes from funded schemes depend strongly on average capital-market returns[2] and the efficiency of annuities markets (Miles,

---

* The author thanks Franklin Allen, Alan Auerbach, Martin Hellwig, Olivia Mitchell, Tsuyoshi Oyama, Mike Orszag and Len Skerratt as well as other participants in a seminar at the Deutsche Bundesbank for help, advice and comments. The paper draws inter alia on Davis (1995a, 1995d, 1999a) and Davis and Steil (2001).

1 An issue arises as to whether the definition should include the mispricing of financial assets. We suggest that, though this may accompany a financial crisis, the failure of payments and of allocation of funds are the defining features. Arguably, mispricing of financial assets is quite common (eg in asset bubbles, exchange-rate misalignments and mispricing of credit risk) without entailing a financial crisis, or even systemic risk, whereas failure of payments and of credit allocation are only seen in a crisis. Mispricing may nonetheless be part of the overall pattern that builds towards a crisis.

2 A low quality of asset managers could also emerge as an issue in some countries, leading to a need for enhanced regulation to protect against risks they may incur.

1999)[3]. The lack of understanding among members of the public of the best approach to financing retirement, including the appropriate level of saving, may in itself aggravate retirement risks in private systems (Mitchell and Bodie, 2000), although some outstanding risks may be ameliorated by feasible financial innovations, such as a reverse-mortgage system.

The article is structured as follows: In the second section, we provide a framework for assessing financial instability. In the third, we outline the projected patterns of ageing, while, in the fourth and fifth we briefly trace the likely macroeconomic and financial-market consequences of ageing. In section six, we go on to consider some of the potential outworkings of the process that may give rise to systemic risks. These may be divided into a number of subcategories, including financial-stability issues arising from the macroeconomic trends anticipated from ageing, the financing difficulties of pay-as-you-go pensions, precautionary saving by the population, as it ages in the presence of unsustainable pay-as-you-go schemes, risks for financial stability arising from the growth of institutional investors and the difficulties that may arise from the pattern of asset accumulation during funding.

There is a considerable element of uncertainty in many of the risks of instability, although we seek, wherever possible, to base them in existing experience of financial turbulences – not least those of Japan which have some links to rapid ageing in that country[4]. More generally, it is important to note at the outset that the effects of ageing on national financial systems are themselves highly uncertain, not least because they will be influenced by as yet unknown future reforms (e.g. reducing the benefits from pay-as-you-go systems), and fiscal or regulatory changes (e.g. taxation of the assets of funded schemes). Due to this strong element of uncertainty to outcomes for individual countries, our main focus is on global issues, and on generic aspects that will occur to differing degrees in different countries. In particular, we distinguish effects likely to arise in countries dependent on pay-as-you-go which finance the burden of ageing by taxation; those arising in countries dependent on pay-as-you-go which use bond issuance and countries which pre-fund pensions in private markets[5].

Meanwhile, in setting out the potential pitfalls for financial stability that could arise specifically from ageing, it is important to add that other causes of financial instability are unlikely to fade from view. Among the most important may be

---

3 Generally, the differing risks to which pay-as-you-go and funding are subject suggest a need for retaining elements of both systems as a means of diversification.

4 Brooks (2000) points to the exceptional ageing of the Japanese population since 1950, related to the decline in the birth rate after WW2. This has meant, for example, that the youth-dependency rate has declined from 60% in 1950, the highest in the OECD, to 23% in 2000, which is the lowest.

5 Note that we do not assess separately the effects of publicly-managed funding initiatives (see Davis, 1998).

those arising from intense competition in liberalized and securitized financial markets, where agency problems are important and there are mispriced government guarantees (see Davis, 1995a).

2 Indicators of financial instability

As a preliminary to analysis of possible systemic risks linked to ageing, it is useful to set out some elements of a framework for analysing and seeking to predict periods of financial instability. This is set out in detail in Davis (1999a). We suggest that many of the strands of the theory of financial instability have a contribution to make to our understanding of financial crises (see also Mayer (2000) and the articles summarised therein), but that the explanations are in most cases partial. In our view, a selective synthesis drawing on the evidence of actual crises is the correct approach to adopt. The relevant theories include those of:

- "debt and financial fragility", which suggests that over-indebtedness, asset-price booms and banking crises are a normal feature of the cycle (Fisher, 1933; Kindelberger; 1978; Minsky, 1977);
- "monetarist", that bank failures impact on the economy via a reduction in the supply of money (Friedman and Schwartz, 1963);
- "uncertainty", as opposed to risk, as a key feature of financial instability, linked closely to confidence, and helping to explain the at times disproportionate responses of financial markets in times of stress (Shafer, 1986);
- "disaster myopia", that competitive, incentive-based and psychological mechanisms lead financial institutions and regulators to underestimate the risk of financial instability (Guttentag and Herring, 1984; Herring, 1999);
- "asymmetric information and agency costs", that these well-known market failures of the debt contract help to explain the nature of financial instability, e.g. credit tightening as interest rates rise and asset prices fall (Mishkin, 1991), or the tendency of lenders to make high-risk loans owing to the shifting of risk linked to agency problems (Allen and Gale, 1999, 2000a);

and complementing these,

- "bank runs", that panic runs on banks (which may follow the various stimuli identified by the above theories) link to the maturity transformation they undertake, and the relatively lesser liquidity of their assets (Diamond and

Dybvig, 1983);[6] such theory can also be applied to failures of securities-market liquidity (Davis, 1994, 1999b);

- "herding" among institutional investors, as a potential cause for price volatility in asset markets, driven e.g. by peer-group-performance comparison, that may affect banks and other leveraged institutions (Scharfstein and Stein, 1990; Davis and Steil, 2001);
- "industrial", that effects of changes in entry conditions in financial markets can both encompass, and provide a supplementary set of, underlying factors and transmission mechanisms to those noted above (Davis, 1995a).

It is also recognized that *inadequacies in regulation* may heighten[7] tendencies to take excessive risks. Mispriced "safety net" assistance generates moral hazard, which, if not offset by enhanced prudential regulation, may lead to heightened risk-taking (McKinnon and Pill, 1997). This pattern may be particularly threatening as developments such as deregulation and increased competition reduce franchise values (Keeley, 1991). Moreover, lenders in the interbank market may not have the correct incentives to discriminate between banks (by price or quantity rationing) and discourage risk–takers (Bernard and Bisignano, 2000).[8]

There is also a need for consideration of the *role of international capital flows*. Traditionally, the focus of the literature on exchange-rate crises (Krugman, 1991) has been on the possible gains from speculation against a depreciation of a fixed parity, given the size of the nation's foreign-exchange reserves and various fundamental factors, notably the net external asset position and the balance of payments. Some models suggest that such a process is akin to bank runs described above. The contribution of international capital flows to recent crises and their international transmission introduce a number of additional elements:

- the issue of exchange-rate pressure, resisted by the authorities via interest-rate increases, which may trigger or aggravate financial instability;
- complications introduced by the financing of the public or private sector in foreign currency, which makes balance-sheet positions sensitive to exchange rates, and leads to a potential link from depreciation in the context of a currency crisis to more general financial instability;

---

6 Note that such "runs" lead to a contraction in the money supply, in line with the monetarist view, if the depositors seek cash, but not if they "run" to "safer" banks.

7 Risks can arise from agency problems independently of the safety net (Allen and Gale, 1999).

8 In the case of excessive regulatory protection, "excessive competition" can arise despite knowledge by lenders of the true probabilities of cyclical shocks and even the distribution of disastrous outcomes. Institutions may, as a consequence, hold inadequate capital on a risk-adjusted basis.

- the increasing role of institutional investors as a conduit for capital flows. As they are under severe pressure to perform in line with peers, they are particularly likely to "herd" into rising markets, and to seek rapid withdrawal from falling markets, destabilising domestic financial markets and exchange rates (Davis, 1995c);
- a possible link of contagion where there are cross-country similarities in trade patterns (Glick and Rose, 1998).

In Davis (1999a) we assess the indicators of financial instability that can be derived from these theories (so-called "*macroprudential indicators*" or MPIs) in the light of experience of financial crises both in the banking sector and in securities markets (see also IMF, 2000). In advance of crises, common generic patterns include:

- Unanticipated regime shifts towards laxity on the part of monetary, fiscal or regulatory authorities (including "financial liberalisation")
- Easing of conditions for new entry of intermediaries to the relevant market
- Debt accumulation, leading to heightened leverage (economy-wide, by individual sectors or in individual markets)
- Asset-price booms (be it property or security prices)
- Financial innovation (and rapid growth of the markets concerned)
- Concentration of risk on the part of financial institutions (implying excessive optimism in respect of potential "correlations")
- Declining capital adequacy of financial institutions
- Monetary tightening or unanticipated regime shifts towards rigour on the part of monetary, fiscal and regulatory authorities.

Of course, many of these features have occurred separately without entailing a crisis, and indeed are part of the normal functioning of a market economy. It is their combination and acuteness that is crucial to the occurrence of financial instability.

A number of econometric studies have examined shorter-term quantitative macroeconomic and financial developments prior to banking crises, offering complementary MPIs. For example, Kaminsky and Reinhart (1999) examined 20 small, open economies from 1970-95, to assess macroeconomic variables whose behaviour is systematically different in the period prior to banking and currency crises. Whereas in the 1970s, before widespread financial liberalisation, there was no close link between currency and banking crises, in the 1980s and 1990s there were clear interrelations, with banking-sector problems often beginning before currency crises, while the latter often deepened the banking crisis. As regards indicators, both currency and banking crises were preceded by recession, declines in the terms of trade, stock-market crashes, real exchange-rate appreciation, prior lending booms, increases in the money multiplier, and increases in real interest rates.

Demirguc-Kunt and Detragiache (1998a), estimated a multivariate logit model of banking crises only, using data from a sample of 53 developing and developed countries over 1980-94. They again pinpointed low economic growth as well as high inflation as key macroeconomic indicators of the probability of financial crises. High real short-term-interest rates – often implemented in the context of a need to bring inflation under control - were also associated with systemic banking problems, as well as vulnerability to balance-of-payments problems (proxied by an adverse terms-of-trade shock), and to sharp capital outflows (ratio of M2 to foreign-exchange reserves). Explicit deposit-insurance schemes made crises more likely, pointing to a link to moral hazard in the context of the safety net. Finally, Hardy and Pazarbasioglu (1998) sought to identify only leading indicators by looking only at lagged variables in their logit estimates, and also seeing the pre-crisis year as a separate event from the crisis, focusing on data from 38 countries from 1980-97. They found that banking distress is associated with a sharp fall in GDP growth; boom-bust cycles of inflation, credit expansion and capital inflows; rising real interest rates and an increasing incremental capital-output ratio; declining bank deposits; a sharp fall in the real exchange rate, declining imports and an adverse terms-of trade shock.

In the rest of the article, we shall seek evidence and hypotheses for a potential effect of ageing on instability in the light of these mechanisms and indicators, as well as considering whether systemic risk might take a different form from that observed in the past.

3 The ageing problem

OECD countries have all witnessed an increase in life expectancy and a decline in the birth rate in recent decades. These have already given rise to an ageing population, with a high proportion of the population currently in the high-saving age groups (around 45-65) and also an increasing burden of dependents relative to the population of working age. The higher life expectancy is, the longer individuals expect to live after retirement and the greater the need for retirement income. As Table 1 shows, the life expectancy at birth in the G-7 countries has risen from around 72 to 78 between 1970 and 2000. Life expectancy in Japan is now 81. Underlying these patterns are better health care, medical advances, and improved overall living standards.[9]

Except in the US, there has also been a decline in birth rates since 1970, which has reduced the size of the younger generations, who would otherwise borrow and offset the saving of their older counterparts (see Table 2). In 2000, there were exceptionally low fertility rates[10] (of below 1.5) in Germany, Italy and Japan, while the rate in France, Canada and the UK was around 1.7 and that in the US was 2.1. Only in the US is the fertility rate sufficient alone (i.e., without immigration) to generate a stable population. Underlying the decline in fertility is a pattern of later marriage and greater activity of women in the labour market, which has increased the opportunity cost of having children, as well as more general social and attitudinal changes.[11] Reflecting the decline in fertility, the generation born in the EU in the 1970s is 17% smaller than that of the 1960s, and the 1980s generation is 25% smaller.

In all of the G-7, as well as the rest of the EU, the retirement of this baby-boom generation offers a general problem for the macroeconomy, and a particular challenge to systems of retirement-income provision. For all demographic projections for OECD countries show a continuation and intensification of the ongoing process of ageing in the future. Table 3 shows that the demographic shift will be particularly marked in the years from now onward. Whereas in 2000, the average G-7 elderly[12]-dependency ratio was around 26%, it is expected to rise to 48% in 2035 and 52% in 2050. In Japan and Italy, the elderly-dependency ratio will be over 60% in 2050. The ageing of the population is also anticipated in the US, but the level expected in 2050 remains somewhat lower than that in the rest of the G-7 and the EU. There is also expected to be an increasing proportion of very old individuals, who may need additional, and costly, health care as well as pensions. The share of young dependents is expected to be flat, but they tend to be less costly than the old.[13] The total-dependency ratio (including those under age 20 and over age 65 in the numerator) will be over 80% in 2050 in Germany and Italy, according to these projections.

---

9 Accompanying these is a pattern of early retirement, thereby also lengthening the potential retirement period (Davis, 1997).

10 Fertility rates indicate the number of children born to an average woman over her lifetime.

11 Davis (1997) notes that the highest fertility rates among EU countries today are in Scandinavian countries, which provide comprehensive and subsidized child-care facilities, thus spreading the burden of childcare from the family to the economy as a whole.

12 The elderly-dependency ratio is the ratio of those over 65 to those aged 20-64.

13 Heller et al. (1986) accordingly estimate that social expenditures will rise in the major industrial countries even if savings in education and family benefits are taken into account.

4 Consequences of ageing for the macroeconomy and pension systems

4.1 The principal effects of ageing on the macroeconomy

A useful starting point in assessing the potential links from the above patterns of ageing to financial stability is to consider the potential impact of ageing on growth, saving and investment at a macroeconomic level. Our aim in this section is not to provide any novel results, but rather to offer a brief survey of the relevant literature, so as to provide a benchmark for the assessment of financial effects.

As regards *growth*, it is widely considered that it will decelerate as ageing proceeds, principally because of lower labour-force growth (see, for example, Turner et al. (1998)). There will also be lower growth in living standards (i.e. GDP per head) than has been the case in recent decades, reflecting the accompanying increase in the dependency ratio. Effects on growth of a fall in labour-force growth are unlikely to be offset by higher *investment*. Indeed, investment is likely to decline, given a lesser need for capital widening, while capital deepening is likely to be limited by diminishing returns. Moreover, slower growth will tend to reduce returns on capital directly, thus again putting downward pressure on investment – the example of slower Japanese output and investment growth since 1970, and accompanying lower returns to capital, may be cited in this context. On the other hand, higher labour-force participation could help to underpin growth. Another unknown factor in this context is the response of *productivity* to ageing. It is suggested by Disney (1996) that there is no link detectable from ageing to productivity, implying investment rates will be crucial to growth[14], not least given possible "endogenous" links from investment to productivity.

An important element in assessing *private saving* is the view taken of the life-cycle hypothesis, which postulates initial dissaving in young adulthood, followed by saving for retirement and finally dissaving in retirement. Most estimates of the determination of saving based on macro evidence imply that this pattern holds, implying that an older population will save less. For example, Masson and Tryon (1990) find an elasticity of 1 from the dependency ratio to the savings rate, using pooled cross-section and time-series data for industrial countries, although later work by Masson et al. (1995) reduced this estimate to 0.14. Pure cross-section studies, such as Horioka (1991), find estimates of around 0.76 for this elasticity. The implication is that ageing will sharply reduce saving. An example may be the

---

[14] Some authors suggest ageing slows technical progress as innovation becomes less profitable with a shrinking market for capital goods, and owing to the lesser dynamism of an ageing population (Wattenberg, 1987). In contrast, Cutler et al. (1990) suggest that innovation increases as labour gets scarce.

decline of saving in Japan in the 1980s accompanying ageing (although this was reversed in the 1990s, as discussed below). On the other hand, the micro evidence based on household-survey data is much more equivocal, suggesting that older people continue to save, albeit at a lower level. Such comparisons are made more difficult by the existence of behaviour that is peculiar to cohorts (Disney, 1996). Accordingly, Börsch-Supan (1996), for example, finds the net effects of the dependency ratio on saving are close to zero.

Public saving is largely driven by the scale of the public pension system in the light of ageing and the means of financing adopted (e.g. taxation versus debt finance). As the population ages, the public sector will tend to lower its saving, ceteris paribus. Rapid increases in the proportion of the population over 65 (the dependency ratio) combined with generous social-security-pension schemes are particularly threatening. The issue is well summarized in Tables 4-5 from Dang et al. (2001) and Chand et al. (1996), which show projections of expenditures for pay-as-you-go pension schemes. Such pressures will be enhanced by increased public health expenditures (Dang et al., 2001), albeit partly offset by lower expenditure needs on education (Turner et al., 1998). If unsustainable increases in contributions or in public debts are to be avoided, there is a need for reform of social-security-pension schemes in many countries, which may involve a combination of lower benefit rates and higher retirement ages in social security systems.

The *system of retirement-income provision* may impact on growth and private saving as well as on public saving. Concerning growth, the argument is that pay-as-you-go, if contributions are perceived as a tax, tends to distort labour markets. On the other hand, funding, by removing such distortions as well as increasing the efficiency in allocation of capital funds, and long-term saving, may enhance prospects for economic growth (see for example Börsch-Supan and Tinios, 2001). It may also boost saving itself, although there are likely to be high levels of substitution from discretionary saving which mitigate the overall effect (see the survey in Kohl and O'Brien (1998)). An element of funding of pensions will also allow diversification of risk by investment outside the domestic economy.

Concerning private saving, there is some evidence for the US and in international cross-section (Feldstein, 1974, 1977, 1995) that unfunded social security pensions reduce private saving. This can be justified theoretically by a life-cycle framework, whereby a guarantee of income to maintain consumption after retirement gives rise to a form of implicit wealth accumulation, and the need to save during the working life is lessened.[15] Underlying this approach is a view that workers see contributions as a form of saving, and not as a tax, and that

15 A further mechanism inducing lower saving under pay-as-you-go social security is that those who are myopic and would otherwise have continued working till they die are now able to retire.

confidence is maintained in the promises of pay-as-you-go. As the population ages and the size of unfunded liabilities increases, the negative effect on private saving could increase sharply (unless the effect is offset by increasing uncertainty over whether pension promises will be kept). Feldstein's results have been disputed (for a review, see Munnell, 1987), and other evidence suggests that the effect, even if negative, may be small, for example, because social security induces early retirement, which gives incentives to save more to cover the longer retirement period, or because changes in intra-family transfers (e.g. bequests) may have offset the increase in public-sector transfers, thus leaving the need for old-age saving identical (Barro, 1974).

What happens to private saving when there is a reform and funding replaces pay-as-you-go? Countries with generous pay-as-you-go social security tend to have low levels of private pension provision (see Table 6, which uses total institutional investment as a proxy for private pensions). This suggests potential substitution. Disney (2000) offers some evidence that reform of public pension systems may have the effect of boosting private saving, also highlighting a study of Italy (Attanasio and Brugiavini, 1999), which shows strong results in support of the hypothesis. This supports the case made by World Bank (1994), the conditions under which funding will have a positive effect on private saving - namely, myopia, limited access to credit, and lack of credibility of the pensions scheme - are precisely those whose absence will lead pay-as-you-go to reduce private saving. So a switch from pay-as-you-go to funding is unambiguously likely to raise private saving.

Even in the absence of reform, one positive effect on private saving that may accompany pay-as-you-go where confidence in future benefits is weak is a high level of precautionary household saving. Analysts suggest that this has already been typical of Japan (see Nakagawa, 1999), where surveys suggest that the recovery in the savings propensity in the 1990s partly reflected an increase in uncertainty[16] about pension systems among young households, while the old households felt anxiety about nursing care and hence sought to save more than any other age groups even in the last stages of their life cycle.[17] There is also anecdotal evidence of such effects in EU countries. Moreover, if there is debt finance of pension liabilities, there may be interactions between private and public saving

---

16 The issue of uncertainty is verified in the Nakagawa results by survey responses. Oyama and Yoshida (1999) find a similar result – albeit for different kinds of uncertainty - including the error on a first-order GARCH model in income in the consumption function.

17 The rise in saving was not seen as solely age-related. The middle-aged and elderly low-income households felt anxiety about employment conditions, given the economic situation (Nakagawa op cit), and in particular the downward revision to growth after the collapse of the financial bubble of the 1980s. Of course, the view that the pension system is unsustainable may also have strengthened as economic growth fell in the 1990s.

owing to "Ricardian" effects whereby private saving responds positively to public dissaving, as individuals look ahead to the taxes needed to repay government debt.

4.2 Projections of growth, saving and investment

A number of projections of growth, saving and investment patterns in response to ageing have been produced in recent years by international organizations and analysts. Viewed with caution, these help to give some benchmarks of likely financial flows that experience has often shown to be closely linked to the likelihood of instability. A key detail usually omitted from such calculations, but crucial for the purposes of the current exercise, is the composition of the flows in terms of instruments. We will comment on this issue further in Section 5.

Turner et al. (1998) provide a simulation of the global effects of population ageing (focusing both on changing population growth and on age structure) using the OECD's international dynamic general-equilibrium macromodel MINILINK. Reflecting the declining labour supply with ageing, economic growth is forecast to decline to 0.25% per annum in Japan, 1% in Europe and 1.4% in the United States by around 2030. Pressures on saving and investment arise, first, from the slowdown in growth, which reduces investment needs directly. Second, a decline in the weight of the OECD in the world economy tends to improve OECD current accounts (and hence saving-investment balances) as non-OECD imports rise faster than OECD import demand. The US, Europe and Japan all generate balance-of-payments surpluses of 2-3% of GDP up to 2025, thus building up net external assets that help to buttress GNP. On the other hand, eventual downwards pressures on public and private saving are greater in the OECD than elsewhere, generating – in combination with exchange-rate appreciation – deficits for the three OECD regions after 2025. As world investment in this simulation falls less than saving, world real interest rates are expected to rise slightly, reinforcing the decline in investment. Reflecting differing returns on capital, interest rates are higher in EMEs than in the OECD. The authors note that higher saving in OECD countries could generate quite different results, with lower real interest rates and consequently higher investment and capital-labour ratios. There would also be greater net external assets, boosting OECD GNP via inflows of interest, profits and dividends[18]. As we noted above, some limited boost to saving could be anticipated from a switch from pay-as-you-go to funding.

Masson and Tryon (1990) use the IMF's global econometric model MULTIMOD in a similar manner to assess the combined effect of ageing

---

[18] The return on such investments will depend on factors such as labour and product-market reforms in the EMEs, as well as the overall size of such flows from the OECD (if the flows are sufficiently sizeable, they will depress the return on capital in the EMEs).

(measured by changes in the population-age structure) on private saving, public deficits and overall production. Production is again assumed to link to the labour supply, i.e. the size of population of working age times the participation ratio. The main cost to the economy from ageing arises from lower labour-force growth, and consequent declines in output relative to base, notably in Japan and Germany, while real interest rates in the OECD rise by 3-5% in 2020, as investment exceeds saving. Their model generates large falls in national saving in Germany and Japan from 2000 onwards, as both private and public sectors reduce their saving, while in the US, France, Italy and the UK the net effect is positive for some decades, with increased private saving more than compensating for a rise in the fiscal deficit. The difference in private saving links to the differences in demographic profiles, with more high-saving individuals remaining in the population of the latter countries. As is the case for Turner et al. (1998), the model includes endogenous tax rises, which rise with the social security burden, rather than assuming fixed contribution rates to social security.

Börsch-Supan (1996) concludes that, until around 2010, demographic effects will increase private sector saving, given positive saving of the elderly and the baby-boom generation being in the high-saving period. He envisages flat investment/GDP, implying changes in the net balance of supply and demand for funds will be driven largely by the government sector. In a "worst-case scenario", where governments run sizeable deficits to cover pension expenditures, substantial shortfalls are envisaged in the overall balance after 2005.

Unlike the above, most other studies have tended to be partial and focus on individual countries or private or public saving only. For example, Roseveare et al. (1996) assess two scenarios for saving patterns, which differ in the size of the assumed negative effect of the dependency ratio on saving, and on the degree of Ricardian equivalence when government deficits increase. They see private saving as a proportion of GDP across all OECD countries falling 3-6 percentage points between 2000 and 2030, depending on the scenario, and national saving declining by 8 to 16 percentage points, given debt financing of pension expenditures (i.e. fixed contribution rates) and assuming a partial Ricardian response of private saving to government dissaving. In France, Austria, Denmark and Finland, net national saving is forecast to be negative in 2030 in both scenarios. As noted by Turner et al. (1998), negative national saving seems unlikely in the light of positive national investment, given the large balance-of-payments deficits that would be implied. Changes in interest rates and exchange rates – omitted from the simulation - would be likely to occur, helping to equilibrate saving and investment.

Looking at private saving in the EU, Miles and Patel (1996) suggests that, as long as the 'baby-boom generation' remains in the labour force, an increase in private saving should be expected, building to a maximum of 2.5 percentage points in 2020, after which saving declines as individuals retire. Over this period, the rise in private saving would be more than enough to offset changes in government saving. Their projection is based on estimates suggesting a life-cycle view of saving is warranted, whereby assets are accumulated at different rates over the working life, rather than a simple coefficient on the proportion of elderly people, as for Masson's work quoted above.

Bikker (1996) focuses on balance-of-payments effects of ageing and concurs with the general equilibrium studies that the effects in OECD countries may be towards a surplus, as long as national saving is boosted by ageing, which seems possible as long as the 'baby-boom' generation remains at work. But once people in this generation retire and begin to dissave, there could be balance-of-payments problems.

Besides saving, the amplitude of shifts in the balance of payments will also depend on the path of investment; in line with the simulations noted above, most studies suggest that investment rates will fall, which would temper the increase in external deficits. For example, Cutler et al. (1990) suggest that total investment may fall with ageing, given the reduced need for capital widening with a smaller workforce; they also envisage a fall in the rate of return on capital from 6.7% in 1990 to 3.5% in 2025. Disney (1996) shows a significant negative relationship between the dependency ratio and fixed-capital growth over 1977-92 in 24 OECD countries. Blommestein (1998) again sees falling investment as likely to occur as the labour force shrinks and the capital-labour ratio rises, depressing returns to new investment.

5 Financial-market effects of savings patterns

In the context of these growth and saving/investment patterns, the ageing of the population will have an important impact on financial markets, which we seek in this section to trace in terms of supply and demand for individual instruments. We defer discussion of shifts in prices and yields to Section 6.

In the case of pay-as-you-go, we have seen that confidence in the system may reduce private saving. More likely in the medium term, and consistent with the simulations cited above, is that private saving may, rather, be boosted by deteriorating confidence in social security. In countries where private pension funds are not extensive, these savings may be invested in either life insurance contracts and mutual funds (as in Europe) or mainly in bank deposits (as in Japan). They may also be sunk into real estate. As in Japan at present, such savings may well be reinforced by Ricardian effects as and when governments run large deficits (at which point saving may also flow into government bonds). We shall argue below that such precautionary saving could trigger financial instability independently of the flows occurring either directly via pay-as-you-go or pension funding.

The simulations suggest that there will indeed be a sharp increase in the supply of government bonds in due course, as governments start to run deficits to cushion the effects of ageing on contribution rates. Initially, private saving may be sufficient to finance such deficits. But even if precautionary and Ricardian effects continue to operate, the retirement of the "baby-boom" generation implies that public deficits will eventually be accompanied by balance-of-payments deficits. Hence, part or all of the demand for government debt has to be found offshore. Given that the pattern of ageing, and hence lowering of government saving, is anticipated to be common across OECD countries, a significant part of such demand would have to emanate from emerging-market economies (EMEs). Alternatively, if pay-as-you-go is financed in a strict balanced-budget manner, the implication for financial instruments will be less direct. In effect, it will depend on (possibly negative) effects on personal saving arising from a high level of taxation, as well as from the decline in private saving likely to accompany ageing more generally.

There will also be effects on financial markets when pensions are largely funded. During the transition phase, as the working population ages while accumulating for retirement, there will be considerable demand for securities, notably in the form of equities (where regulations permit) and bonds. Given the contrasting portfolios of institutional investors and households, and the evidence of a lack of offsetting shifts in portfolios when institutional investment increases (King and Dicks-Mireaux, 1988), relative demand for deposits is likely to decline. Over time, there will be a shift within demand for securities from equity to bond-related instruments because the growing maturity of pension schemes, and the increased demand for annuities per se, would necessitate the holding of shorter-duration assets.

Such flows arising from funding will again not be purely domestic, to the extent that ageing occurs at different rates in different countries. Reflecting a desire for diversification, it seems likely that there will be considerable gross-capital flows between OECD countries and from OECD to EME countries during this phase, in the form of bond and equity finance. These are likely to considerably exceed the amplitude of net flows (i.e. arising from saving-investment imbalances and consequent balance-of-payments disequilibria), see Reisen (1998)[19].

When an increasing proportion of the population retires and begins to live on the accumulated assets, domestic demand for securities in OECD countries could fall sharply. There would also be withdrawals of financing from EMEs. Schieber and

---

19 Reisen (1998) points to both the offsetting patterns of saving and the diversification benefits arising from EMEs as helpful in ensuring adequate returns on OECD pension funds that invest in EMEs, not least in the light of the tendency for returns in OECD countries to fall in coming decades. Blommestein (1998), on the other hand, points to the low returns and high risks of EMEs, in combination with increasing correlations with the OECD markets, which suggest that EME investment is not a panacea.

Shoven (1994) point out that decumulation is an ineluctable process for defined-benefit pension funds, and suggest that they will cease to contribute to US net saving around 2024. They note, however, that this effect is unlikely to occur for defined-contribution funds in the foreseeable future. Given the need to finance annuities, demand for equities would fall more than demand for bonds (Brooks, 2000). Poterba (1998) focuses on extant information on age-specific asset holdings (excluding defined-benefit pension funds), corrected for cohort effects in order to evaluate this issue. He concludes that asset demand may indeed rise as households age, and notes that surveys suggest that there is a decline in risk tolerance at ages over 65, but suggests that there is less evidence of a downturn in asset holdings at the end of the life cycle. He thus considers that a sharp fall in demand for securities is unlikely to arise in coming decades[20].

Even if there were to be net decumulation of securities by OECD investors, global demand will also depend on the degree to which other countries, e.g. in the Far East or Latin America, experience slower demographic ageing, and thus provide a countervailing factor in the context of globalized financial markets. Note, however, that maintaining global demand for securities would require them not only to substitute for capital inflows from OECD countries, but also to generate substantial surpluses to cover declines in demand for securities in OECD countries themselves.

6 Effects on systemic risks

There are a number of potential financing issues that may be envisaged in the wake of the above-mentioned trends that may give rise to financial turbulence. These may be divided into, respectively, risks arising from the overall macroeconomic development, the difficulties of pay-as-you-go, the risks arising from the manner of funding via institutional investors, and the risks arising from the above-mentioned pattern of asset accumulation during funding. We consider that the dangers arising from "no reform" are much more severe than those likely to arise from funding, where the latter largely require a readjustment on the part of regulators to a change in the locus of risk, rather than a dramatic increase in its level. Note, however, that, given the need to finance the transition, a country reforming a generous social security system and shifting to funding will not entirely escape some of the difficulties outlined for pay-as-you-go.

---

[20] He admits, however, that his analysis fails to cover defined-benefit pension funds.

6.1 Risks arising from the overall macroeconomic development

If the predictions of an initial rise in private and national saving relative to investment are correct, then, from the perspective of a country that faces relatively rapid[21] ageing, risks may initially arise from a balance-of-payments surplus which leads to an appreciating currency and loss of competitiveness by domestic industry, generating credit losses by lending institutions (Japan in the 1980s being a possible example). Such losses could be especially marked if ageing is at the same time putting downward pressure on economic growth rates and returns on physical capital.

Exchange-rate appreciation in this period may be aggravated by the degree of home bias among investors. Uncertainty over the future course of the investment-saving balance can lead to heightened perceptions of currency risk, which enhances such home bias and puts further upward pressure on the exchange rate. This case is particularly made for Japan during the past two decades, in which the linkage between currency risk and home bias was strong, due to the following perceptions by domestic savers and institutional investors: (a) the heterogeneity of its society from other countries -- this is partly reflected in limited labour movements into and out of Japan (b) the size of savings inflows is considered too big to be absorbed in a diversified way in foreign markets.

Later there could be balance-of-payments deficits as national saving diminishes, with the possibility of currency crises occurring if exchange-rate fixity is sought to be maintained. The scope of such problems will also depend on the asset-liability position of the country – arguably a deficit is less likely to cause a currency crisis when there remain net foreign assets, and, correspondingly, the private sector is not exposed to currency movements via foreign-currency borrowing[22]. While currency crises are not themselves of systemic concern, they tend to be closely linked to banking crises, as noted by Kaminsky and Reinhart (1999).

In this context, note that the macroprudential indicators cited in Section 2 do not include balance-of-payments deficits per se, suggesting that the period with an appreciating exchange rate may be just as threatening to financial stability, especially if savings flows generate abundant liquidity for bank lending, as was the case for Japan in the 1980s. Indeed, credit expansion was one of the MPIs cited in Section 2. Furthermore, a combination of an investment-saving imbalance and an imperfect international capital market - i.e. ample liquidity in the home

---

21 Balance-of-payments effects would be attenuated if all OECD countries were to "age" simultaneously, but projections suggest this will not be the case.

22 Note that both the Asian and Latin American crises as well as the Swedish banking crisis were closely linked to foreign-currency exposure of the private or public sectors (on Sweden see Englund, 2000).

market – may in turn encourage the loose conduct of fiscal and monetary policies, if the imbalance is mistakenly seen as a cyclical rather than a structural one. Again, the Japanese experience during the 1980s may be cited, when the fiscal surplus (including the surplus of public pension funds) was viewed externally as a source of the long-lasting Japanese current-account surplus. This situation - i.e. when fiscal and monetary measures are taken in an attempt to correct the structural imbalance due to the demographic factor - is also likely to lead to the creation of a financial bubble, though this case has not yet been observed in many other countries than Japan.

The experience of the Asian, Latin-American and Mexican crises suggests, however, that balance-of-payments deficits did often accompany both currency crises and systemic risks, with the withdrawal of external credit flows being the triggering event. Furthermore, a number of the MPIs mentioned in Section 2 may move adversely during the advanced portion of the process of ageing sketched in Sections 4 and 5, notably lower economic growth, stock-market volatility and/or increases in real interest rates ensuing when "high-savers" retire. These again point to potential risks for the financial system.

Macroeconomic consequences of ageing in the OECD may spill over to countries outside the OECD, because an acceleration and reversal of capital flows to EMEs may well be a marked feature of the process of ageing (corresponding to surpluses and deficits in OECD countries). It was seen in the Asian crisis that a rapid inflow of saving can generate wasteful investment, while the outflow can cause marked macroeconomic difficulties, asset-price volatility, foreign-exchange-market turbulence and banking crises (Stiglitz, 2000).

6.2 Risks arising from the difficulties of pay-as-you-go

If current social security pension systems remain unchanged, there will need to be sharp increases either in contributions or in fiscal deficits. Owing to the foreseeable and long-term nature of the changes ahead, it seems likely that reform-reducing benefits of pay-as-you-go will diminish,[23] if not necessarily eliminate,[24] the amplitude of the problems that would otherwise arise. It this light, and also to emphasize the need for reform and provide a polar case, it remains useful to trace

---

23 Moreover, there are likely to be significant non-linearities, implying that reform will entail a more-than-proportional reduction in risks, despite leading to an increased fiscal burden (via higher taxes or issue of "recognition bonds" for accrued pension rights) in the short run.

24 It can be argued that, even if reform implies a wholesale switch to funding, some fiscal problems may remain, given the need to issue bonds or raise taxes in order to finance accrued benefits. The crucial element of certainty about future borrowing needs would reduce adverse effects, however.

out the extreme cases where such reform is not forthcoming (Cremer and Pestieau, 2000) cite some of the political difficulties that reforms face).

Before considering effects on public finances, one can trace the potential effects of increased precautionary saving by the household sector as and when confidence in social security promises ebbs. We have noted this is already a feature of Japan. Since countries with generous pay-as-you-go have underdeveloped private pension systems, it is possible that such flows of precautionary saving may go directly or indirectly[25] to banks, as in Japan in the last two decades. In certain conditions, again as observed in Japan, this may give rise to banking sector risk. The conditions include, first, inadequate credit-risk assessment, second, moral hazard from the safety net, and third, domestic or international institutional investors offering attractive securities financing to the more highly-rated companies, facing banks with adverse selection (on disintermedaition in Japan, see Hoshi et al., 1993). Funds may then be lent on to high-risk borrowers at inadequate spreads, including commercial property and property-related claims. The consequence could be a greater incidence of property-linked bubbles. Particularly where there is a balance-of-payments surplus, the banking flows could also be diverted to the international interbank market, as was the case prior to the Asian crisis (Bernard and Bisignano, 2000). Given the weakness of credit assessment in the international interbank market, the consequence may be to unduly ease the credit constraints on the banking systems where the funds become available, leading to misallocation of resources. Systemic crises in the past three decades have frequently been directly linked to the ebbs and flows of international interbank funding (see also Davis, 1995a).

Even if funds are invested in life-insurance companies, avoidance of systemic risk is not guaranteed. Owing to the nature of their liabilities, as well as regulations (Davis 2001b), life insurers tend to invest more in domestic bonds than pension funds do (pension funds diversify internationally and invest heavily in equities). A shortage of government bonds that may continue for some time ahead, as well as competition in asset management driving life insurers to increase returns, may prompt more investment in higher-risk assets, such as high-yield bonds and low-rated securitized loans. Besides their general effect on credit expansion, which could generate fragility in the non-financial sectors, such funds may again feed a property boom, leaving the insurers as well as banks vulnerable to a downturn in the property cycle.

Turning to the public-finance issues as the population ages, tax financing of public pensions is likely to give rise to economic problems, beginning with a decline in inward investment, slower economic growth and growing evasion of

---

25 Hargraves et al. (1993) show that in the 1980s Japanese households sought to draw down their bank deposits and invest in life and pension claims, but life insurers and pension funds bought corporate bonds and companies deposited the proceeds with banks.

contributions as tax rates rise (World Bank, 1994). The process may culminate in translocation of domestic industry and productive labour. Driven to an extreme, such "factor-flight" effects could denude a country of a significant proportion of its productive resources. Property, being non-tradable, could be hardest hit among asset markets. Even a modicum of such shifts could have severe recessionary effects, impacting on the solvency of non-financial companies as well as on employment, thus in each case sharply increasing loan defaults. Banks could consequently suffer major credit losses on their domestic business if, as seems likely, the scale of economic problems arising from tax finance is poorly anticipated. Credit losses would threaten banks' access to the international interbank market. There is the possibility of debt crises for the private sector, and deficits for the public sector, as the country faces a vicious cycle of diminishing tax revenues and rising tax rates to seek to finance pension payments. A likely further outcome is default of the pension payments themselves, as in Eastern Europe in the 1990s and some Latin American countries in the 1980s (World Bank op. cit.).

Bond financing of unreformed pay-as-you-go may give rise to sharply increasing interest rates, as credit ratings deteriorate and governments sell to increasingly unwilling holders of their debt offshore. Economic growth would decline sharply, given effects on investment of higher real interest rates, while private-sector borrowers may be "crowded out" of domestic bond markets by the government. Experience of countries, such as Italy, Sweden, Ireland and Denmark, that have faced fiscal crises suggests that, besides direct effects of higher interest rates and crowding out, uncertainty generated by a fiscal crisis, as well as Ricardian effects, act to dampen expenditure at a macro level, leading to recessionary tendencies, as well as inflation (Giavazzi and Pagano, 1995). Credit ratings would be likely to fall lower for private firms than the government (since private credit ratings are usually lower than those of the government).

Together, these consequences of bond financing may lead to financing difficulties for the corporate and household sectors, and would increase the default risk faced by banks and life insurers. Again, credit losses and lower ratings would threaten banks' access to the international interbank market. Higher real rates and a recession would also hit equity prices. Falling and volatile bond and equity prices accompanying debt finance could impinge on leveraged investors taking positions, as well as on the viability of any funded pension arrangements (showing that the risks to pay-as-you-go and funding are not entirely independent in this case). Again, real estate, being less internationally tradable than securities, would be particularly exposed to rising long-term interest rates, and prices could fall sharply, threatening banks' collateral. Note that most OECD fiscal crises in the past (e.g. in Italy, Ireland and Denmark) have occurred in the context of financial systems with significant structural regulation and excess profitability of banks. Thus, the effects of these past events on banks may underestimate the likely consequences in a liberalized financial system. Equally, fiscal difficulties for the government will sharply limit its ability to finance recapitalization of the banking sector in the wake of a banking crisis, which would also impinge on banks' "support" credit ratings in the absence of such a crisis, raising their cost of funds.

Ultimately, the public sector itself could face a solvency crisis, owing to a "snowball" increase in debt and interest payments, which would have major consequences for private financial markets. Governments might be tempted as a last resort to monetise the debt and thereby cause hyperinflation, wiping out the value of money fixed assets held e.g. by banks and life insurers. The EMU context would make such deficits and monetisation particularly problematic, given spillovers within the Single Currency area, although global spillovers would also be present if a significant number of countries were affected. In this context, it may be added that contagion effects on other government bond markets might be anticipated, whereby the occurrence of a fiscal problem due to ageing in a single significant OECD country could generate sharply higher financing costs in other countries with similar potential problems. This could threaten to generate self-fulfilling fiscal crises in a number of countries, parallel to the contagious currency crises in Asia following trade patterns, as cited in Glick and Rose (1998).

6.3 Risks arising from funding via institutional investors

We now turn to issues arising from the funding of pensions. In this section, we discuss the consequences of the differences in financial structure that accompany private funding, separately from the saving flows examined in the next section.

In general, a financial system characterized by institutional investors and extensive capital-market financing may be more stable than a bank-based one, especially if there is mispriced safety-net protection in the latter and low values of banking charters. For, in normal times, institutional investors, having good information[26] and low transactions costs, are likely to speed the adjustment of asset prices to fundamentals; this should entail price volatility only to the extent that fundamentals are themselves volatile. Moreover, the diversity in types and sizes of institutional investors - in their liabilities, incentives and consequent attitudes to risk - should be stabilizing to financial markets. The liquidity that institutional activity generates may dampen volatility, as is suggested by lower share-price volatility in countries with large institutional sectors (Davis and Steil, 2001). In a global context, enhanced cross-border portfolio investment undertaken by institutional investors should enhance the efficiency of global capital markets by equalizing total real returns (and hence the cost of capital) between markets.

[26] The concept of superior information of institutions is underpinned by studies showing that initial public offerings that are largely subscribed by institutions tend to do well, while those that are largely purchased by the general public tend to do badly (Trzcinka, 1998).

It can, moreover, be argued that securitized financial systems have important stabilizing features, such as ease of marking to market, matched assets and liabilities – notably for mutual funds and defined-contribution pension funds – and distance from the safety net. There are wider opportunities to diversify and spread risk. And the "multiple channels of intermediation" available to the corporate sector in securitized financial systems will reduce the impact of any crises which affect either banks or securities markets (see Greenspan, 1999; Davis, 2001a).

Institutional investors may, however, also sporadically give rise to unfamiliar types of financial instability from the point of view of regulators and market players, which would be accentuated as they grew during the expansion of precautionary saving and/or funding. Already in existing experience of financial instability, one can distinguish two particular types of financial turbulence they give rise to:

A first type involves extreme market-price volatility after a shift in expectations and consequent changes in institutional investors' asset allocations. Such crises are focused mainly on the consequences for financial institutions of sharp price changes, which result from institutional "herding" as groups of such institutions imitate one anothers' strategies for reasons related to the information and agency problems between the investors and the asset managers (see the summary in Davis and Steil (2001)). Whereas misaligned asset prices and sharp price movements during corrections may not in themselves have systemic implications,[27] these may emerge when such movements threaten e.g. institutions that have taken leveraged positions on the current levels of asset prices. Examples are the stock-market crash of 1987, the ERM crisis, the 1994 bond-market reversal and the Mexican crisis (Davis, 1995c). There were also elements of this in the Asian crisis. Indeed, as discussed in Section 6.4, EME securities markets, as well as those for forex, derivatives, bonds and equities in OECD countries, could be vulnerable to these shifts.

A second type involves protracted[28] collapse of market liquidity and issuance (Davis, 1994). Again, often involving one-way-selling by institutional investors as they seek to shift asset allocations simultaneously, the distinction is often largely one of whether markets are sufficiently resilient, and whether market-maker structures are suitably robust. Also, such crises tend to characterize debt markets rather than equity or foreign exchange. The risks are acute not only for those holding positions in the market but also for those relying on the market for debt finance or liquidity – which increasingly include banks. Examples in the past have tended typically to be rather specific and idiosyncratic markets, which by nature relied on a narrow investor base, market-maker structure and/or issuer base (junk

---

27 They may, however, lead to resource misallocation.

28 It is not denied that all sharp price changes will tend to affect market liquidity to a greater or lesser degree

bonds, floating-rate notes, Swedish commercial paper, ECU bonds). However, the events following the Russian default and the rescue of the hedge fund LTCM were much more serious, as liquidity failure was threatened in markets such as the US securities-repurchase (repo), swaps, commercial paper (CP), corporate and Treasury-bond markets (see IMF, 1998; Davis, 1999b). In this context, note that real-estate crises, which could also follow institutional herding, have elements of illiquidity as well as price declines.

Price-volatility and market-liquidity-based crises may threaten EMEs, banks and the non-financial sector more than institutional investors themselves. The latter can "sit out" such crises, given their long-term liabilities. But, as noted above, institutional investors may tend to focus increasingly on debt claims as members approach retirement, while competition in asset management may lead investment managers also to be willing to take heightened credit risks in order to maximize their return on assets (Bishop, 1998; Allen and Gale, 1999). In countries such as the US, defined-benefit pension funds have pension-benefit insurance that may generate significant moral hazard (Bodie and Merton, 1992). This suggests that institutional investors could increasingly invest in high-yield bonds and EME sovereign and corporate debt, and thus be vulnerable to credit risk to a greater extent than has been the case in the past. Credit cycles could, in other words, affect institutional investors as well as banks. Solvency could be threatened directly for life-insurance companies and defined-benefit pension funds if a significant proportion of their assets defaulted.

Besides market and credit risk, errors on the asset or liability side (e.g. in terms of guaranteed returns or mortality projections) could also threaten the solvency of a range of institutional investors, as has been seen already in Japan, and with certain UK companies such as Equitable Life.

Three points may mitigate related systemic concerns, first, that insurance companies and defined-benefit pension funds are not easily subject to runs on suspicion of insolvency, given they have matched and long-term assets and liabilities, while mutual funds and defined-contribution pension funds are not themselves subject to solvency risks – credit risks are passed directly to the household sector. Second, most institutional-investor claims are not insured, or the insurance is mutual, thus generating incentives for interfirm monitoring. Third, given the ease of adopting market-value accounting for securitized claims, it can be argued that debt crises are much less likely in corporate-bond markets than for banks, where the deterioration of credit quality is hidden from view in the balance sheet (Davis, 2001a). Markets can still make mistakes, however, as witness the repeated bond-based debt crises of the late Victorian period; and US experience suggests that bond markets generally find rescheduling after financial distress difficult (Gilson et al., 1990).

Beyond the consequences for asset-price volatility and securities-market liquidity, further risks may arise for the banking sector in an institutionalized financial system. A lesser proportion of saving being channelled via banks, given lower deposit inflows and greater competitiveness of capital-market financing, may give rise to banking crises of the familiar type, where banks take increased risks so as to boost their profitability in a highly competitive market situation, while higher-quality credits seek capital-market financing. It can be argued that the banking crises in a number of countries in the late 1980s were linked to the heightened competition banks faced from the capital markets (Davis and Steil, 2001). In this context, note that a number of authors such as Demirguc-Kunt and Detragiache (1998b) have looked at the effect of financial liberalization on systemic risks, and found that banking crises were more likely to occur in liberalized financial systems. Crises tended to occur a few years after liberalization, and were linked to a decline in bank-franchise value, because monopoly power is eroded (see Hellman et al., 2000). Securities-market competition can arguably have a similar effect on franchise values and risk taking.

6.4 Risks arising from the pattern of asset accumulation during funding

The asset flows from pension saving are popularly considered to be already having an effect on asset prices, in driving the bull market in equities in countries such as the US. Certainly, work on savings flows such as Poterba (1998) does imply that current high levels of retirement saving in the form of securities may have driven up market prices (although their effect on returns has been much smaller). It may be the case that the institutionalization of saving has boosted equity prices via a fall in the risk premium (Blanchard, 1993), reflecting better diversification and lower risk-aversion of collective investment vehicles. Such tendencies can overshoot, leading to bubbles. On the other hand, most work on the relation of equity prices to mutual fund flows suggests that the direct effects of inflows on prices are hard to detect (Fortune, 1998; Engen et al., 2000).

While the above discussion is largely focused on equity, asset bubbles could also accompany the pension-asset build-up, to the extent that retirement investment focuses on debt or real-estate claims. The property market might be particularly susceptible. Finance of property development by bond issue or direct institutional investment are two possible channels. Following Allen and Gale (1999, 2000), if institutions become significantly involved in property investment and related credit finance, the massive inflows to pension funds and insurance companies could generate expectations of ongoing boosts to credit flows into property. There may also be increasing uncertainty about future such flows as ageing progresses. These in the past have tended to foreshadow asset bubbles generated by such credit flows. Since banks would remain exposed to property prices via collateral, and would probably also lend to higher-risk projects, they could be badly hit when decumulation occurs, or earlier if institutional investors seek to adjust their asset allocation away from property for any reason.

A large share of OECD-pension saving in EMEs can also lead to bubbles and financial stability risks in the latter owing to institutional behaviour, even before an eventual repatriation of funds, as foreseen in Section 6.1. Owing, for example, to autonomous shocks affecting profitability and creditworthiness, there may be a flight of investable funds back to the OECD or to other EMEs. In this context, note that securities are in principle much easier to repatriate than bank loans. Indeed, the behaviour of OECD institutional investors is already widely considered to destabilize EMEs, not least owing to the tendency to invest in EMEs as a bloc, rather than focusing closely on individual countries' fundamentals (Buckberg, 1996).

These patterns of volatility are underpinned by the sharp difference in the relative size of EMEs and OECD institutional investors, whereby, according to 1995 data, 1% of domestic equity holdings by institutional investors in the G-7 countries is equivalent to slightly over 1% of global stock-market capitalization but would be equivalent to 27% of the market capitalization of the emerging Asian economies and 66% of Latin American ones. Hence, if portfolio shifts on the scale of only 1% of domestic equities were to be focused solely on the emerging markets, the effects would be very disruptive. The relative growth of EME market capitalization and OECD institutional investors suggests that this size differential is likely to be maintained for some decades ahead.

Beyond the issue of changing asset-allocation decisions, which can drive price falls even if pension saving is still rising, there are also arguments suggesting that asset prices will fall during the process of ageing itself, as it interacts with pension funding. A fundamental aspect is that the return on capital may fall as a consequence of a lower labour supply and lower growth during ageing. As noted, Cutler et al. (1990) see the real return on capital halving. An issue for funding is whether it can generate offsetting increases in efficiency and resource allocation, e.g. via growth of capital markets, that may help to offset this by boosting productivity and investment.

Besides the return on physical capital, a key issue is whether asset prices will also be put under downward pressure in coming decades by declining saving in OECD countries implicitly affecting the real interest rate or the risk premium. Schieber and Shoven (1994) note that, given the correlation of ageing in OECD countries, and the likely decumulation of defined-benefit pension-fund assets, there could be widespread falls in asset prices, linked to high real interest rates. Supporting this, Erb et al. (1997) find a positive correlation in the US between the fraction of the population 25-45 and 65+ and stock returns,[29] while those 45-65 have a negative effect. Looking at a range of OECD and EME countries, they find a positive relation of stock returns to the average age of the population. Brooks (2000) focuses on the relation between ageing and the demand for equities and

29 And thus a negative effect on prices.

bonds, and suggests that there will be excess demand for bonds and excess supply of equities in coming decades, with a marked decline in the returns on the retirement savings of baby boomers. In this context, solvency could be threatened for life-insurance companies and defined-benefit pension funds that had made undertakings based on expected returns on assets formed during the bull period (as has been the case recently for the Japanese life insurers and pension funds, and for UK insurers such as Equitable Life).

Poterba (1998), on the other hand, looks at past data and finds virtually no link of demographic patterns to asset returns, and particularly no positive results were found for equities. His conclusion is that there is no strong likelihood of a "baby bust" effect depressing asset returns markedly around 2020-2030,[30] even if the recent rise in baby-boom savings may have had a positive effect. Neuberger (1999) argues that the increase and subsequent decrease in flows will be balanced by rises and falls in equity issues, with little effect on prices and returns. Also, the increase in the ratio of pensioners to workers is already under way, and will continue steadily rather than abruptly, again casting doubt on the idea of a cycle. Or, at least, the market will take on board such gradual future shifts without major and abrupt adjustments in prices. Furthermore, OECD countries are ageing at different rates, and there may be offsetting demand for securities from EMEs.

In general, we are sympathetic to the view that, even if there are price changes with the decumulation of pension funds, they will not necessarily be precipitate, and hence the issue is one affecting the level of retirement income and the welfare of pensioners, rather than systemic financial-market stability. On the other hand, the tendency of markets to overreact in both directions is apparent from the recent boom in IT stock prices, and hence vigilance is needed. If there are indeed such sharp changes in market prices during ageing, they may give rise to losses threatening solvency on the part of financial institutions or investors taking leveraged positions, along the lines set out in Section 6.3. It would also impact on the adequacy of funding and replacement rates offered by pension funds and life-insurance companies. Such a sharp shift away from securities could extend to other asset markets, notably those for commercial property and possibly to residential property. These effects would be compounded if, for example, the older population had a lower demand for housing services. Mankiw and Weil (1989), for example, made controversial[31] projections of lower house prices in the US as a consequence of ageing. The banks, which hold such property as security, would find credit quality declining as a consequence.

---

30 Commenting on Poterba's work, Abel (2001) points out that an elastic supply of capital as the population ages could generate a fall in asset prices, even if the demand for assets falls little as baby boomers retire.

31 Engelhardt and Poterba (1991) and Hendershott (1991) provide counter-evidence to this suggestion.

7 A re-evaluation of the theories and indicators

Tracing the relevance of the theories outlined in Section 2 in the light of the risks assessed in Sections 3–6, it is suggested that financial instability peculiar to the ageing of the population would only partly resemble that most typical of the past, namely banking crises following stimuli to credit expansion such as financial liberalization, as outlined in the *financial fragility* paradigm. On the other hand, reflecting a shift from equity to debt and from government bonds to private bonds, credit cycles could affect institutional investors, also impacting upon banks. The effects of financial instability arising via a corporate debt crisis on bonds held by institutional investors have not yet been widely seen – although the threatened bankruptcy of Japanese life insurers may give some clues. The *monetarist* impact of bank failures on the economy via the money supply may also be attenuated if banks are a less important component of the financial system (as well as the safety net now being well-developed).

On the other hand, there will clearly be heightened *uncertainty* about the likely financing flows that will occur as the population ages, especially when pay-as-you-go is dominant, and possible *disaster myopia* by institutions – and regulators - that have not made appropriate provision for the effects of ageing on the economy and the financial system. *Agency costs* will be most crucial in the context of the relation between the asset manager and the ultimate investor, and managers may indeed seek high-risk placements, especially in the debt market, to gain business. The more traditional forms of agency cost in banking will still be present, however (whereby banks with low franchise values and/or mispriced safety-net protection will have incentives to take risks).

Runs from institutional investors are not feasible in respect of the bulk of their liabilities, namely long-term contracts, but life insurers – and money-market funds - have proved vulnerable to liquidity crises in respect of their short-term debt. With lower credit quality, banks may become even more vulnerable, despite the lesser role they are likely to play in the financial system. *Herding* by institutional investors will meanwhile become a much more important cause of financial instability than hitherto, as outlined in Sections 6.3 and 6.4. The *industrial competition* between asset managers will equally become as important in generating financial instability (e.g. by generating herding) as that between banks.

As regards *macroprudential indicators*, the size and composition of saving flows of the private and public sectors would seem to be the most crucial. Both gross and net flows would need to be taken into account, given their relevance for credit, liquidity and market risk, while the evolving portfolios of institutional investors will come to the fore, particularly where funding plays a key role. Debt accumulation would still be of particular interest, with the focus being on public rather than private debt, given the likely increase in government debt in pay-as-you-go, as well as the increased appetite of institutional investors for debt, given the growing maturity of pension schemes and the provision of annuities. On the other hand, one would need caution in using fiscal data because the meaning of a given level of debt changes with shifts in population structure, because of the ratio of explicit to implicit debt. High-frequency movements will still be dominated by explicit debt, but financial markets will be increasingly aware of – and react to – the implicit debt also. Moreover, changes in debt levels might provide little information in the context of a pension reform.

Asset-price booms may well be driven by the accumulation of either precautionary savings (diverted via banks) or funding per se – and reversed sharply as the saving for retirement unwinds, or when the asset allocations of institutional investors shift. Banks may find they have to concentrate on high-risk assets (including property loans) if institutional investors are providing extensive bond finance for higher-rated firms. The flood of retirement funds likely to be generated in the coming decades will be fruitful areas for financial innovations, which experience shows are often mispriced in the first instance.

As regards regime shifts when funded sectors become very large, small changes in the details of regulation may have unanticipated effects at a macroeconomic level, comparable for example to the "capital crunch" that is widely considered to have followed the Basle Capital Adequacy Accord of 1988. Policies which act to encourage or discourage institutional investment in real estate could be examples. And shifts in fiscal policy, whether or not related to pay-as-you-go pensions, could have regime-shift effects on asset markets, even if the main regime shift is likely to be that of ageing itself, and its impact on growth and returns.

Looking at the shorter-term indicators of instability, we can highlight that fiscal policies accompanying pay-as-you-go are likely to raise real interest rates and/or lead to a slump. Meanwhile, real exchange-rate appreciation and credit expansion may accompany the accumulation phase in OECD countries, while recession, stock-market crashes and currency crises may accompany decumulation. EMEs may experience sizeable capital inflows that are later unwound.

8 Conclusions

Although the precise details are subject to considerable uncertainty, it seems likely that the process of population ageing will involve major shifts in financing, which may give rise to financial turbulence and systemic risk. The locus and scale of these effects will also depend on the predominant approach to retirement-income provision. It is argued that the financial-stability risks arising from continuing with unsustainable pay-as-you-go systems would be more substantial than those arising from funding. Fiscal crises can have incalculable consequences for private financial markets, also generating transnational spillovers, while pension funding involves more an adaptation by regulatory authorities to a more securitized and institutionalized financial system. Concerning policy, for social security the key issue is reform, so that the fiscal difficulties and their consequences for financial stability foreshadowed above do not arise. Central bankers have a strong interest in ensuring that such reform occurs, given the difficulties for the policy-mix and inflation that could otherwise occur. For institutional investors involved in funding, policy issues arising include the need for prudent personal asset regulation, the absence of guarantees generating moral hazard and the international diversification of institutional portfolios, so that they are less dependent on the performance of the domestic economy than would otherwise be the case. Bank regulators need to be aware of the side-effects that patterns of ageing can have for banks' assets and liabilities, and to ensure that capital adequacy and sound risk assessment are maintained.

It is important not to see these effects on financial stability in isolation. The chosen approach to retirement-income provision is likely to have major effects on economic performance as well as on financial stability. In particular, retention of pay-as-you-go would be likely to accentuate the existing distortions that it engenders for labour and product markets, while switching to funding both reduces these distortions and gives rise to benefits in the fields of investment and technical progress. The economy will accordingly be stronger and more resilient to any financial turbulence that may arise. Government finances would be in better shape for any bank recapitalization that may be needed.

Furthermore, some of the risks outlined above may arise independently of ageing. Banking-sector risks arising from financial liberalization may be an ongoing, rather than one-off, phenomenon. Also, it can be argued that the process of institutionalization is likely to occur independently of ageing, owing to technical progress which boosts the comparative advantage of institutions, relative to banks[32] and the growing wealth of the household sector, which leads them to seek professional management for high-risk-high-return assets (Davis and Steil 2001). Adaptation of regulatory authorities to this shift will thus be required in any case.

---

[32] Despite pressures, such as institutionalization, leading to convergence in financial structures, some differences are likely to remain for some time, and their impact on the vulnerability of the financial sector to the types of instability highlighted here is worthy of further study (for background see Allen and Gale, 2000a).

References

Abel, A. B. 2001. *Will Bequests Attenuate the Predicted Meltdown in Stock Prices when Baby Boomers Retire?*. NBER Working Paper No. 8131.

Allen, F. and D. Gale. 1999. Bubbles, Crises and Policy, *Oxford Review of Economic Policy*, 15/3, pp. 9-18.

Allen, F. and D. Gale. 2000a. Bubbles and Crises, *Economic Journal*, 110.

Allen, F. and D. Gale. 2000b. *Comparing Financial Systems, MIT Press.*

Attanasio, O. and A. Brugiavini. 1999. *Social Security and Household Saving,* mimeo, UCL and University of Venice.

Auerbach, A. J., L. J. Kotlikoff, R. P. Hagemann and G. Nicoletti. 1989. *The Economic Dynamics of an Ageing Population:* The Case of four OECD Countries, Department of Economics and Statistics Working Paper No. 62, OECD, Paris.

Barro, R. J. 1974. Are Government Bonds Net Wealth?. *Journal of Political Economy*, 82, pp. 1095-1117.

Bikker, J. A. 1996. National Savings, the Current Account and Ageing Populations, a Pension Fund Model. *Economic and Financial Modelling*. Summer 1996, pp. 1-20.

Bernard, H and J. Bisignano. 2000. *Information, Liquidity and Risk in the International Interbank Market; Implicit Guarantees and Private Credit Market Failure.* BIS Working Paper No. 86.

Bishop, G. 1998. *Securitising European* Saving, Salomon Smith Barney, London, 1st December 1998.

Blanchard, O. J. 1993. The Vanishing Equity Premium. In eds. R. O'Brien, *Finance and the International Economy.* 7, Oxford University Press.

Blommestein, H. 1998. Ageing induced Capital Flows to Emerging Markets do not solve the OECD's Basic Pension Problem. In eds. H. Blommestein and N. Funke, *Institutional Investors in the New Financial Landscape,* OECD, Paris.

Bodie, Z. and R. C. Merton. 1992. Pension Benefit Guarantees in the United States; A Functional Analysis. In ed. R. Schmitt, *The Future of Pensions in the United States,* University of Pennsylvania Press.

Börsch-Supan, A. 1996. The Impact of Population Ageing on Savings, Investment and Growth in the OECD Area. In *Future Global Capital Shortages - Real Threat or Pure Fiction?,* OECD, Paris.

Börsch-Supan, A. and P. Tinios. 2000. *The Greek Pension System, Strategic Framework for Reform.* Presented at the Bank of Greece/Brookings Institute conference on Greece's Economic Performance and Prospects, Athens, December 7-8, 2000.

Brooks, R. 2000. *Life Cycle Portfolio Choices and Asset Market Effects of the Baby Boom,* IMF Working Paper WP/00/18.

Buckberg, E. 1996. *Institutional Investors and Asset Pricing in Emerging Market.* Working Paper WP/96/2, International Monetary Fund.

Chand, S. K. and A. Jaeger. 1996. *Ageing Populations and Public Pension Schemes,* IMF Occasional Paper No. 147.

Cremer, H. and P. Pestieau. 2000. Reforming our Pension System; is it a Demographic, Financial or Political Problem?. *European Economic Review,* 44, pp. 974-983.

Cutler, D., J. Poterba., L. Sheiner and L. Summers. 1990. *An Ageing Society, Opportunity or Challenge?.* Brookings Papers on Economic Activity, p. 1.

Dang T. T., P. Anoli and H. Oxley. 2001. *Fiscal Implications of Ageing Projections of Age-Related Spending,* OECD Economic Department Working Paper No. 305

Davis, E. P. 1994. Market Liquidity Risk. In D. Fair and R. Raymond (eds)., *The Competitiveness of Financial Institutions and Centres in Europe,* Kluwer Academic Publishers.

Davis, E. P. 1995a. *Debt, Financial Fragility and Systemic Risk,* revised and extended version, Oxford University Press.

Davis, E. P. 1995b. *Financial Fragility in the Early 1990s, what can be learnt from International Experience?.* LSE Financial Markets Group Special Paper No. 76.

Davis, E. P. 1995c. Institutional Investors, unstable Financial Markets and Monetary policy. In F. Bruni, D. Fair and R. O'Brien eds., *Risk Management in Volatile Financial Markets,* Kluwer, Amsterdam.

Davis, E. P. 1995d. *Pension Funds, Retirement-Income Security and Capital Markets, an International Perspective,* Oxford University Press.

Davis, E. P. 1997. Population Ageing and Retirement income provision in the European Union. In Ed. B. Bosworth and G. Burtless, *Ageing Societies, the Global Dimension.* Brookings Institution Press, Washington DC, (also Special Paper, Royal Institute of International Affairs, London).

Davis, E. P. 1998. *Investment of Mandatory funded Pension Schemes.* Discussion Paper No. PI-9908, the Pensions Institute, Birkbeck College.

Davis, E. P. 1999a. *Financial Data needs for Macroprudential Surveillance,* Lecture Series No 2, Centre for Central Banking Studies, Bank of England.

Davis, E. P. 1999b. *Russia/LTCM and Market Liquidity Risk.* The Financial Regulator, 4/2, Summer 1999, pp. 23-28.

Davis, E. P. 2001a. *Multiple Channels of Intermediation, Corporate Finance and Financial Stability,* Working Paper No. 01115, International Monetary Fund.

Davis, E. P. 2001. Portfolio Regulation of Life Insurance Companies and Pension Funds. Working Paper No. PI-0101, The Pensions Institute, Birkbeck College, London.

Davis, E. P. and B. Steil . 2001. *Institutional Investors, MIT Press.*

Demirguc-Kunt, A and E. Detragiache. . 1998a. *The Determinants of Banking Crises* in *Developing and Developed Countries.* IMF Staff Papers, 45, pp. 81-109.

Demirguc-Kunt, A and E. Detragiache. 1998b. *Financial Liberalisation and Financial Fragility*. IMF Working Paper No WP/98/83.

Diamond, D. and P. Dybvig P. 1983. Bank runs, Deposit Insurance and Liquidity. *Journal of Political Economy*, 91, pp. 401-19.

Disney, R. 1996. *Can we Afford to grow older?,* MIT Press.

Disney, R. 2000. Declining public pensions in an era of Demographic Ageing: will private pensions fill the gap?. *European Economic Review,* 44, pp. 957-973.

Engelhardt, G and J. M. Poterba. 1991. Demographics and House Prices; The Canadian Evidence. *Regional Science and Urban Economics,* 21, pp. 539-46.

Engen, E. M and A. Lehnert. 2000. Mutual Funds and the US Equity Market. *Federal Reserve Bulletin,* December 2000, pp. 797-812.

Englund, P. 1999. The Swedish Banking Crisis, Roots and Consequences. *Oxford Review of Economic Policy,* 15/3, pp. 80-97.

Erb, C. B., C. R., Harvey and T.E. Viskanta. 1997. Demographics and International Investment, *Financial Analysts Journal*, 53, July/August, pp. 14-28.

Feldstein, M. 1974. Social Security, Induced Retirement and Aggregate Capital Formation, *Journal of Political Economy*, 82, pp. 905-6.

Feldstein, M. 1977. Social Security and Private Saving; International Evidence in an Extended Life Cycle Model. In M. Feldstein and R. Inman. *The Economics of Public Services,* International Economic Association.

Feldstein, M. 1995. *Social Security and Saving, new Time Series Evidence.* NBER Working Paper No. 5054.

Fisher, I. 1933. The Debt Deflation Theory of Great Depressions, *Econometrica,* pp. 337-57.

Fortune, P. 1998. Mutual Funds, Part II; Fund flows and security returns, *New England Economic Review,* Jan/Feb, pp. 4-22.

Friedman, M. and A. J. Schwartz. 1963. *A Monetary History of the US 1867-1960,* NBER, New York.

Froot, K.A., D.S. Scharfstein and J.C. Stein. 1992. Herd on the Street: Informational Inefficiencies in a Market with Short-Term Speculation. *The Journal of Finance*, 47, pp. 1461-84.

Giavazzi, F. and M. Pagano. 1995. *Non Keynesian Effects of Fiscal Policy Changes; International Evidence and the Swedish Experience,* Working Paper No 1284, CEPR, London.

Gilson, S, J. Kose and L. Lang. 1990. Troubled Debt Restructurings, *Journal of Financial Economics*, 27, pp. 315-53.

Glick, R. and A. K. Rose. 1998. *Contagion and Trade; why are Currency Crises Regional?.* mimeo, Federal Reserve Bank of San Francisco.

Greenspan, A. 1999. *Do Efficient Financial Markets Mitigate Financial Crises?.* Speech to the Financial Markets Conference of the Federal Reserve Bank of Atlanta, 19 October.

Guttentag, J. M and R.J. Herring. 1984. Credit Rationing and Financial Disorder. *Journal of Finance*, 39: pp. 1359-82.

Hagemann, R. P. and G. Nicoletti. 1989. *Ageing Populations; Economic Effects and Implications for Public Finance.* Department of Economics and Statistics, Working Paper No. 61, OECD, Paris.

Hardy, D. C. and C. Pazarbasioglu. 1998. *Leading Indicators of Banking Crises; was Asia Different?.* IMF Working Paper No. WP/98/91

Hargraves, M., G.J. Schinasi and S. R. Weisbrod. 1993. *Asset Price Inflation. In the 1980s: a Flow of Funds Perspective,* IMF Working Paper No. WP/93/77.

Heller, P., R. Hemming and P.W. Kohnert. 1986. Ageing and Social Expenditures in Major Industrial Countries 1980-2025, Occasional Paper No. 47, International Monetary Fund, Washington DC.

Hellman, T. F., K. Murdock and J.E. Stiglitz. 2000. Liberalisation, Moral Hazard in Banking and Prudential Regulation; are Capital Requirements Enough?, *American Economic Review*, 90, pp. 147-165.

Hendershott, P. H. 1991. Are Real House Prices Likely to Decline by 47%?, *Regional Science and Urban Economics,* 21, pp. 553-65.

Herring, J. 1999. Credit Risk and Financial Instability. *Oxford, Review of Economic Policy,* 15/3, pp. 63-79.

Horioka, C. Y. 1991. The Determinants of Japans Private Saving Rate, the Impact of the Age Structure of the Population and other Factors. *Bank of Japan Economic Studies Quarterly,* 42, pp. 237-53.

Hoshi, T, A. Kashyap and D. Scharfstein. 1993. *The Choice between Public and Private Debt; an Analysis of Post Deregulation Corporate Financing in Japan.* Working Paper No. 4211, National Bureau of Economic Research.

IMF.1998. *World Economic Outlook and International Capital Markets,* Interim *Assessment December 1998 - Financial Turbulence and the World Economy.* IMF, Washington D.C.

IMF. 2000. *Macroprudential Indicators of Financial System Soundness.* Occasional Paper No. 192, IMF, Washington D.C.

Kaminsky, L. G. and C.M. Reinhart. 1999. The Twin Crises; the Causes of Banking and Balance-of-Payments Problems, *American Economic Review*, 89, pp. 473-500.

Keeley, M. C. 1990. Deposit Insurance, Risk and Market Power in Banking. *American Economic Review,* 80: pp. 1138-99.

Kindleberger, C. P. 1978. *Manias, Panics and Crashes,* A History of Financial Crises, Basic Books, New York.

King, M. A and L. DicksMireaux . 1988. Portfolio Composition and Pension Wealth: an Econometric Study. In eds. Z. Bodie , J.B. Shoven and D.A. Wise, *Pensions in the US Economy,* University of Chicago Press.

Knight, F. H. 1921. *Risk, Uncertainty and Profit, Boston;* No. 16 in a Series of Rare Texts in Economics, Republished by the LSE.

Kohl, R. and P. O'Brien. 1998. *The Macroeconomics of Ageing, Pensions and Savings;* a Survey. Economics Department Working Paper No. 200, OECD, Paris

Krugman, P. 1991. Financial Crises in the International Economy. In M. Feldstein, ed. The *Risk of Economic Crisis,* Univ. of Chicago Press.

Leibfritz, W., D. Roseveare, D. Fore. and E. Wurzel. 1996. Ageing Populations, Pension Systems and Government Budgets - how do they Affect Saving?. In: *Future Global Capital Shortages - Real Threat or Pure Fiction?,* OECD, Paris.

Mankiw, N. G and D.N. Weil . 1989. The Baby Boom, the Baby Bust and the Housing Market. *Regional Science and Urban Economics,* 19, pp. 238-258.

Masson, P. and R.W. Tryon. 1990. *Macroeconomic Effects of Projected Population Ageing in Industrial Countries.* IMF Staff Papers, 37, pp. 453-485.

Masson, P., T. Bayoumi. and H. Samiei. 1995. *International Evidence on the Determinants of Private Saving.* IMF Working Paper No. W95/51.

Mayer, C. P. 1999. The Assessment; Financial Instability, *Oxford Review of Economic Policy,* 15/3, 1-8.

Miles, D. and B. Patel. 1996. *Saving and Wealth Accumulation in Europe, the Outlook into the next Century.* Merrill Lynch, Financial Research, 3 June 1996.

Miles, D. K. 1999. *Funded and Unfunded Pension Schemes; Risk, Return and Welfare.* Mimeo, Imperial College, London.

Minsky, H. P. 1977. A Theory of Systemic Fragility. In: E.I. Altman and A.W. Sametz, ed. *Financial Crises,* Wiley, New York..

Mishkin, F. S. 1991. Asymmetric Information and Financial Crises: A Historical Perspective, in R.G. Hubbard ed. *Financial Markets and Financial Crises,* University of Chicago Press, Chicago.

Mitchell, O. S and Z. Bodie. 2000. *A Framework for Analysing and Managing Retirement Risks,* Working Paper No. PRC WP 2000-4, Pension Research Council, The Wharton School, University of Pennsylvania.

Munnell, A. H. 1987. The Impact of Public and Private Saving Schemes on Saving and Capital Formation. In: *Conjugating Public and Private, the Case of Pensions.* International Social Security Association, Studies and Research, 24, pp. 219-36.

Nakagawa, S. 1999. Why has Japans Household Savings Rate Remained High during the 1990s?. Reprinted from Bank of Japan Monthly Bulletin, April 1999

Neuberger, A. 1999. *Long Term Savings Flows and the Capital Market,* Morgan Stanley Dean Witter Global Pensions Quarterly, January 1999.

Oyama, T. and K. Yoshida. 1999. *Does Japan Save too Little or do Other Countries save too much?.* Working Paper No. 99-5, Research and Statistics Department, Bank of Japan.

Poterba, J. M. 1998. *Population Age Structure and Asset Returns - An Empirical Investigation.* NBER Working Paper No. 6774.

Reisen, H. 1998. Warning; Past Pension Fund Performance is no Guarantee for Future Performance. In eds. H. Blommestein and N. Funke, *Institutional Investors in the New Financial Landscape,* OECD, Paris.

Roseveare, D., W. Leibfritz., D. Fore. and E. Wurzel. 1996. *Ageing Populations, Pension Systems and Government Budgets: Simulation for 20 OECD Countries.* Economics Department Working Paper No. 168, OECD, Paris.

Scharfstein, D. S. and J. C. Stein, 1990. Herd Behaviour and Investment, *American Economic Review,* 80, pp. 465-79.

Schieber, S. J. and J. Shoven. 1994. *The Consequences of Population Ageing on Private Pension Fund Saving and Asset Markets.* NBER Working Paper No. 4665.

Shafer, J. R. 1986. Managing Crises in the Emerging Financial Landscape, *OECD Economic Studies,* 8, pp. 56-77.

Stiglitz, J. E. 2000. Capital Market Liberalisation, Economic Growth and Instability, *World Development,* 28, pp. 1075-1086.

Trzcinka, C. 1998. The Conflicting Views and Management Practices of Institutional Equity Investing, *Financial Markets,* Institutions and Instruments, 7, pp. 20-53

Turner, D., C. Giorno, A. De Serres, A. Vourch and P. Richardson. 1998. *The Macroeconomic Implications of Ageing* in a Global Context, Economics Department Working Paper No. 193, OECD, Paris.

Wattenburg, B. 1987. *The Birth Dearth,* Pharos Books, New York.

World Bank. 1994. *Averting the old Age Crisis;* Policies to Protect the old and Promote Growth, The World Bank, Washington DC.

Table 1. Life Expectancy at Birth

Years	1970–1975	1980–1985	1990–1995	2000
United Kingdom	72	74	76	78
United States	73	75	77	77
Germany	71	73	76	77
Japan	74	77	79	81
Canada	73	76	78	79
France	72	75	78	79
Italy	72	75	78	79

Source: World Bank (1996), U.S. Department of the Census.

Table 2. Fertility Rates

Number of Children per Female	1970–1975	1980–1985	1990–1995	2000
United Kingdom	1.8	1.8	1.8	1.7
United States	1.8	1.8	2.0	2.1
Germany	1.5	1.4	1.2	1.4
Japan	1.9	1.8	1.5	1.4
Canada	1.8	1.7	1.9	1.6
France	1.9	1.8	1.7	1.7
Italy	2.2	1.4	1.3	1.2

Source: World Bank (1996), U.S. Department of the Census website.

Table 3. Projections of Elderly-Dependency Ratio to 2050

Population 65 and over as a Percentage of Population Aged 20–64	1960	2000	2035	2050
United Kingdom	17.9	26.6	44.6	45.3
United States	15.4	21.7	38.2	37.9
Germany	16.0	26.6	54.1	53.2
Japan	9.5	27.7	53.9	64.6
Canada	13.0	20.4	42.2	45.9
France	18.8	27.2	47.5	50.8
Italy	13.3	28.8	56.8	66.8

Source: Dang et al. (2001).

Table 4. Projections of Pension Costs (OECD Estimates)

Pension expenditure/ GDP	1995	2000	2010	2020	2030	2040
United Kingdom	4.3	4.3	4.1	3.9	4.3	4.1
United States	4.9	4.4	4.3	5.4	6.2	6.3
Germany	11.7	11.8	11.2	12.6	15.5	16.6
Japan	6.8	7.4	8.6	8.5	7.5	8.2
Canada	5.3	5.1	6.0	7.6	9.6	10.4
France	12.2	12.1	13.1	15.0	16.0	15.8
Italy	13.4	14.2	14.3	14.9	15.9	15.7

Source: Dang et al. (2001).

Table 5. Present Value of net Pension Liability, 1995–2050 (IMF Estimates)

As % of 1994 GDP	Net Pension Liability	Memo: Contribution Gap
United Kingdom	5	0.1
Germany	111	3.4
France	114	3.3
Italy	76	2.5
United States	26	0.8

Source: Chand and Jaeger (1996). The contribution gap is the difference between the contribution rate that is needed to reduce the net asset position to zero and the current contribution rate.

Table 6. Social Security Benefits and Institutional Investment

	Social Security Replacement Ratios	Institutional Investment/ GDP (%), 1998
United Kingdom	60–33	197
United States	71–45	176
Germany	45–43	32
Japan	Na	63
Canada	57–26	105
France	67–51	90
Italy	78–75	54

Note: Replacement ratios—pensions as a percentage of final salary—are for final salaries of $20,000 and $50,000. The data are for 1997.

Source: Watson Wyatt (1997), Table 1.6.

Discussion

Franklin Allen

Philip is clearly one of the world's leading experts on pensions and he's also an expert on crises. He is the ideal person to write this paper. It is a thought-provoking and interesting analysis of the problem.

The starting point is a discussion of financial stability. *Systemic risk* is defined as a sequence of events entailing heightened risk of a financial crisis. A *financial crisis* is defined as being a major and contagious collapse of the financial system, entailing inability to provide payment services or to allocate funds for investment. These are fairly narrow definitions. With ageing, it may well be that any resulting financial crisis will take a long time to develop. The crisis in Japan that resulted from the aftermath of the bubble in asset prices in the late 1980-'s is an example of a long-drawn-out crisis. The impact of aging may be similar, in that it develops over several years and its impacts are damaging without any discontinuous change.

The second step of the analysis is to outline the aging problem. This is well-known, but the paper does a nice job of summarizing the literature succinctly. The categorization of pension systems developed is a useful one. Philip distinguishes between three different types.

1. Pay-as-you-go.
2. Bond issuance.
3. Pre-funded pensions in private markets.

The final part of the paper develops the consequences of the aging problem and the type of pension scheme adopted for the financial system.

The main difference that I have with Philip is that I would use a different categorization of crises. I would divide the causes of crises into five different types. Only some of these (distinguished below by **) are likely to be associated with aging.

1 Financial panic

Panics are crises that are *random events* unrelated to changes in the real economy. The classical form of this view suggests that panics are the result of some kind of "mob psychology" or "mass hysteria" (see, e.g., Kindleberger (1978)). The modern version, developed by Diamond and Dybvig (1983) and others, is that bank runs are self-fulfilling prophecies. Given their assumptions of first-come, first-served liabilities and costly liquidation of some assets, there are multiple equilibria. If everyone believes that a banking panic is about to occur, it is optimal for each individual to try and withdraw her funds. Since each bank has insufficient liquid assets to meet all of its commitments, it will have to liquidate some of its assets at a loss. Given the first-come, first-served nature of deposit contracts, those depositors who withdraw initially will receive more than those who wait. On the one hand, anticipating this, all depositors have an incentive to withdraw immediately. On the other hand, if no one believes a banking panic is about to occur, only those with immediate needs for liquidity will withdraw their funds. Assuming that banks have sufficient liquid assets to meet these genuine liquidity demands, there will be no panic.

An important issue within the Diamond and Dybvig framework is that of equilibrium selection. One simple way of modelling which of these equilibria occurs is to assume it depends on extraneous variables or "sunspots". For example, if a sunspot occurs, people believe the bad equilibrium will prevail, and this will be self-fulfilling. If a sunspot does not occur, people will believe that the good equilibrium will prevail, and this will also be self-fulfilling. Another, more sophisticated way is to assume informational imperfections, as in Morris and Shin (1998). They show how introducing a small degree of informational imperfection allows a unique equilibrium to be determined.

With regard to the issue of aging, there seems to be no particular reason why this should lead to more financial panics than before. This would happen if aging was expected to change the equilibrium selection mechanism, but this seems unlikely.

2 Business cycle**

An alternative to the "sunspot" view is that banking panics are a natural outgrowth of the *business cycle*. An economic downturn will reduce the value of bank assets, raising the possibility that banks will be unable to meet their commitments. If depositors receive information about an impending downturn in the cycle, they will anticipate financial difficulties in the banking sector and try to withdraw their funds. This attempt will precipitate the crisis. According to this interpretation, panics are not random events but a response to unfolding economic circumstances. Building on the empirical work of Gorton (1988) that nineteenth-century banking crises were predicted by leading economic indicators, Allen and Gale (1998) develop a model that is consistent with the business-cycle view of the origins of banking panics[1]

In contrast to Diamond and Dybvig (1983), the basic problem is not one of multiple equilibria. Rather the problem is that banks use deposit contracts that involve a fixed promise. If the returns on a bank's assets are low, the bank will be unable to satisfy its promise. Depositors will be able to deduce this. Early consumers with urgent liquidity needs and late consumers who do not require liquidity to consume will try and withdraw simultaneously. Since the bank has only limited liquid assets, it will be unable to meet everybody's demands and there will be a run.

Crises that result from banks being unable to meet their fixed promises to depositors may well result from the aging problem. The heavy requirements of a large cohort of elderly people requiring liquidity simultaneously may well increase the frequency of this type of crisis. The types of financial system that are likely to be particularly susceptible are those that are bank-based. Market-based financial systems, where there is widespread use of non-contingent contracts in the provision of pensions, may also suffer from an increased occurrence of financial crises.

3 Inconsistent government macro policies**

The third type of crisis is where the government follows financial and exchange-rate policies that are inconsistent with stability. The first-generation models of currency crises (e.g., Krugman (1979)) and the second-generation models (e.g., Obstfeld (1986; 1994) and Calvo (1988)) have these types of features. For example, in first-generation models the government may be running a deficit that is financed partially by expanding the money supply. The resulting inflation acts as a tax. This will be inconsistent with a fixed or pegged exchange rate and will trigger a currency crisis.

Generous pay-as-you-go and bond-issuance pension systems may well lead to an increase in this type of crisis. In addition to runs on the currency, anticipation of funding pension systems through an inflation tax may lead to a collapse in the government bond market, and it may not be possible to roll over short-term debt.

---

1 See also Chari and Jagannathan (1988), Jacklin and Bhattacharya (1988), Hellwig (1994), and Alonso (1996)).

4 Bubble collapse**

Crises often follow what appear to be bubbles in asset prices. Examples are the crises in Norway, Finland and Sweden in the late 1980-'s and early 1990-'s, the Japanese bubble of the 1980's and the subsequent lost decade of the 1990's and the Asian crises of 1997. Allen and Gale (2000a) provide a theory of bubbles and ensuing crises based on the existence of an agency problem. Many investors in real estate and stock markets obtain their investment funds from external sources. If the ultimate providers of funds are unable to observe the characteristics of the investment, there is a classic risk-shifting problem. Risk-shifting increases the return to investment in the assets and causes investors to bid up the asset price above its fundamental value. A crucial determinant of asset prices is the amount of credit that is provided for speculative investment. Financial liberalization, by expanding the volume of credit for speculative investments, can interact with the agency problem and lead to a bubble in asset prices.

This type of crisis is again likely to be exacerbated by the aging problem. The types of financial system that are likely to be particularly susceptible are those with significant market funding of pensions and institutional investors.

5 Contagion and financial fragility

The prevalence of financial crises has led many to conclude that the financial sector is unusually susceptible to shocks. One theory is that small shocks can have a large impact. A shock that initially affects only a particular region or sector or perhaps even a few institutions can spread by contagion to the rest of the financial sector and then infect the larger economy. There are a number of different types of contagion that have been suggested in the literature.

Banks are linked in several ways including payment systems and interbank markets. These linkages can lead to a problem of contagion. We start by considering models of payment-system contagion. Building on a locational model of payment systems developed by McAndrews and Roberds (1995), Freixas and Parigi (1998) have considered contagion in net and gross payment systems. In a net payment system, banks extend credit to each other within the day and at the end of the day settle their net position. This exposes banks to the possibility of contagion if the failure of one institution triggers a chain reaction. In a gross system, transactions are settled on a one-to-one basis with central bank money. There is no risk of contagion, but banks have to hold large reserve balances. A net payment system is preferred when the probability of banks having low returns is small, the opportunity cost of holding central bank money reserves is high, and the proportion of consumers that have to consume at another location is high. Freixas,

Parigi and Rochet (1999) use this model to examine the conditions under which gridlock occurs. They show that there can be gridlock when the depositors in one bank withdraw their funds, anticipating that other banks cannot meet their netting obligations if all their depositors have also withdrawn their funds. Rochet and Tirole (1996a) consider the role of the too-big-to-fail policy in preventing contagion.

Allen and Gale (2000b) focus on a channel of contagion that arises from the overlapping claims that different regions or sectors of the banking system have on one another through interbank markets. When one region suffers a banking crisis, the other regions suffer a loss because their claims on the troubled region fall in value. If this spillover effect is strong enough, it can cause a crisis in the adjacent regions. In extreme cases, the crisis passes from region to region and becomes a contagion. Aghion, Bolton and Dewatripont (1999) also consider a model of contagion through interbank markets. In their model, there are multiple equilibria. In one equilibrium there are self-confirming beliefs that a bank failure is an idiosyncratic event and, in the other, there are self-fulfilling beliefs that a bank failure signals a global shortage of liquidity. Lagunoff and Schreft (1998) study the spread of crises in a probabilistic model. Financial linkages are modelled by assuming that each project requires two participants, and that each participant requires two projects. When the probability that one's partner will withdraw becomes too large, all participants simultaneously withdraw, and this is interpreted as a financial crisis. Van Rijckeghem and Weber (2000) document linkages through banking centres empirically. Rochet and Tirole (1996b) use monitoring as a means of triggering correlated crises: if one bank fails, it is assumed that other banks have not been properly monitored, and a general collapse occurs.

None of these theories of contagion and financial fragility suggest that the aging problem will lead to an increase in this type of crisis. They are much more due to the structure of the payment system and the form of the interbank market. These are independent of the pensions system.

In conclusion, there does seem to be some likelihood of an interaction between the aging problem and financial stability. The types of crisis that are most likely to be triggered by the ageing problem are those due to the business cycle, where fixed promises are made and there are insufficient resources to meet them, those due to inconsistent government macro policies and those due to bubble collapse.

References

Aghion, P., P. Bolton and M. Dewatripont. 1999. *Contagious Bank Failures.* Princeton University, Working Paper.

Allen, F. and D. Gale. 1998. Optimal Financial Crises. *Journal of Finance,* 53, pp. 1245-1284.

Allen, F. and D. Gale. 2000a. Bubbles and Crises. *Economic Journal,* 110, pp, 236-255.

Allen, F. and D. Gale 2000b.Financial Contagion. *Journal of Political Economy,* 108, pp. 1-33.

Alonso, I. 1996. On Avoiding Bank Runs. *Journal of Monetary Economics,* 37, pp. 73-87.

Calvo, G. 1988. Servicing the Public Debt: The Role of Expectations. *American Economic Review,* 78, pp. 1411-1428.

Chari, V. and R. Jagannathan. 1988. Banking Panics, Information and Rational Expectations Equilibrium, *Journal of Finance,* 43, pp. 749-60.

Diamond, D. and P. Dybvig. 1983. Bank Runs, Deposit Insurance and Liquidity. *Journal of Political Economy,* 91, pp. 401-419.

Freixas, X. and B. Parigi. 1998. Contagion and Efficiency in Gross and Net Interbank Payment Systems., *Journal of Financial Intermediation,* 7, pp. 3-31.

Freixas, X., B. Parigi and J. Rochet. 2000). Systemic Risk, Interbank Relations and Liquidity Provision by the Central Bank. *Journal of Money, Credit & Banking,* 32, pp. 611-38.

Gorton, G. 1988. Banking Panics and Business Cycles. *Oxford Economic Papers,* 40, pp. 751-781.

Hellwig, M. 1994. Liquidity Provision, Banking and the Allocation of Interest Rate Risk. *European Economic Review,* 38, pp. 1363-1389.

Jacklin, C. and S. Bhattacharya. 1988. Distinguishing Panics and Information-Based Bank Runs: Welfare and Policy Implications. *Journal of Political Economy,* 96, pp. 568-592.

Kindleberger, C. 1978. *Manias, Panics, and Crashes: A History of Financial Crises*, New York: Basic Books.

Krugman, P. 1979. A Model of Balance of Payments Crises. *Journal of Money, Credit and Banking,* 11, pp. 311-25.

Lagunoff, R. and S. Schreft. 1998. A Model of Financial Fragility. forthcoming, *Journal of Economic Theory.*

McAndrews, J. and W. Roberds. 1995. Banks, Payments and Coordination. *Journal of Financial Intermediation,* 4, pp. 305-27.

Morris, S. and H. Shin. 1998. Unique Equilibrium in a Model of Self-Fulfilling Currency Attacks. *American Economic Review,* 88, pp. 587-597.

Obstfeld, M. 1986. Rational and Self-Fulfilling Balance of Payments Crises. *American Economic Review,* 76, pp. 72-81.

Obstfeld, M. 1994. Risk-Taking, Global Diversification, and Growth. *American Economic Review,* 84, pp. 10-29.

Rochet, J. and J. Tirole. 1996a. Interbank Lending and Systemic Risk. *Journal of Money, Credit and Banking,* 28, pp. 733-762.

Rochet, J. and J. Tirole. 1996a. Controlling Risk in Payment Systems. *Journal of Money, Credit and Banking,* 28, pp. 832-862.

Van Rijckeghem, C. and B. Weber. 2000. *Spillovers Through Banking Centers: A Panel Data Analysis.* IMF Working Paper,WP/00/88, Washington D.C.

Discussion

Martin Hellwig

The paper by Davis on "Ageing and Financial Stability" makes three main points:

- Whereas anybody concerned with financial stability is aware of risks inherent in old-age provision based on financial (and real) investments, it is important to appreciate that in an ageing society a pay-as-you-go system may involve even more risk. Such risk concerns (i) the claimants themselves as the pension system may be increasingly unable to service their claims and (ii) the overall financial and *economic* system as the fiscal pressures associated with a pay-as-you-go system under stress affect the overall macroeconomy.
- A shift towards funding requires an adaptation of prudential regulation and supervision. Regulation should aim to eliminate existing home country biases and use international diversification of fund portfolios to reduce exposure to country-specific shocks, including country-specific demographic shocks. Regulation and supervision must also take care to eliminate or at least reduce moral hazard on the side of financial institutions.
- Financial institutions as well as regulators need to be aware of the implications that the demographic shift will have for financial systems under different modes of old-age provision. In particular, they need to be aware of the changes in asset price levels and asset price structures that may arise from the fiscal pressures associated with a pay-as-you-go system under stress and even from the changes in demographic structures themselves.

On the implications of pay-as-you-go systems for financial stability in ageing societies, I am, at once, more optimistic and more pessimistic than Davis. I am more optimistic because pension debt is *not* fixed in terms of well-defined claims. There is no issue of "solvency" in any literal sense. It is always possible to change the rules of these systems so as to reduce the claims of retirees, if this is politically convenient. This will harm the retirees and undermine confidence in the state, but at the same time it will reduce the fiscal pressures that may arise from the pay-as-you-go system.

At the same time, I am more pessimistic than Davis about the implications of pay-as-you-go systems for financial stability in ageing societies. Considerations of political economy suggest that in order to "preserve social symmetry", expropriations of pension wealth is likely to be accompanied by expropriations of other forms of wealth as well. Indeed the two forms of expropriations may well be coupled, e.g., by the introduction of rules that reduce claims on the pay-as-you-go system in proportion to whatever incomes are available from other sources, in particular, from private savings for old-age provision. An example of such a rule is provided by the treatment of private insurance claims in the new German law on provisions for surviving dependents.

From the perspective of today, such prospects work directly against the need to enhance real saving and real investment as a way to smooth the effects of the demographic shift. We are used to thinking about people's skepticism in the viability of the pay-as-you-go system as inducing them to make some old-age provisions on their own, through private saving. This view may be inappropriate if skepticism about the ability of the political system to deal with the crisis of the pay-as-you-go system extends to the treatment of returns on private savings by the political system at a time of crisis.

In thinking about "ageing and financial stability", we need to distinguish between the implications of the demographic shift on the stability of financial systems and the implications of potential financial instability for the system of old-age provision that we want to have in place as the demographic shift occurs. Concerning the former, I am not convinced that the effects of the demographic shift on financial systems are any more serious than the shocks that financial systems had to deal with over the past three decades, such as: the oil price shocks of the seventies, the twin deficits that were associated with American Reaganomics in the eighties, the real-estate boom and bust that took place in many countries between the mid eighties and the mid nineties, and, most recently, the boom and bust in IT ventures. These developments have subjected financial systems and financial institutions all over the world to tremendous - and mostly unforeseen - stress, inducing failures of financial institutions and indeed crises of entire financial systems in many countries. The question of how financial systems deal with such stress and how statutory rules can be designed to make financial systems more robust is by and large independent of the demographic shift.

Indeed, being more gradual and coming less as a surprise, the impact of the demographic shift may provide less of a shock than some of the other developments. To be sure, asset prices are likely to be depressed when there are many retirees trying to cash in on their savings and few young people willing to acquire assets. But this is an effect that we can prepare for. As of today, we know that between say 2020 and 2040, this effect will be at work. This is quite different from the abrupt swings in monetary and fiscal policy that underlay the interest shocks of the early and the late eighties or the currency crises of the nineties. Markets and the price mechanism are there to guide economic activity, unceasingly inducing people to adjust to new circumstances. Financial stability is the more at risk the less scope there is for anticipating such change. Therefore I have difficulties seeing the demographic shift as a major factor threatening the stability of financial systems.

Concerning the implications of financial instability for the system of old-age provision, we need to ask to what extent it is desirable or even feasible to immunize old-age provision from the shocks to which financial systems are regularly exposed. Existence of these shocks is a fact of life. Different financial systems have different ways of dealing with them, resulting in different allocations of the associated risks. We need to assess these risk allocations in order to see where the institutions governing them should be improved.

What standard do we have for assessing the allocation of risks in a given system of old-age provision? Political discussions tend to emphasize the safety and predictability of retirement incomes, and clearly this is a major concern. The observation that the concern for safety is sometimes no more than a rhetorical device designed to justify the erection of barriers to entry against anybody who doesn't meet some presumed standard of safety[1] should not divert attention away from the fact that underlying this concern there is a genuine need. People would not want to have their retirement incomes fluctuate as wildly and unpredictably as corporate profits or stock market valuations.

This being said, it is not clear that it makes sense to insulate retirement incomes from all the shocks to which the macroeconomy is exposed. As macroeconomic risks cannot be diversified away, the question is not how to get rid of them, but how to allocate them over people in the economy. Any suggestion that retirees ought to be immunized against the risks of an economic downturn in 2030 must be based on the notion that somebody else is better qualified to bear these risks. This claim would be at odds with the usual presumption that an efficient risk allocation involves *all* people in the economy sharing undiversifiable risks in proportion to their respective degrees of risk tolerance. Unless we take retirees to be infinitely more risk averse than other people, a complete insulation or retirement incomes from such risks seems undesirable.

From the perspective of the individual, some sacrifice of safety and predictability of retirement incomes is desirable if the greater exposure to undiversifiable risks is accompanied by sufficiently larger mean returns. This depends on the individual's risk preferences as well as the risk premia that emerge from the interaction of people in market and non-market institutions. An *a priori* assessment by a politician, a regulator - or an academic economist - is hardly appropriate. The problem of prudential regulation is all the more difficult.

Why is this a matter of public policy at all? Why shouldn't the provision for old age be left to the people concerned, to make their choices about saving, asset allocation, and patterns of retirement incomes as they see fit in view of what the

1 Thus in the German context, concerns about the reliability of annuity plans providing longevity insurance provide turf for political infighting between the banking and insurance sectors.

market has to offer? Wouldn't they know best what to make of a given risk-return tradeoff according to their preferences?

To anybody who is used to thinking about old-age provision in terms of a statutory pay-as-you-go system or about financial institutions in terms of prudential regulation, such questions will be ludicrous. After all, these have "always" been domains of the government, occupied in response to evident "failures" of the laissez-faire economy as seen, e.g., in the financial crises of the early thirties.

It is instructive to frame this discussion in terms of standard considerations of "market failures", i.e. reasons as to why market outcomes may be inefficient. I see two main concerns:

- On the side of final investors, there may be moral hazard as they neglect some of the repercussions of their choices on the rest of the system.
- Information asymmetries and moral hazard may impede relations between final investors and financial institutions; statutory prudential regulation and supervision may be a way of handling the public-good aspects of monitoring and control of financial institutions.

A precise analysis of these concerns is needed to provide guidance as to what precisely regulatory intervention should be trying to do.

In the context of old-age provision, the behaviour of final investors is often considered to be subject to moral hazard arising from the prospect of government subsidies being available to provide a minimal standard of living. The compulsory character of retirement contributions is justified by the need to avoid people not saving anything while they earn and then being a burden on the public purse in old age.

A very similar consideration suggests that governmental support of a minimal standard of living may induce moral hazard in the form of excessive risk taking as well as moral hazard in the form of insufficient saving. Investment strategies that provide very high payoffs with some positive, not very high probabilities may be attractive if failures are cushioned by the availability of public support. Some statutory regulation of risk taking by people saving for old age may therefore be justified by the very same concern that also justifies the compulsory character of retirement contributions.[2]

[2] But note that both, the concern for insufficient saving and the concern for excessive risk taking resulting from government support of minimal living standards, will support regulations ensuring saving and safety levels *only* to the extent needed to avoid reliance on the public purse. This is rather less than the levels currently prevailing in continental Europe.

The underlying incentive problem is the same as the well known asset substitution problem at depository institutions, but here it concerns the final investor rather than the banker working with the final investor's money. In either case, the lack of penalties in the event of failure has adverse effects on attitudes towards the tradeoff between the probability of success and the size of the prize in the event of success.

In principle, incentives for private investors could be improved if the government committed to not providing subsidies to support a minimal standard of living. However this is not a realistic alternative. Ethical values and concerns for social stability make any such commitment incredible. As long as the state has some commitment to social policy, people can count on being provided with some minimum standard of living. In a sense, some elements of a pay-as-you-go system are always going to be with us.

The very same consideration applies to Davis's call for an absence of government guarantees of financial institutions. Ever since the savings & loans crisis in the United States, this has been on the agenda of regulatory reform. However I am skeptical about the feasibility of such a reform. Even if governments abandon any outright commitment to the preservation of financial institutions, they will still come to the rescue if circumstances are such that the alternatives look worse. If there is a prospect of a wholesale crisis of a substantial part of the financial system, a government will prefer to step in, using the resources of the central bank and the fiscal authorities to prop up the system rather than run the risk of the kind of deep and protracted macroeconomic crisis that can be induced by financial crises. Indeed, in some circumstances, such intervention by the government may well be the only sensible thing to do.

Therefore the call for an absence of guarantees generating moral hazard needs to be somewhat refined so as to take account of the rationale for government intervention *ex post* as well as the problems caused by moral hazard *ex ante.* A policy of non-intervention, which avoids all moral-hazard effects may simply be infeasible. Given the severity of potential consequences of financial crises, it may not even be desirable *ex ante*.

Davis rightly emphasizes the role of agency problems in financial relations and the need to strengthen prudential regulation and supervision in order to eliminate or at least reduce these problems. We should however be careful to distinguish between different kinds of problems. The literature on agency problems in financial relations distinguishes between problems concerning effort choice and problems concerning risk choice. Insufficient effort is sometimes also referred to as the reaping of (excessive) "private benefits"; excessive risk taking is just the asset substitution problem mentioned above. The words "insufficient" and "excessive" involve a comparison to whatever would be chosen if the parties in question could fully commit their future actions as part of the financing contract rather than first have the financing contract and then have the party in charge of management choose whatever mode of behaviour it likes best.

Theoretical analysis suggests that effort choice is mainly a problem for outside share finance and risk choice mainly a problem for outside debt finance. Outside share finance induces the moral hazard that the manager neglects the benefits to outside shareholders of his taking more effort. Outside debt finance induces the moral hazard that the manager neglects the harm to outside creditors of his choosing a strategy that involves a high bankruptcy probability. Given this relation between types of moral hazard and modes of finance, the literature studies problems of corporate control in terms of moral hazard with respect to effort choice or private benefits and problems of banking regulation in terms of moral hazard with respect to risk choice. Problems of corporate control tend to concern relations between management and dispersed outside shareholders; banking regulation is concerned with the protection of dispersed outside depositors.

In the context of old-age provision, these considerations are relevant for the assessment of different contractual forms in funded systems. The distinction between plans involving defined benefits and plans involving defined contributions is closely related to the distinction between debt and equity. A plan involving defined benefits is debt-like in the sense of there being a given obligation to one's clients. By the reasoning given above, I should expect such plans to be subject to moral hazard in the form of excessive risk taking. A plan involving defined contributions is equity-like in the sense that retirement income depends on whatever returns have been achieved. I should expect such a plan to be subject to moral hazard in the form of insufficient effort taken by fund managers (or excessive private benefits being taken by them).

Presumably one needs different modes of regulation and supervision to deal with these different types of moral hazard. As we think about regulatory intervention, we should therefore be careful to distinguish between the different types of plans, taking account of the fact that the nature of the contract determines what type of moral hazard we need to worry about most.

This being said, I have the impression that, at least in Europe, regulatory thinking is dominated by the defined-benefits/debt-finance paradigm, with a focus on solvency and the almost complete avoidance of risk. As mentioned above, I am not convinced that a complete immunization of retirement incomes from undiversifiable risks is feasible. (Or even desirable: Avoidance of excessive risk taking is not necessarily the same as avoidance of all risk taking!) To the extent that it isn't, an excessive reliance on the defined-benefits paradigm may entail significant dangers to financial stability.

Enthusiasts of funded pension plans, in particular funded plans in the defined-contributions mode, point to the role of competition among fund managers as an incentive mechanism ensuring that there is no problem of insufficient effort. I have some difficulties with this view.

A first difficulty concerns the assessment of performance. The person who saves and invests for his or her retirement is interested in long-term returns. To what extent are long-term return prospects captured by whatever information is available from day to day or from year to year? The empirical literature suggests that investment fund quality is quite difficult to assess, and moreover may not exhibit any particular stability properties over time. Given these difficulties of assessment, the incentive implications of competition among funds are quite unclear.

A second difficulty concerns the mobility of investors. For competition to work, investors must be free to move from fund to fund, from pension plan to pension plan. In fact there is an impediment, which has to do with the insurance component of a pension plan. The scope for offering a lifetime annuity depends on the assessed longevity of the beneficiary. Given that longevity is not observable and verifiable, this gives rise to the usual problems of asymmetric information impeding the functioning of competitive markets. People who want to change pension plans at age sixty may find that there is no market because competitors are afraid of a winner's curse problem, i.e., afraid that people who want to change plans at this age are people whom their current plans are treating badly because they suspect an above average longevity. Lack of effort or excessive reaping of private benefits on the side of the pension fund's management may simply be a way to worsen conditions for people who, over the course of the years, have given rise to the suspicion that they may be "bad risks" for an annuity plan.

To avoid this difficulty, we may want to uncouple the investment and the annuity portions of a pension plan. For instance, a contract concluded at the age of twenty might give the investor the right to convert any sum of capital into an annuity at a given rate at the age of sixty-five. Between the ages of twenty and sixty-five, the same investor might have lots of contracts with investment funds manageing the allocation of savings across assets. If these funds have nothing to do with the provision of annuity insurance, there should be no impediment to the investor's moving from one fund to another if he or she is unhappy about the first fund's performance. But then we must ask what is the viability of the contract providing for the annuity. How do we deal with the possibility of insolvency there? Or equivalently, how do we deal with the risk that as of the age of sixty-five, when it is time to convert a sum of capital into an annuity, the agent in question may be considered to have a high life expectancy and therefore to be a "bad risk" for an annuity plan?

As shown by the preceding considerations, the implications of financial instability for (funded) old-age provision involve serious open questions. To study these questions and understand where policy intervention of some sort or other may be called for, we have to get down to a detailed analysis of relations between the different parties in terms of information and incentives at the microeconomic level. This is the more urgent because in relations between people who save and invest in order to provide for their old age and the financial institutions that serve them, there is less scope for learning by experimentation than in ordinary financial relations. If it takes until retirement to find out about flaws in the financial relations that one has entered into, it is too late to do anything about it. This gives the agency problem in financial relations concerning retirement provision a special pungency.

Should Monetary Policy be Different in a Greyer World?

David Miles

"The older we grow the greater become the ordeals"

Goethe 'Maxims and Reflections'

"At 50 you begin to be tired of the world, and at 60 the world is tired of you"

Count Oxenstierna 'Reflections and Maxims'

1 Introduction

Goethe thought life got more difficult the more of it one had experienced. Will monetary policy be more difficult, or at least different, in a world in which the population is older? This is a relevant question because in the absence of some ghastly catastrophe then it is almost certain that the population in the developed world will, on average, be substantially older in the future. Within Europe, the proportion of the population aged over 65 is likely to be about 50% greater 30 years from now. In some countries the rise in the proportion of those over 65 is particularly dramatic. In Germany population might fall as well as age (although migration may more than offset this). If there are to be a far greater proportion of elderly people will the way in which monetary policy is operated need to be different? This is the question that this paper seeks to investigate.

What I do is take a series of different models, all of which are types of overlapping-generation models but with different degrees of complexity, and try to calibrate some of the possible impacts of ageing upon monetary policy. I focus particularly on how the transmission mechanism of monetary policy might change. In doing so I will rely upon central projections of future population structure made by the United Nations. At the outset it is important to stress just how uncertain future demographic structure is. This reflects enormous uncertainty about changes in longevity, about future fertility rates and about the scale of immigration. I will not return to this point and will consistently use central United Nations estimates. But I do note that forecasts of life expectancy made over the last 20 years have consistently underestimated the rise in longevity, and by substantial margins. The scale and duration of the decline in fertility in the post-

Second-World War-period has also consistently exceeded forecasts. We simply do not understand trends in fertility well at all. Forecasts of the scale of migration, which depend both upon political decisions within European countries and on developments outside, are also subject to enormous errors.

2 Some Possible Effects

It is not difficult to think of several ways in which demographic change could alter the way in which monetary policy should be conducted. Consider the following:

1. Changes in interest rates have a different impact upon consumption and saving for people of different ages. The elderly have done most of their saving and are generally long on assets and short on human capital. The young are in the opposite position. Changes in interest rates are likely to influence the behaviour of the young by changing the cost of consumption at different points over the life cycle. For the old, the major impact might be the change in asset prices induced by a shift in short term interest rates. The two impacts - on the young and the old - are very unlikely to be equal. Because of this as the relative numbers of old and young change in the population, the impact of interest-rate changes varies.

2. The stock of financial assets held by households and the relative importance of those assets to human capital varies over the life cycle; this is likely to mean credit restrictions matter more to the young than to the middle-aged and the elderly. The more people there are in the later parts of their working lives and in early retirement, and the fewer young people there are, then the less important are credit (or liquidity) constraints. This alters the impact of changes in interest rates upon expenditure.

3. The composition of the stock of wealth held within the economy may well be different in an older society, as will be the aggregate stock of wealth. With changes in the relative importance of equities, bonds and property in portfolios, the way in which changes in interest rates affect the economy will change.

4. Changes in demographic structure are also likely to have a substantial impact upon the capital-labour ratio and the overall real return on capital within the economy. Movements in what one might call the equilibrium real interest rate clearly have an impact upon monetary policy. For example, the simple rule John Taylor has proposed (see for example Taylor (1993)) has the short-term nominal interest rate set by the monetary authorities equal to a constant plus terms reflecting the deviation of inflation and output from target and trend, respectively. The constant is the real interest rate (in a closed economy the rate that matches saving and investment at trend output) plus target inflation.

5. Differences in the pace of ageing across countries are likely to generate significant cross-border flows of capital. This may influence exchange rates.

All of the above suggest that the transmission mechanism of monetary policy may change significantly as demographic structure evolves. One of the main purposes of this paper is to try and get a quantitative handle on some of the magnitudes involved here. But there is another important way in which ageing might affect monetary policy. Within Europe now monetary policy is conducted with the aim of keeping inflation low. The target range for inflation and the rate at which inflation should be brought back within that target range reflect perceived trade-offs (between inflation stabilization and output stabilization) and preferences. As the transmission mechanism changes the trade-off between inflation stability and output (unemployment) stability may alter. But preferences between the two may also change. The old may attach higher importance to low and stable inflation, and less importance to low and stable unemployment, than do the young. The old may also attach more importance to asset-price stability than to either consumer-price or output stability - and they may do so to an increasing extent if funded pensions become more important. One political implication of ageing may, therefore, be a change in the goals of policy. To assess that interesting question one needs to know something about the portfolios of assets held by the elderly (for example, how many of those assets have real values that are adversely influenced by unanticipated inflation) and, also, about political decision-making. This paper does not really focus on that second issue and concentrates more on assessing some of the mechanisms listed under 1-5 above and which have a direct bearing on the transmission mechanism. Nonetheless, the political dimension and possible changes in the goals of policy are, I believe, important. I return to them at the end.

Plan of the paper

In the first section of this paper I use a fairly standard overlapping-generations, general equilibrium model of the European economy to assess what the scale of change in savings rates, the capital stock, capital-labour-ratios, and real rates of return on investment might be as a result of ageing. I then take a somewhat simplified version of that basic overlapping-generations model to assess whether the impact of changes in monetary policy upon expenditure is likely to be different with an older population. Both these models abstract from portfolio choice in that they assume a single asset (a claim on corporate capital). But some of the important impacts of ageing arise as a result of shifts in portfolio allocation. To analyse the portfolio allocation issue we obviously need to consider several assets, and so we have to introduce uncertainty. At least two sorts of uncertainty matter in an environment where the age structure of the population is changing: rate-of-return uncertainty (which matters for both the young and the old, but may be relatively more important for the latter); labour-income uncertainty (which is more important for the relatively young). Portfolio allocation for the old and young will differ, because the relative importance of these sources of uncertainty is very different. To analyse the differences we need multiple sources of risk and this makes the task of solving models substantially more difficult. In the third section of the paper I use a model with multiple sources of uncertainty to try to assess how portfolios of assets, degrees of credit restrictions, and stocks of wealth, might change as a result of ageing. I then consider the implications of that for monetary policy.

One obvious, but important, point emerges from all these simulations. The way in which ageing might influence monetary policy depends critically upon the way in which ageing changes the structure of pension systems. If substantial rises in the old-age-dependency ratio (the ratio of those of retirement age to those of working age, the inverse of the support ratio) generates a substantial scaling-back in the generosity of unfunded state pensions, then the impact upon savings rates, real interest rates, portfolio allocation and the stocks of wealth will be very different from a situation where unfunded pensions continue to be generous and contribution rates on workers rise. It is impossible to reach conclusions about how monetary policy might be different without specifying how the pension arrangements themselves evolve. As a result I consistently consider various pension reform scenarios.

3 A Simple Overlapping-Generations Model

Consider a standard Taylor type rule for monetary policy of the form

$$(1) \quad i_t = \delta_t + \alpha(\pi_t - \pi^*) + \beta(y_t - y_t^T)$$

i_t is the nominal (short-term) rate of interest at time t

π_t is inflation at time t

π^* is the inflation target

y_t is log output

y_t^T is the log of potential (or trend) output

α, β are (positive) coefficients.

Sevensson (2001) defines the first term in the equation (which we denote δ_t) as "the sum of the average short real interest rate and the inflation target". Because the short real rate may change over time, we add a subscript to δ.

So δ can be interpreted as the real rate of interest plus target inflation. Whether or not Taylor-type rules really are desirable, let alone optimal, is a moot point. (For a detailed discussion of the issue see Svensson (2001) and Woodford (2001). But even if monetary policy is more discretionary than following a rule, such as equation 1, it will still be necessary for the monetary authorities to estimate the equilibrium real rate of return in the economy.

The aim in this section is to assess how that real rate might itself evolve as the population of Europe ages. I use an overlapping-generations model of the sort pioneered by Auerbach and Kotlikoff (1987). The model is similar to that described in detail in Miles (1999) where more detail of the structure is provided. Briefly, the model has many cohorts alive at any point in time. The relative numbers of people at different ages at each point is set equal to estimated past, and projected future, numbers for Europe. (The United Nations' central projections for European population structure in the period to 2050 are used).

Individuals make decisions on consumption, saving and labour supply in order to maximise a lifetime utility function that depends upon consumption and leisure consumed in each period. We assume individuals begin to make economic decisions aged 20, retire aged 64, and receive pension benefits until they die. The labour-supply decision depends on real wages and real interest rates, both now and into the future. Consumption depends upon the same factors. Expectations are rational, which in the context of the non-stochastic version of the model implies perfect foresight.

We assume that Europe is a closed economy, so that the real rate of return on capital (which is the real interest rate in this model) evolves in order to balance the capital used in the productive sector to desired holdings of assets by households. The evolution of the capital stock, labour supply - and therefore real wages and real interest rates - depends upon demographic structure and upon the way in which public policy towards pensions evolves.

A short appendix (annex 1) lists the key equations. Here I simply sketch the structure of the model. Agents maximize a utility function that depends only upon the discounted value of the utility they receive from consumption of goods and of leisure in each period of their lives; this is a "pure" life-cycle model and there are no bequests. We assume agents have a within-period constant elasticity of substitution utility function defined in terms of consumption and of leisure. The evolution of consumption and of labour supply over time for each agent depends on all future expected interest rates, after-tax wages and the profile of (agent-specific) labour productivity; choices depend on the intertemporal elasticity of substitution of consumption and leisure, as well as the within-period degree of substitutability. Agents face a single budget constraint which makes the present discounted value of consumption equal to the present value of post-tax labour incomes and of receipts of the state pension.

As a result of the uneven pattern of labour income over life, agents will chose to save during much of their working life and dissave in retirement. The pattern of labour income reflects several factors, one of which is exogenous: the number of units of effective labour that an agent is endowed with moves over time exogenously. This exogenous profile of effective labour units supplied reflects, in turn, two factors. First, we assume that there is labour-augmenting technical progress that increases the effective units of labour supplied by workers of all ages at a constant rate per period. There is a second, age-specific, form of productivity growth which reflects an agent's changing productivity over his/her life cycle. We use information on the age-wage relation to generate an age-related productivity factor for all working cohorts.

Labour income for a particular cohort is the product of the number of units of effective labour supplied and the post-tax real wage per unit. The latter is endogenous, and depends on the tax rate needed to balance the state pension scheme and on the aggregate capital-to-labour ratio, which, via a Cobb-Douglas production function, determines the marginal productivity of labour.

How the model is solved depends crucially on whether we assume that the interest rate is exogenous (making a small, open-economy assumption) or is determined by the domestic capital stock, which in a closed-economy version equals the sum of the stocks of wealth of each cohort at each point in time. In an open-economy version with an exogenous real interest rate, it is fairly easy to solve the model; but, since we are focusing on Europe as a whole, it is better to make a closed-economy assumption so that capital used by companies matches wealth held by households. (Borsh-Supan, Ludwig and Winter (2001), in this volume, consider the open-economy issues and the scale of capital flows in detail).

The model is solved in the following way: we first interpolate from historical data, and from UN projections, the size of age cohorts to construct a series running from 1901 to 2060 on the relative numbers of people of ages from 20 to 80 alive at each point. We then take an initial guess at the interest-rate vector for the 100-year period 1960-2060. (In fact we need to solve the model back to 1901, so as to work out the optimal lifetime strategy of someone born then who is in the last year of his life in 1960). That initial guess implies a capital-to-labour ratio which generates a time series for the gross (of labour taxes) real wage per unit of effective labour. For a given replacement rate (i.e. the ratio between the state pension paid and the average gross earnings of workers), we can calculate the tax rate on labour income required to balance the PAYG state scheme. This rate varies over time, as the ratio of workers to retired varies.

Given the initial guess at interest rates and the implied time series of after tax real wages per unit of effective labour, we can now solve the dynamic programming problem of every agent alive at every period. Faced with a profile of real wages, of tax rates, of real interest rates, and given the (exogenous) profile of labour productivity over life, each representative agent works out optimal labour supply, optimal consumption and optimal saving. The aggregate level of wealth at any point is simply the sum of the net wealths of each cohort then alive, weighted by the size of that generation. In any period, the net wealth of each generation is different, depending on where they are in the life cycle.

The total stock of wealth and the aggregate supply of labour forthcoming at the initial guess at the time-path of interest rates generate a new time series for the capital-labour-ratio, which implies a new pattern of interest rates. This new value provides the start value for the second iteration. The procedure is repeated until a fixed point is reached; at that stage the actions of every agent alive at any point between 1960 and 2060 are optimal given the time series of interest rates and post-tax real wages they face over their lives; and how those interest rates and post tax wages evolve are themselves consistent with the decisions of all individuals. At the equilibrium the government pension scheme is also balanced. Since in 2060 there is an agent who will live to 2120, and in 1960 there is someone born in 1901, we have to solve the model 60 years either side of the period for which we actually present results.

We consider three different regimes. In the first, unfunded state pensions continue to provide the larger part of resources for the retired. More specifically, we assume that the gross replacement rate (the ratio between the state pension and the, gross of contributions, average level of wages) stays at 50% in the future. This requires a substantial increase in the contribution rate which we assume is levied on the wages of all employees. (The contribution rate would need almost to double, on the United Nations' demographic projections, to keep such an unfunded pension system in balance).

In the second regime we hold the contribution rate at its year-2000 level. In the year 2000, we assume that the unfunded state pension is worth 50% of average wages. The replacement rate then needs to decline smoothly in order to preserve balance in the system at an unchanged contribution rate. Ultimately, the replacement rate would need to fall from 50% to around 27% if the tax on labour income to fund the system were to be constant.

In the third pension regime, we assume that public pensions become steadily less generous as the result of an explicit policy to phase out state, unfunded, pensions. We assume the government announces now (actually in the year 2000) a policy of reducing the replacement rate from its current level, assumed to be 50%, to zero in a steady transition which begins in the year 2020. We delay the implementation of the phasing-out of the state pension for an obvious reason; we need to allow those in or near retirement a period of unchanged pensions to avoid a large and unexpected decline in resources, with insufficient time to adjust. Those who will receive lower state pensions have at least 20 years to adjust.

The model is solved in a standard way – we seek a set of real interest rates and real wages which clear the goods market and the labour market in every period, and such that each individual generation is maximizing its utility whilst having rational expectations. The key parameters reflect substitutability between consumption and labour, substitutability between consumption at different points in time (the intertemporal substitutability, which in this model is the inverse of the coefficient of risk aversion), the rate of pure time preference, the substitutability between capital labour in the production function, and the process governing the evolution of labour productivity over the life cycle. There is great uncertainty about the key parameters. I present the results of simulations showing one set of parameters, but I discuss the robustness of the findings and briefly describe the main features of the results with a quite different set of parameters.

For the main simulations, we set the inter-temporal elasticity parameter to one third (implying a coefficient of relative risk aversion of 3). We set the rate of pure time preference to *minus* 1.5% per annum – a negative rate might seem unusual but it is consistent with the work of Hurd (1989). We use a simple Cobb-Douglas production function and set the share of capital income to 25%. We assume labour productivity for all workers rises at 2% a year. We also assume that workers have an age-related element to labour productivity. We assume productivity is a quadratic function of age, such that (abstracting from general productivity growth) earnings peak in the mid-40s at a level around one-third higher than earnings at age 20. Adding to this age profile the exogenous 2%-a-year general productivity growth implies that the wage of a typical 40-year-old are about double that person's wage at age 20.

Charts 1-3 show how the savings rate, real interest rate, and capital-output ratio evolve when we solve the model assuming a constant replacement rate for the unfunded state pension of 50%. Chart 1 shows that the savings rate declines dramatically over the next 50 years. Despite this, the equilibrium real interest rate falls for about 30 years before rising slightly to a level still substantially lower than in the year 2000. It may seem strange that a falling savings rate generates a falling real rate of return. But the decline in the labour force, and the resulting rise in the capital-to-labour-ratio, more than offsets the reduction in new savings and with a much higher capital labour ratio the marginal rate of return on capital falls. Chart 2 shows that the scale of this impact is significant. The real interest rate in the year 2000 in the model is about 4.6 %. This falls to around 4.1% by 2030.

If the contribution rate to the state pension is held constant at its 2000 level, the generosity of state pensions has then to fall and the reduction in the savings rate is much less (chart 4). The stock of wealth, and therefore capital, in the economy would be significantly higher if state pensions were to become significantly less generous. Individuals respond by saving more. The real interest rate now falls by even more from 4.6% in 2000 to about 3.9% by 2035. The capital-output ratio rises more.

Not surprisingly, in this forward-looking model with rational agents, if state pensions were to be phased out entirely, people respond by saving substantially more. As a result, the capital stock is larger, the capital labour ratio greater, and the real interest rate lower, than if state pensions continued to be worth 50% of average earnings. Chart 5 shows that the fall in the savings rate is now much gentler; the contrast with chart 1 is stark. Chart 6 shows that with state pensions slowly phased out, the real interest rate falls from 4.6% in year 2000 to 3.4% by 2050; this is a very substantial reduction in the real rate of return. The following table summarizes the evolution of the real interest rate under the alternative scenarios.

Table 1: Real Interest Rate: European Economy Simulations

	Base Case: constant replacement rate	Constant Contribution Rate	Pensions Phased out from 2020
1970	4.90	4.90	4.90
1980	4.80	4.80	4.80
1990	4.78	4.78	4.78
2000	4.68	4.60	4.60
2010	4.49	4.37	4.30
2020	4.28	4.14	3.95
2030	4.14	3.97	3.84
2040	4.16	3.87	3.71
2050	4.28	3.83	3.42
2060	4.39	3.79	3.21

Different sets of parameters generate different results. For example, if we set the coefficient of relative risk aversion at 2 (so that the intertemporal elasticity of substitution is 0.5), the rate of time preference at +0.5% a year and productivity growth at the lower level of 1% a year, we generate a lower overall level of saving and a somewhat smaller reduction in rates of return as the population ages. But the decline in the savings rate and the path followed by the rate of return, and particularly their sensitivity to the nature of pension reform, are similar to the base case. In fact, these features of the simulations are robust to quite large variations in parameter estimates.

Overall two points emerge from the results. First, within the context of a model assuming consumption-smoothing over time and forward-looking behaviour, the savings rate is likely to decline substantially within Europe. This is neither a surprising nor a new result. It reflects the fact that the sorts of overlapping-generations models that are often used, and that are used here, imply life-cycle-type smoothing, where individuals may save substantially during their working years and then reduce their stock of assets in retirement. But how large that reduction in the stock of assets in retirement might be in practice depends upon whether a life-cycle-type model of behaviour is realistic. It also depends on whether bequests matter. In the model presented here there is no bequest motive. If bequests are important and deviations from life-cycle behaviour significant, the decline in the saving rate illustrated in charts 1, 4 and 5 is likely to be overstated. But this means the falls in real interest rates shown in Table 1 are probably too *small*. What the simulations show is that even though savings might fall substantially, the real interest rate is likely to be lower. If life-cycle type behaviour where bequest motives are not important, *is* misleading, and savings rates will not fall by as much as implied by the model used here, then real interest rates could fall even more. (Brugiavini (2001), in this volume, reviews the literature on the link between saving and demographic structure and discusses evidence which is hard to square with a pure life-cycle-view).

What really drives interest rates down in this model is that with a falling aggregate labour supply, the capital-labour ratio increases and drives down the marginal return on capital, which is the real interest rate. Obviously if savings fall by less the capital stock is even larger, the capital-labour ratio greater, and the decline in the real interest rate even more marked. If monetary policy is successful in keeping the average inflation rate the same in the future as it has been recently in Europe, then the implication of this is obvious and straightforward – nominal interest rates will, on average, be lower in the future than they have been in the recent past. The magnitude of the effects is not small. The final simulation (where unfunded pensions are phased out) shows a decline in the real interest rate from 4.6% to 3.2% over a 50-year-horizon. The constraint of a zero floor on the short-term nominal interest rate is faced by all central banks but has rarely been binding in recent decades (with the notable exception of Japan in recent years). But a substantial decline in the equilibrium real rate would mean that the chances of that constraint binding would be higher. Our simulation results suggest that this could be a factor over the next few decades, especially if inflation stays low.

4 Effects on the Transmission Mechanism

But will the transmission mechanism of monetary policy itself be different? To assess this, we need to go beyond the model presented in the previous section. That was a purely real model, in which people's wealth is accumulated in the form of claims on the capital stock. Furthermore, it made no sense to talk of an exogenous change in the interest rate in that model, since the real interest rate was determined endogenously to clear the capital market. To think about the transmission-mechanism issue, it is actually helpful to move to a simpler model than that of the previous section, since now we want to treat the interest rate as exogenous, and consider the impact of movements in that exogenous rate. Here I consider a particularly simple case. I take an economy in which the rate of return on assets is given exogenously. Individuals accumulate wealth in the form of financial assets that pay a given stream of income. We can think of these as perpetual bonds that pay one unit each year, or as equities that pay a constant stream of dividends. People buy these assets during years when they are net savers, and sell them when they reduce the stock of assets. The set-up is pretty much as above, and we assume individuals maximize the same sort of lifetime-utility function. I simplify here and assume that individuals supply one unit of labour inelastically for a working life that lasts 44 periods (which we can think of as running from age 20 to age 64). Other parameters of the utility function (the rate of time preference; the intertemporal substitutability of consumption) are as in the main simulations reported above. We set the real wage exogenously (which is consistent with the exogenous-real-interest-rate assumption; we must think of the capital-labour ratio as set independently of domestic private sector behaviour). Once again we assume an age-productivity profile that has real wages peaking in the early 40's.

What we do is, first, calculate the equilibrium flow of consumption, stock of savings and national income for an economy whose population structure matches that of Europe in the year 2000. One could think of this as an open-economy-version of the previous model, where we exogenously set the interest rate, although obviously the small open-economy assumption is not really applicable to Europe. I assume a 50% replacement rate on a state, unfunded pension financed from contributions from current workers and in balance. I assume the real interest rate is 4%, *and is expected to stay there*. I then calculate what happens when that interest rate is changed in a way that was not anticipated. (For simplicity, I assume the change is perceived to be permanent). This change has several impacts. First, it has an immediate impact upon the price of financial assets (remember these are simple, perpetual securities that pay one unit each period, and their value is inversely related to the interest rate). Second, the price of moving consumption from one period to another changes. This price effect has both an income effect and a substitution effect.

These effects - the wealth effect and the price-change effect - have different relative strengths for the young and the old. For the young, the impact due to the change in the price of consumption is significant, since they have a long time to live and the cost of consuming more now and less in the distant future changes substantially. For the relatively old, there is likely to be a larger impact from the change in the value of their existing assets than from a change in the price of shifting consumption from now to slightly later in their (short) remaining lives. We work out the impacts of the change on people of different ages, and then aggregate across generations to show the effect on total consumption and saving. Having calculated the aggregate impact of a change in interest rates in a model whose demographic structure matched Europe in 2000, I then conduct exactly the same experiment with the population structure matching that predicted for Europe in the year 2050. In thinking about the impact in 2050, I consider several different scenarios for state pensions. Table 2 shows the results.

Table 2: The impact of a permanent, but unanticipated, change in interest rates on aggregate consumption

	Generosity of state pensions – replacement rate	Real rate from 4% to 3%	Real rate from 4% to 5%
2000 population	50%	+12.7%	-7.6%
2050 population	50%	+14.8%	-8.9%
2050 population	40%	+15.7%	-9.4%
2050 population	30%	+16.6%	-9.9%
2050 population	20%	+17.5%	-10.3%
2050 population	10%	+18.3%	-10.8%
2050 population	0%	+19.0%	-11.1%

When there is an unanticipated and permanent change in interest rates, there is a very substantial impact upon consumption. A 1% unanticipated, but permanent, rise in real interest rates reduces consumption expenditure by about 7.6% with the 2000 population structure. But that impact is larger with the 2050 population structure; expenditure now falls by about 9%. The elasticity of expenditure in the economy with respect to a change of interest rates is about 18% greater when there are more old people (holding the generosity of state, unfunded pensions constant). The reason is pretty straightforward. An unanticipated increase in interest rates that is expected to be persistent has a very substantial impact on asset values. The reduction in asset values from a rise in interest rates has a larger negative impact on the consumption of the relatively elderly (who hold much of the stock of assets) than the impact upon the relatively young. With more old people, the resulting decline in consumption is greater when interest rates rise. The table also shows that the rise in spending following a cut in interest rates is, not surprisingly, rather significantly greater with the year 2050 population structure than with the year 2000 population structure. Once again, the elasticity of consumption is about 17% larger, with the same pension arrangements. Notice that the impact of a rate cut is very much larger than the effect of a rate rise; this is a reflection of the convexity of the bond price-interest rate relation. The size of these effects is very large and much greater than most empirical studies suggest. To a large extent, this is because the impact here is, by construction, that of an unexpected change in interest rates. Empirical studies find it very hard to isolate the impact of the unexpected component. The impact of an anticipated change in interest rates is very much smaller in the simple model we have used here.

We can also ask another question. How much will the transmission mechanism be different if, as a result of ageing, the generosity of unfunded state pensions is scaled back and people accumulate more wealth to finance retirement and consumption? We now compare the impact of interest-rate changes in the 2000 economy, with a replacement rate of the state pension of 50%, with the impact of interest-rate changes in an economy, with pensions worth only 40%, 30%, 20% or 10% of average wages, and also with no state pensions. Table 2 shows the results. The impact of monetary policy is substantially greater the less generous state pensions are. With state pensions worth only 20% of average wages, the impact of a change in interest rates in 2050 is about 35% greater than in 2000, with pensions worth 50% of average wages. If there were to be no state, unfunded, pensions in 2050 the elasticity of consumer spending with respect to interest-rate changes would be about 50% greater than in 2000. The reason is obvious; less generous state pensions will almost certainly mean higher private saving and a greater stock of assets, particularly held by the relatively elderly. Changes in monetary policy that impact asset prices have a greater effect when the stocks of assets are larger.

So there are two reasons why monetary policy might be more powerful in the future. First, because there are more old people and their consumption might be more sensitive to interest-rate changes, at least to the extent that interest-rate changes affect asset prices. Second, the ageing of the population may generate changes in pension arrangements, which itself increases the stock of financial assets. Both mechanisms, at least in the context of the very simple and stylised model used here, seem to be important. But it is important to stress that what we have calculated are the impacts of a permanent change in interest-rates. Yet most changes in interest rates engineered by central banks are far from permanent. Were our aim to assess the overall size of the effect of interest-rate changes, this would be a huge drawback. But our focus is more on how the impact of monetary policy *changes* as the population ages, and the convenient short-cut of focusing on permanent changes in rates is less serious here.

But all this is in the context of an extremely simple model, where individuals can only save in one asset - effectively a perpetual bond. In fact, people hold a substantial part of wealth in equities, and a larger part in housing; cash is also important. These assets are not perpetual bonds. But because equities and houses are very long-lived assets, their price response to an unanticipated change in interest rates might be captured in a reasonable way by treating them like perpetual assets generating flat cash flows. Yet this is clearly not appropriate for cash. The fact is that people face a huge menu of financial assets to pick from, and this suggests an important policy question; what might be the effect on portfolio allocation, as well as upon the overall stock of financial assets, of ageing, and how might this affect monetary policy? It is to this difficult question that we turn in the next section.

5 Uncertainty, Portfolio Allocation and Ageing

In this section, we present the results of a model where there are many sources of uncertainty. In order for there to be any portfolio choice facing individuals, there has to be uncertainty. In the absence of uncertainty, all assets will have to generate the same return, which means there can only be one asset. Here we allow for individuals to choose between a safe asset with a known real return and one with a volatile return. We also allow for labour-income uncertainty and for the operation of credit restrictions, and we show how savings and portfolio allocation evolve over time. Once again we look at different scenarios, where the generosity of state pensions moves. Our aim in this section is to get some idea of how sensitive the level of saving, and the allocation of that saving between different assets, is, both to changes in demographics and to (induced) changes in the generosity of unfunded state pensions. We also introduce another important element into the discussion. We allow for financial markets to be less-than-perfectly efficient. More specifically, we allow for differences in the efficiency of annuity contracts. In an environment where people may be required to provide much more of their retirement resources from their own savings, the efficiency with which annuities markets work becomes particularly important, both for individual welfare and for aggregate outcomes (because it can significantly affect the overall level of savings in the economy).

The model we use takes a step away from the general equilibrium setting of Section 1. Solving models with multiple sources of uncertainty is hard, and doing so in a general equilibrium setting (which involves solving the individual optimization problems many times) is computationally extremely burdensome. For that reason, we assume exogenous rates of return on safe and risky assets, and ask how savings and portfolio allocation evolve as demographics change and as pension arrangements vary. The model is described in detail in Miles and Cerny (2001) and here I give only the briefest of sketches. (Some technical detail is given in Appendix 2 of this paper).

Briefly, individuals face persistent shocks to labour income through their working life; face uncertainty over the time of death; can buy safe or risky financial assets; cannot borrow against labour income; and face the potential to buy annuities, but at varying degrees of fairness (that is, actuarial fairness). We calculate optimal decisions for individuals facing multiple sources of uncertainty and then undertake Monte-Carlo simulations to calculate the overall stocks of savings, portfolio allocation and consumption. We assume there is a safe asset with a return of 2% real; risky assets have an expected return of 6% but a standard deviation (of log returns) of 17.5%.

In looking to the future, we assume that life expectancy has stabilized at levels predicted for Japan in the year 2050. It may seem bizarre to use Japanese predicted life expectancies 50 years ahead to assess future European demographics. Japanese life expectancy is already greater than is typical in Europe, and it is projected to rise substantially. We assume that, in 50 years from now in Europe, life expectancy might have caught up with Japanese life expectancy, which itself will be substantially higher than today. We use these life expectancies to calculate a steady-state population structure (i.e. assuming constant population) and then solve the model based on required contribution rates to balance pay-as-you-go state-funded pension schemes of varying levels of generosity. We calculate optimal individual paths and simulate 7000 life histories based on random draws from independent, (normal) distributions of shocks to log incomes and log rates of return. Some of the key findings are described in Tables 3 to 6. Each of these tables shows aggregate stocks of wealth, portfolio allocation, savings rates and wealth-to-national-income ratios for various levels of the state, unfunded pension. We consider 3 very different levels of efficiency in the annuities market: perfect (actuarially fair) annuities; annuities with a money's worth ratio of around 90% (which might be typical levels for countries with mature annuities markets, like the US and UK (see Brown, Mitchell and Poterba (1999) and Mitchell (2001), in this volume); and complete inefficiency - that is, the absence of annuities. In this model with uncertain lifetimes (at the individual level) and no bequest motive, how annuities work is essential for portfolio allocation and saving. We assume that, where annuities are available, people can choose the safe/risky asset split within annuities contracts (see Miles and Cerny (2001) for details).

We focus, first, on the overall stock of wealth. The tables reveal, once again, just how sensitive stocks of wealth in Europe are likely to be to pension arrangements. With state, unfunded pensions worth 20% of average wages, the stock of aggregate wealth is around double what it would be with state pensions worth 50% of average earnings. This impact of changing the generosity of state pensions is much greater than the effect of different degrees of efficiency of annuities contracts. If state pensions are low, there does tend to be higher saving when annuity-market efficiency is reduced, as agents take precautions against longevity risk. But the effect is absent when state pensions are generous. The reason for the much lower sensitivity of saving to annuity-market conditions than to the level of state pensions is that variations in annuity-market efficiency have two, offsetting, effects. Lower annuity-market efficiency makes risk-averse agents want to accumulate more to guard better against the risks of living too long (a precautionary effect). But lower annuity-market efficiency makes saving less attractive for those with no bequest motive, since it raises the price of moving consumption to the future. This is the substitution effect. Contrast that with the impact of a reduction in the generosity of the state pension, which has a straightforward positive impact upon the demand for accumulating funds.

We also calculate the fraction of the population that is likely to be liquidity-constrained. Chart 7 shows the percentage of the population that is credit-constrained by age, assuming a state pension with a 50% replacement rate and semi-perfect annuities. This proportion is substantial amongst the relatively young and amongst the very old. A significant fraction of those who live to be very old find that, when there are generous state pensions and relatively unfair annuities contracts, they eventually want to consume everything. This is because, when annuities rates really do not reflect very well survival probabilities, then as survival rates really dip - as they do for the very old - then holding wealth, for people with no bequest motive, becomes increasingly undesirable. So long as there is a generous state pension, the strategy of falling back completely upon that pension becomes optimal for a high proportion of the less-well-off-elderly.

But the less generous state pensions are, the lower is the proportion of the population - particularly amongst the elderly - that are credit-constrained. The mechanism here is clear; with low state pensions, the rate of saving in the economy is higher, and the stock of financial assets is larger. This gives most people more flexibility about spending even in old age, and so credit restrictions become less common. Table 6 shows that as we vary both the generosity of state pensions and the efficiency of annuity contracts, the average proportion of people at different ages that are credit-constrained varies. With pensions worth 40% of average earnings about 47% of 20-year-olds are credit-constrained and about 30% of 25-year-olds. Those fractions fall to 41% and 23%, respectively, if pensions are worth only 20% of average wages. Among the retired, credit restrictions become insignificant amongst all but the very old (90 plus) once state pensions fall beneath 40% of average earnings.

One implication of this is that the proportion of the population that is credit-constrained may fall gradually as the European population ages. This is partly because credit constraints are probably more prevalent amongst the young, but also because reform of state-pension systems may mean greater incentives for the accumulation of savings, which itself reduces credit constraints. Lower credit constraints means that the price impact of monetary policy changes is likely to be greater. Of course, it also means that other aspects of the transmission mechanism - most obviously the credit channel - may become less powerful.

Tables 3-5 also show how portfolio allocation may alter as pension arrangements evolve. What we find is that the proportion of overall wealth that is invested in risky assets, which we think of as the stock market, as opposed to safe assets (which we think of as bank deposits, or treasury bills) depends very sensitively on pension arrangements. The results might seem to be counter-intuitive. What they say is that the *less* generous state pensions are - and the greater the reliance upon accumulated assets is - the lower the proportion of those assets which should be in risky assets. The scale of this impact is large, and it depends to some extent upon how efficient annuity markets are. With perfectly efficient annuity markets (that is, annuities available at actuarially fair rates), then the proportion of portfolios held in risky assets varies from 94% (with generous state pensions with a replacement rate of 50%) to 60% (with no state pensions). With completely inefficient annuities markets (effectively, annuities are not offered), the percentage of assets held in risky form varies from 97% (the 50% replacement ratio of the state pension) to 59% (with no state pensions). The sensitivity of portfolio allocation to the generosity of state pensions is clearly very significant.[1] State pensions, in this simple model, are safe assets, whereas the value of a defined-contribution pension, which is invested in risky assets, is uncertain. The lower the state pension is, the more people are exposed to rate-of-return uncertainty, and the greater is the share of their assets that they put in safe form (which once again we think of as bank deposits). But note that there are two responses to a cut back in state pensions: people hold less of their portfolio in risky assets, but they also hold more financial assets. In fact, the overall level of investment in risky assets is generally higher in the simulations with lower state pensions. Consider the case of semi-perfect annuities: with a 40% replacement rate of state pensions, risky assets are equal to 93% of a stock of wealth of 345, ie. 321. With no state pensions, risky assets are only 60%, but of a much larger stock of wealth of 804, ie. 482.

One would expect that the allocation of portfolios between risky and safe assets (equities versus bank deposits) itself has significant implications for the operation of monetary policy. One of the implications of these simulations, albeit based upon a highly stylised model, is that portfolio allocation is likely to be highly sensitive to pension arrangements.

6 Conclusions

A range of different models - all highly stylized - has been used to focus on how the transmission mechanism of monetary policy might change as the population ages. There are reasons to believe that credit-constraints may become less prevalent; that the level of real interest rates will - on average - be lower; that consumer expenditure might become more sensitive to shifts in interest rates and that the impact of movements in asset prices may become more powerful. The nature of these effects depends very much on how state-pension systems evolve.

---

[1] In contrast, the sensitivity of portfolio allocation to annuity-market efficiency for a given level of state pensions is relatively small.

If, as a result of ageing and pressure on state-pension systems, individuals within Europe will be forced to rely much more on personal savings, then the impact of asset-price volatility will become more important. This itself has implications for monetary policy. We now have in Europe monetary arrangements which attach considerable, perhaps overriding, importance to hitting inflation targets. Those inflation targets are based upon measures of consumer prices and do not reflect asset prices. One implication of ageing may be greater concern amongst populations in Europe about asset-price volatility. The ECB may be required (or may itself decide) to attach much more weight to asset-price volatility than it now does. I have not in this paper tried to quantify or assess how the aims of monetary policy should change in an older Europe, and have focused more on how the implications of a given change in policy might be different. In other words, I have focused more on the transmission mechanism than on the objectives of monetary policy. But those objectives may well need to be different in the future, and this might have as significant an impact (perhaps a more significant impact) upon monetary policy than shifts in the transmission mechanism.

This brings problems with it. Pressure on the central bank to keep asset prices up can obviously conflict with the aim of controlling consumer-price inflation. In countries where the welfare of a significant proportion of the population depends very sensitively on asset prices, these pressures will be greater. Arguably, they are markedly greater now in the US than is the case in Europe. The impact of ageing on the preferences of the central bank has often been thought to work through a greater pressure for price stability, or at least low inflation. But ageing is likely to bring with it greater reliance of the elderly on asset accumulation than in the past. It is less obvious that the old prefer low and stable inflation than that they prefer high asset prices. If the elderly invest a significant part of their financial wealth in equities and indexed debt, then they have some protection against even unexpected inflation. But a generalised fall in asset prices is harder to hide from. A robust conclusion from life-cycle models is that the old are relatively short on human capital and relatively long on marketable assets. They are, as a consequence, likely to be relatively more concerned about asset-price levels. This will become increasingly so if a substantial switch towards funded pensions is undertaken in Europe.

Table 3: Mean return on risky assets 6%; Standard deviation 17.5%; Safe return = 2%; Flat-rate Pensions; Perfect Annuities market

	Replacement rate of state pension			
	0 %	20 %	40 %	50 %
Aggregate Wealth (W)	738	530	350	275
W/National Income	7.54	5.77	4.12	3.89
Saving Rate	16%	13%	10%	8.3%
Risky Assets/Wealth	60%	75%	89%	94%
Share of Capital Income	33%	29%	23%	20%

Table 4: Mean return on risky assets 6%; Standard deviation 17.5%; Safe return = 2%; Flat-rate Pensions; Semi-Perfect annuities (money's worth ratio approximately 90%)

	Replacement rate of state pension			
	0 %	20%	40 %	50 %
Aggregate Wealth (W)	804	550	345	264
W/National Income	7.98	5.88	4.06	3.27
Saving Rate	23%	18%	13%	10%
Risky Assets/Wealth	60%	78%	93%	96%
Share of Capital Income	35%	30%	23%	19%

Table 5: Mean return on risky assets 6%; Standard deviation 17.5%; Safe return = 2%; Flat-rate Pensions; No annuities

	Replacement rate of state pension			
	0 %	20%	40 %	50 %
Aggregate Wealth (W)	915	566	335	250
W/National Income	8.68	5.95	3.95	3.12
Saving Rate	32%	23%	15%	11%
Risky Assets/Wealth	59%	82%	95%	9%
Share of Capital Income	38%	31%	23%	18%

Table 6: % of Population Credit-Constrained by Age

A: average rate of return on risky assets = 6%; flat-rate pension worth 40% of average income at retirement					
Age	20	25	35	80	90
$\beta = 1$	47	29	10	0	0
$\beta = 0.5$	47	29	10	2	15
$\beta = 0$	46	20	10	4	33
B: average rate of return on risky assets = 6%; flat-rate pension worth 20% of average income at retirement					
Age	20	25	35	80	90
$\beta = 1$	41	25	4	0	0
$\beta = 0.5$	41	23	7	0	0
$\beta = 0$	40	22	8	0	2

Annex 1*: The OLG Model:

Each agent maximizes a lifetime-utility function that depends upon consumption of goods and leisure in each period. We assume an additively separable form of the agent's utility function so that lifetime utility for the cohort born in period t=i, denoted U_i, is:

$$\text{(A 1)}\quad U_i = \sum_{t=i}^{t=T+i} \{ [\, u(c_{it},(1-l_{it}))\,]^{1-1/\zeta} / (1-1/\zeta) \} / (1+\rho)^{t-i}$$

where T is the length of life (60 adult years) and ρ is the rate of time preference; c_{it} is the consumption of the cohort born in period i at time t; l_{it} is the proportion of available time devoted to paid work; $1-l_{it}$ is the consumption of leisure ($0 <= l_{it} <= 1$). The parameter ζ determines the degree of intertemporal substitutability of consumption and leisure.

The within-period utility functions, $u(c_{it},(1-l_{it}))$, are assumed to be of the constant-elasticity-of-substitution type:

$$\text{(A2)}\quad u(c_{it},(1-l_{it})) = [\, c_{it}^{(1-1/\varepsilon)} + \alpha(1-l_{it})^{(1-1/\varepsilon)}\,]^{(1/(1-1/\varepsilon))}$$

* This annex draws heavily on Miles (1999).

α is a parameter which determines the intensity of preference for leisure relative to consumption; ε determines the substitutability of consumption and leisure. α will be set to tie down the average supply of labour for an agent over its life; ε determines how the supply varies over time in response to the changing opportunity cost of labour over the agent's life-cycle.

The period-to-period budget constraint is:

$$\text{(A3)} \quad w_{it} = (1+r_t)w_{i\,t-1} + y_{it} - c_{it} + p_{it}$$

where w_{it} is wealth for cohort i at time t, r_t is the interest rate and y_{it} is post-tax labour income. p_{it} is the state pension; it is zero for the first 44 periods of an agent's life (up to and including age 64 if we take 20 as the age of starting work) and is set at a given proportion of the average labour income of current workers in each of the last 16 periods of life (corresponding to a period from early sixties to late seventies). For a given replacement rate (ratio of state pension to average labour income), the tax rate on labour for each period (tax_t) is set so as to balance aggregate pensions paid against total tax receipts. Clearly, this tax rate will depend upon demographic structure and, for a given replacement rate, will vary over time with the ratio of retired to workers.

Post-tax labour income (y_{it}) is the product of the real wage per effective unit of labour, $(w/p)_t$, the proportion of time spent working (l_{it}), the total endowment of effective units of labour at time t, (h_{it}), and one minus the tax rate.

$$\text{(A4)} \quad y_{it} = (w/p)_t \,.\, l_{it} \,.\, h_{it} \,.\, (1-tax_t)$$

Note that how h_{it} evolves reflects both age-specific productivity (shown in figure 1) and time-related productivity.

The lifetime-budget constraint for generation i (assuming, for notational simplicity, a constant interest rate r) is:

$$\text{(A5)} \quad \sum_{t=i}^{t=T+i} c_{it} / (1+r)^{t-i} = \sum_{t=i}^{t=T+i} (y_{it} + p_{it}) / (1+r)^{t-i}$$

Two sets of first-order conditions, in conjunction with the budget constraint, tie down optimal decisions. First is the optimal labour/leisure choice. The relevant first-order condition is:

$$\text{(A6)} \quad (1-l_{it}) = (\mu_{it}/\alpha)^{-\varepsilon} c_{it}$$

where μ_{it} is the effective cost of leisure. For an interior solution ($0 < l < 1$), then $\mu_{it} = (w/p)_t \cdot h_{it} \cdot (1\text{-}tax_t)$. For corner solutions ($l$=0 or l=1), where the time constraint or the retirement constraint may bite, we define a shadow price of labour to make the corner solution satisfy the first-order condition. This ensures that the constraints are satisfied and that the level of consumption is also consistent with optimization at times when labour supply may be constrained.

The second set of first-order conditions determines the evolution of consumption and leisure over time. These are:

$$\text{(A 7)} \quad (c_{it} / c_{it-1}) = \{(1+r_t)/(1+\rho)\}^{\zeta} \left[\left(1 + \alpha\varepsilon(\mu_{it})^{1-\varepsilon} \right) / \left(1 + \alpha\varepsilon(\mu_{it-1})^{1-\varepsilon} \right) \right]^{(\varepsilon-\zeta)/(1-\varepsilon)}$$

In the base-case simulations we set the discount rate equal to -0.015 and the intertemporal elasticity of substitution (ζ) equal to 1/3. The discount-rate assumption is consistent with the empirical evidence of Hurd (1989), who suggested a negative rate. The intertemporal elasticity of substitution is more controversial; Cooley and Prescott (1995) use unity for their simulations, whereas Auerbach and Kotlikoff (1987) use a coefficient of relative risk aversion of 4, implying the elasticity is only 0.25. Empirical work by Hansen and Singleton (1983) and Mankiw, Rotemberg and Summers (1985) suggests values a little over unity, while Grossman and Shiller (1981), Mankiw (1985) and Hall (1980) found values between 0 and 0.4.

We set α, the relative weight attached to leisure against consumption, to a value which (at a general equilibrium for the other base-level parameters, and using European demographics) generated a plausible optimal labour supply over the working life for someone joining the labour force in 1995. We took a plausible average labour supply over the working life to imply 0.33 of the available endowment of time is spent at work. Assuming 8 hours a day spent asleep gives a total time endowment over a working year of 5840 hours. Assuming a typical working year of 48 weeks and a working week of 40 hours would make the average level of l equal to 0.33. We set the parameter determining substitutability between consumption and leisure (ε) equal to 0.8 (see Auerbach and Kotklikoff (1987) for a justification), though we also experiment with a much lower degree of substitutability (0.25). We describe in the next section what pattern of labour supply over an agent's life this degree of substitutability implies. We also describe in more detail what the combination of ζ and ε imply about the evolution of labour supply for different cohorts over time, as real wages per effective unit of labour change in response to exogenous productivity changes and endogenous shifts in the capital-to-labour-ratio.

Having calculated the savings and wealth of members of each generation, the calculation of aggregate savings and wealth at each period is trivial. Total wealth at time t (W_t) is:

$$\text{(A8)} \quad W_t = \sum_{i=t-60}^{t} \lambda_{it}\, w_{it}$$

where w_{it} is the wealth at time t of the representative agent of the generation that became adults in period i. λ_{it} is the weight of the generation "born" i periods ago in the adult population at time t.

Total labour input in each period is the sum of effective labour input from each cohort then of working age; each cohort's effective labour input at time t is the product of its supply of labour multiplied by its endowment of labour. Total labour input at t

$$\text{(A9)} \quad L_t = \sum_{i=t-60}^{t} \lambda_{it}\, h_{it}\, l_{it}$$

The capital stock used in domestic production is:

$$K_t = W_t$$

The total level of production is given by:

$$\text{(A10)} \qquad Q_t = AK_t^{\beta}\, L_t^{1-\beta}$$

We measure output net of depreciation, which means we should choose β to match the net-of-depreciation share of profits and other capital income in net-of-depreciation aggregate output. For the UK, national accounts data reveal that that proportion has been remarkably stable over the past ten years, at 0.25, which is the value we choose. (See the Blue Book, ONS, Table 16.3). With a constant population and in a steady state L_t rises at the rate of labour-augmenting technical progress. Otherwise, how L_t evolves from one period to the next depends on the changing age composition of the labour force, since labour units supplied by each cohort are age-specific.

Equilibrium in production requires that r_t equals the marginal productivity of capital:

$$\text{(A11)} \qquad r_t = \beta A(L_t/K_t)^{1-\beta}$$

Real wages at time t per unit of effective labour supply are:

$$\text{(A12)} \qquad (w/p)_t = (1-\beta)A(K_t/L_t)^{\beta}$$

Annex 2*: Optimal saving and portfolio allocation under uncertainty:

In this model, given the stochastic processes for labour income and for rates of return (and conditional on pensions arrangements and mortality rates), agents choose consumption (and therefore saving) and portfolio allocation in each period to maximize expected lifetime-utility. We assume an additively separable form of the agent's lifetime-utility function. We also assume a constant coefficient of risk aversion, the inverse of the intertemporal substitution elasticity. Agents are assumed to know the probabilities of surviving to given ages. Agent k who is aged j at time t maximizes:

$$(1) \qquad U_k = E_t \left[\sum_{i=0}^{i=T-j} s_{ij} \{ [c_{kt+i}]^{1-\zeta} / (1-\zeta) \} / (1+\rho)^i \right]$$

where T is the maximum length of life possible (which we take as 120 years of age), and the probability of surviving i more periods conditional on reaching age j is s_{ij} ($s_{0j}=1$); ρ is the rate of pure time preference; c_{kt+i} is the consumption of the agent in period t+i.

ζ is the coefficient of relative risk aversion.

Agents face two constraints:

First, there is a budget constraint, governing the evolution of financial assets taken from one period to the next.

$$(2) \qquad W_{kt+1} = [W_{kt} + \exp(y_{kt})(1-\tau) - c_{kt} + b_{kt}](\lambda \exp(r_{st}) + (1-\lambda)\exp(r_{ft}))$$

W_{kt} is the stock of wealth of agent k in period t

y_{kt} is the log of gross labour income

τ is the tax rate on labour income. Tax paid is simply a proportion of gross income

b_{kt} is the level of the unfunded, state pension received by an agent; this pension is zero until age 65.

λ is the proportion of financial assets invested in risky assets.

---

* This annex draws heavily on Miles and Cerny (2001).

r_{st} is the one-period (log) rate of return on risky financial assets between period t and period t+1.

r_{ft} is the one-period (log) rate of return on safe financial assets between period t and period t+1

For ease of notation, we have not given agent-specific subscripts to asset returns but we will allow for returns to depend on characteristics of the investor; because probabilities of death are specific to agents of a given age, rates of return on assets that might have annuities features will be agent-specific. We will describe how rates of return on financial investments are determined shortly.

Agents also face a borrowing constraint; wealth cannot be negative:

$$W_{kt} \geq 0 \qquad \text{for all k and t.}$$

This constraint may bind in various periods. Whether it does so depends in a complex way upon the profile of the deterministic component of labour income, the realizations of income and rate-of-return shocks, portfolio choices, the degree of risk aversion and the volatility of shocks. It also depends on the tax rate and the generosity of state pensions.

We also assume agents cannot take short positions in either safe or risky assets:

$$0 \leq \lambda \leq 1$$

Agents work from age 20 to the end of their 64th year (if they survive that long) and are retired thereafter. We assume that the profile of gross-of-tax labour income reflects three factors. First, there is a time-related rise in general labour productivity. Second, there is an age-related element to the growth of labour income over an agent's life. This is modelled as quadratic in age.

The age-specific part of the log of labour income is:

$$(3) \qquad \alpha + \gamma \text{age} - \theta \text{age}^2$$

We set γ and θ so that the age-income profile matches patterns that have been typical (we discuss calibration issues in detail in the next section).

There is also an idiosyncratic (agent-specific) stochastic element of labour income. The log of labour income for an agent is the sum of the age-related element, the time-related element and the additive income shock. Denoting the log of gross labour income of agent k who is aged j in period t as y_{kt} we have:

$$(4) \qquad y_{kt} = \alpha + gt + \gamma j - \theta j^2 + u_{kt}$$

$$u_{kt} = \phi\, u_{kt-1} + e_{kt}$$

where $e \sim N(0, \sigma_e)$

α is a constant.

g is the rate of growth of labour productivity over time.

ϕ reflects the degree of persistence in idiosyncratic shocks to labour income; empirical evidence from a range of countries suggests ϕ is high and that idiosyncratic shocks to income typically have a high degree of persistence. We set $\phi = 0.95$, $\sigma_e = 0.125$

We assume that rates of return on risky financial wealth vary across periods due to random shocks that hit stock and bond markets. We assume that rates of return on savings at a particular time - both on safe and risky assets - differ between individuals, because financial institutions take into account the probabilities of the death of agents and offer age-related investment products. More specifically, financial institutions offer the following contracts: for every $ invested in period t, with a given risky/safe split, the investor receives the market return adjusted for a probability of survival to the next period. If markets are perfect, the probability used in making this adjustment is the true survival probability. But we allow for imperfections (stemming from adverse selection or some other types of cost), which mean that the two are not equal. If the agent dies the institution keeps the funds. With no bequest motives, agents will always choose these contracts over ones which just pay the market rate of return. If the insurance element of this contract is offered on actuarially fair terms, the ex-post rate of return on a $ invested in the risky asset during period t by an agent k who is aged j and who survives to the next period is given by:

$$\text{(5)} \quad \exp(r_{st}) / s_{1j}$$

s_{1j} is the probability of surviving one more year, conditional on reaching age j

We assume r_s is the sum of the mean log return and an unpredictable shock.

$$r_{st} = r + v_t$$

r is the mean rate of return on risky assets (we set it at 6%)

v_t is the random element of the rate of return on assets in period t.

We assume v is iid and normal: $v \sim N(0, \sigma_r)$ (we set $\sigma_r = 0.175$)

For a $ invested in the safe asset, the return to an agent aged j is:

$$\text{(6)} \quad \exp(r_f) / s_{1j}$$

We set $r_f = 0.02$

We assume that the distribution of returns on assets is itself unaffected by pension arrangements; the most obvious justification for this is that there is a global capital market, where claims on safe and risky assets can be traded.

We can write the log returns on actuarially fair investments in risky and safe assets, respectively, to an agent aged j at time t as:

(7) $r + v_t - \ln(s_{1j})$

(8) $r_f - \ln(s_{1j})$

If markets are perfect, this financial contract can be offered at no risk by financial institutions because they pass on all the rate-of-return risk to investors, and are assumed to be able to take advantage of the law of large numbers and face no uncertainty about the proportion of agents who will survive. It seems natural to assume that financial firms will offer insurance against risks that are idiosyncratic (individual length-of-life risk), but not offer insurance against systematic risk (rate-of-return risk). The contracts offered by financial institutions can be thought of as highly flexible individual retirement accounts (or personal pension schemes). Effectively, agents have their own pot of assets, into which they pay contributions and make deductions. Contribution rates and drawdowns from the fund are subject only to the constraint that the pot of assets can never fall below zero. The average rates of return on the fund increase with age, since survival probabilities decline with age. Just as the standard flat annuities available for a given sum rise with age, so the average rate of return offered by financial institutions increases with age.

In effect we are assuming that financial institutions offer one-period annuities. These are the vehicles through which agents save for retirement. Agents are able to draw down such accounts in a flexible way in retirement. Individuals may decide to mimic the payments from standard flat annuities by having the "pot" size (ie. W) decline with age at a rate that is offset by rising average rates of return.

We introduce a measure of the efficiency of annuities markets. When this measure, β, is 1, the annuities market works perfectly. When $\beta = 0$, annuities are, effectively, not offered. The survival probability implicit in the contract offered by a financial institution is a weighted average of the true survival probability, s_{1j}, and the rate when no annuity is offered, an effective survival probability of unity. β is the weight placed on the actuarially fair survival probability

The rate of return paid on one-period savings invested with λ in the risky asset and $(1-\lambda)$ in the safe asset for an agent aged j at time t becomes:

(9) $\{\lambda \exp[r + v_t] + (1-\lambda)\exp[r_f]\} / [\beta s_{1j} + (1-\beta)]$

This way of modelling the efficiency of annuity contracts allows the departure from actuarially fair contracts to vary with age. The greater age is, the lower is the probability of surviving, and, for all $\beta < 1$, the greater is the departure from actuarially fair contracts.

The set of first-order conditions from individual k's optimization problem is:

if $\quad c_{kt} < [W_{kt} + \exp(y_{kt})(1-\tau) + b_{kt}]$

then

$$(10) \quad U`(c_{kt}) = E_t\,[s_{1j}\ \{ U`[c_{kt+i}]\{ \lambda \exp(r + v_t) + (1-\lambda)\exp(r_{ft}) \}/ [\beta s_{1j} + (1-\beta)] \} / (1+\rho)]$$

or else:

$$c_{kt} = [W_{kt} + \exp(y_{kt})(1-\tau) + b_{kt}]$$

and

$$(11) \quad U`(c_{kt}) \geq E_t\,[s_{1j}\ \{ U`(c_{kt+I})\ \{ \lambda \exp(r + v_t) + (1-\lambda)\exp(r_{ft}) \} / [\beta s_{1j} + (1-\beta)]\ \} / (1+\rho)]$$

where $\quad U`(c_{kt})$ is $\partial U_k / \partial c_{kt}$

We also require a condition for optimal portfolio allocation:

either:

$$0 = E_t\,[\,U`(c_{kt+I})\{ \exp(r + v_t) - \exp(r_{ft}) \}\,]$$

and $0 \leq \lambda \leq 1$

or else:

$$(12) \quad 0 < E_t\,[\,U`(c_{kt+I})\{ \exp(r + v_t) - \exp(r_{ft}) \}\,]$$

and $\lambda = 1$

or else:

$$0 > E_t\,[\,U`(c_{kt+I})\{ \exp(r + v_t) - \exp(r_{ft}) \}\,]$$

and $\lambda = 0$.

(10) holds when the borrowing constraint is not binding. When the constraint binds complementary, slackness implies that (11) holds. (12) is a standard condition for optimal portfolio allocation. Corner solutions may arise where

agents wish only to invest in the safe asset or in the risky asset; for an internal solution, the first equality at (12) must hold.

Athough characterizing optimal plans is easy enough, solving explicitly for optimal consumption and for the optimal accumulation path for funds is not possible. Instead, we have to turn to numerical methods. We solve the problem backwards in a now-standard way (see Deaton (1991), Zeldes (1989) and Skinner, Hubbard and Zeldes (1995)). This involves constructing large grids in the state space and solving for optimal saving and portfolio allocation at those grid points using numerical integration and interpolation.

References

Auerbach, A. and L. Kotlikoff. 1987. *Dynamic Fiscal Policy*, Cambridge University Press.

Borsch-Supan, A; A. Ludwig and J. Winter. 2001. Ageing and International Capital Flows, this Volume.

Brown, J., O. Mitchell and J. Poterba. 1999. The Role of Real Annuities and Indexed Bonds. In: *an Individual Accounts Retirement Program,* NBER Working Paper No. 7005.

Brugiavini, A. 2001. Ageing and Saving in Europe, this volume.

Cooley, T. and E. Prescott. 1995. Economic Growth and Business Cycles. In *Frontiers of Business Cycle Research*, ed. T. Cooley, Princeton, Princeton University Press.

Deaton, A. 1991. Saving and Liquidity Constraints. *Econometrica,* 59, pp. 1221-48.

Grosssman, S. and R. Shiller. 1981. The Determinants of the Variability of Stock Market Prices. *American Economic Review*, 71, No. 2.

Hall, R. 1980. *Intertemporal Substitution in Consumption,* NBER Working Paper No. 720, July.

Hansen, G. 1993. The Cyclical and Secular Behaviour of the Labour Input: Comparing Efficiency Units and Hours Worked, *Journal of Applied Econometrics*, 8, pp. 71-80.

Hansen, P. and K. Singleton. 1983. Stochastic Consumption, Risk Aversion and the Temporal Behaviour of Asset Returns, *Journal of Political Economy*, 91, No. 2.

Hubbard, G., J. Skinner and S. Zeldes. 1995. Precautionary Saving and Social Insurance, *Journal of Political Economy*, 103, pp. 360-399.

Hurd, M. 1989. Mortality Risks and Bequests. *Econometrica*, 57, pp. 779-813.

Mankiw, G. 1985. Consumer Durables and the Real Interest Rate. *Review of Economics and Statistics*, August.

Mankiw, G., J. Rotemberg and L. Summers. 1985. Intertemporal Substitution in Macroeconomics., *Quarterly Journal of Economics*, 100, pp. 225-59.

Miles, D. 1999. Modelling the Impact of Demographic Change Upon the Economy. 1999, *The Economic Journal*, 109, No. 452, pp. 1-37.

Miles, D. and A. Cerny. 2001. *Risk, Return and Portfolio Allocation under Alternative Pension Systems with Imperfect Financial Markets,* CEPR Discussion Paper (April).

Mitchell, O. (2001). Developments in Decumultaion: The Role of Annuity Products in Financing Retirement, this volume.

Mitchell, O., J. Poterba and M. Warshawsky. 1997. *New Evidence on the Money's Worth of Individual Annuities,* NBER Working Paper No. 6002.

Svensson, L. 2001. What is Wrong with Taylor Rules? Using Judgement in Monetary Policy through Targeting Rules, mimeo.

Taylor, J. 1993 Discretion versus Policy Rules in Practice, Carnegie-Rochester Conference Series on Public Policy, 39, pp. 195-214.

Woodford, M. 2000. The Taylor Rule and Optimal Monetary Policy, mimeo, Princeton University, December.

Zeldes, S. 1989. Optimal Consumption with Stochastic Income. *Quarterly Journal of Economics*, 104, pp. 275-98.

Figure 1:
European Saving Rate: constant replacement rate

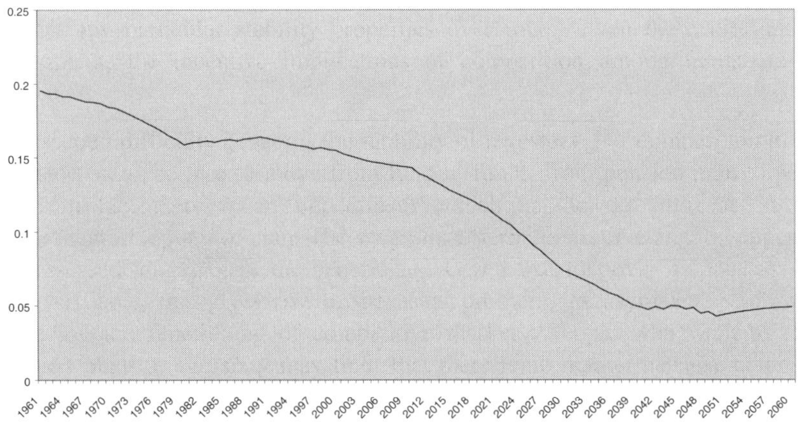

Fugure 2: European Real Interest Rate: constant replacement rate

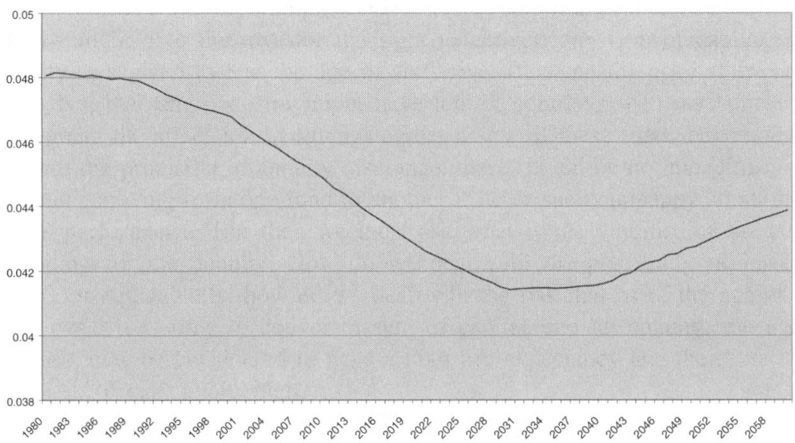

Fugure 3: European Capital Output ratio
: constant replacement rate

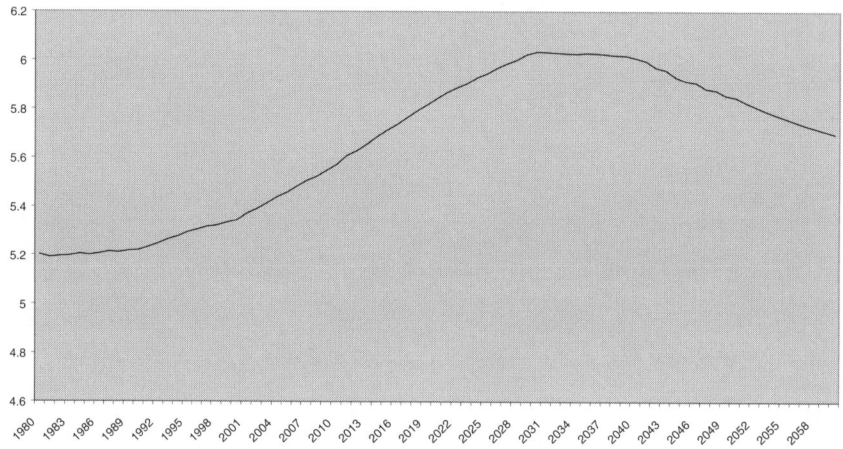

Figure 4:
European Saving Rates: constant contribution rate

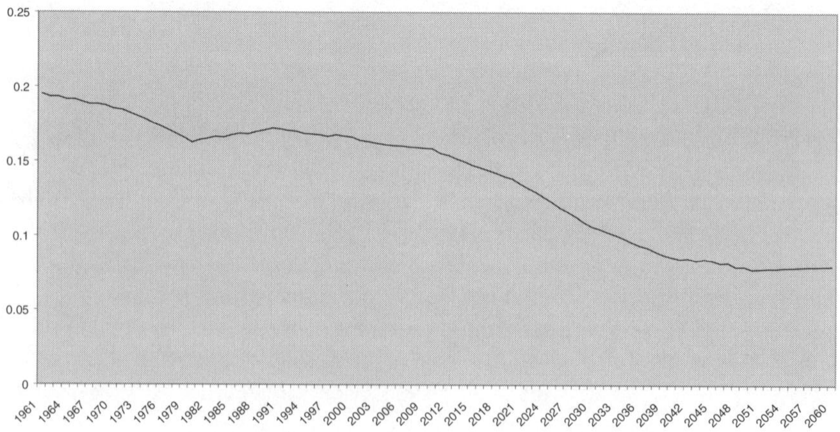

Figure 5:
European Saving Rates: phase out state pension completely

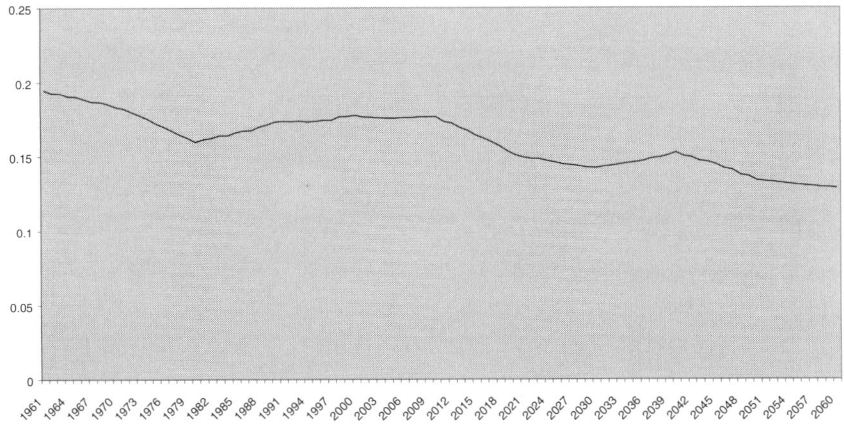

Figure 6: European Real Interest Rate - phase out pension completely

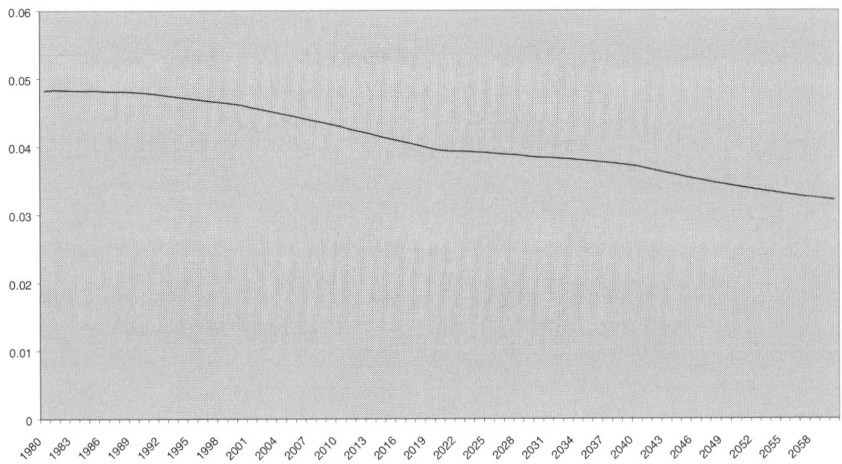

Figure 7: proportion of future population credit constrained by age: 50% replacement rate; semi-perfect annuities.

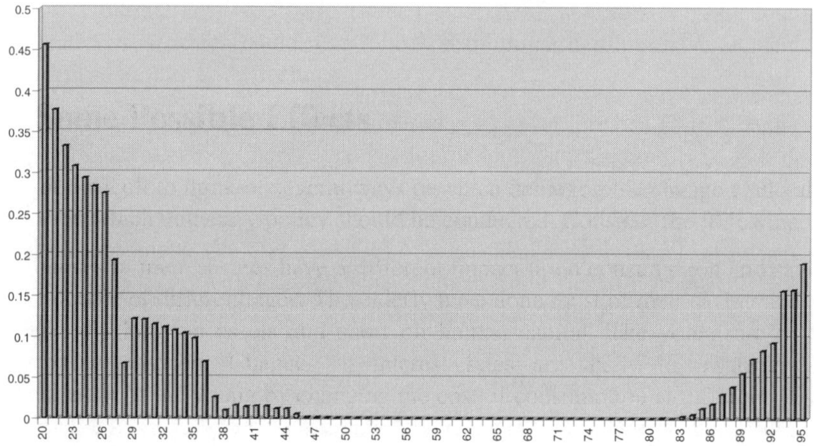

Discussion

Hiroshi Fujiki

1 Introduction

It is my great honor to participate in this very interesting conference as a discussant. Central bank conferences on monetary policy and academic research in this area in the 1990s asked the question: „how to reduce inflation“. Recent studies put more emphasis on „how to maintain low inflation to enhance macroeconomic stability and agents' welfare“ in a context of optimal feedback monetary policy rule based on stochastic forward-looking macroeconomic models. Those studies do pay attention to monetary transmission mechanism, asset prices, and the level of government debt (for example, the literature on the fiscal theory of price levels or optimal taxation, such as Walsh (1998)). However, those papers are not specific about the details of dynamic fiscal policy, particularly the role of a public pension system with regard to changing demographic structure, hence I believe that the topic of this conference is very important.

I enjoyed reading the paper by Professor Miles. I think this paper will become a nice benchmark study of the monetary transmission mechanism in the context of a changing demographic structure. As I understand it, this paper says that demographic changes and the generosity of the public pension system influence economic agents' expectations and optimization over time, as well as the future monetary transmission mechanism. Therefore, this issue, albeit long-term and highly political, should be considered a very important current issue for central bankers. Technical difficulties inherent in the simulation study of dynamic fiscal policy employed in Professor Miles' paper make it difficult for him to consider the relationship between demographic changes and pension systems in terms of optimal monetary policy rules. Such technical difficulties might frustrate central bankers who need more concrete advice: for example, a feedback rule relating some specific target variables (maybe income and wealth distributions, which will depend on demographic structure?) to their operational variables, such as short-term interest rates. However, I think that such frustrations should not lead to the conclusion that Professor Miles' paper is not helpful to central bankers. Rather, Professor Miles' paper addresses one of the most important and fruitful areas of research in monetary policy. We should have an acceptable class of macroeconomic models with which central bankers can talk about the relevant policy targets, policy instruments and the transmission mechanism under changing demographic structures and pension reforms, namely, the details of the supply side of an economy.

In Japan, demographic changes and pension reforms are two of the most serious structural issues relating to monetary policy (See Yashiro (1999), Fujiki, Nakada and Tachibanaki (2001) for example). Japanese monetary policy is facing other serious structural issues: the banks' non-performing loan problem, corporate debt problems, slowly but steadily increasing unemployment, and the rapid accumulation of government debt. The Bank of Japan has been fighting against a liquidity trap under conditions of a small but negative inflation rate since the collapse of the bubble economy in the late 1990s. I hope to contribute to this conference by adding some lessons from Japan. In the following sections, first, I would like to summarize the main message of this paper, as far as I understand it. Second, I want to make several comments and pose questions on the models in the paper. Finally, I will make some comments regarding the conclusion of this paper.

2 Summary of Professor Miles' paper

Professor Miles responds to this question by simulating three overlapping-generation models and considers the effects of demographic changes and the generosity of unfunded public pensions on the transmission mechanism of monetary policy.

The first is a closed-economy overlapping-generation model that enables us to understand how the intercept term of the Taylor rule, that is, the real interest rate, depends on the demographic structure. In this model, an agent maximizes his or her lifetime utility by the choice between labour and leisure, subject to a single budget constraint which makes the present discounted value of consumption equal to the present value of post-tax labour incomes and of receipts of the state pension. The state pension is unfunded, and labour income is taxed. Since the evolution of the capital stock and labour supply depends upon demographic structure and the public pension system, real wages and real interest rates implied by the aggregate production function depend on demographic structure. Supposing an exogenous labour-augmenting technical progress, age-specific rate of productivity growth, as well as a population-growth rate obtained from UN projections, simulations of real interest rates are carried out on the basis of three unfunded public pension regimes whose generosities differ.

The results in Table 1 reflect two opposite effects on the real interest rate. First, saving rates and aggregate savings decline, reflecting life-cycle-saving behavior with an increasing older-generation cohort. This means there will be a higher real interest rate. Second, a reduction in aggregate labour supply increases the capital-labour ratio, which means lower marginal productivity of capital and a lower real interest rate. Table 1 shows that, even under the most generous pension regime that discourages private savings, the second effect dominates the first effect, and the real interest rate will be lower.

The second model is a small open-economy version of the first model, which is useful for understanding the impact of an exogenous change in real interest rates. In theory, we expect that unanticipated permanent rises in the real interest rate work in two ways. First, they reduce the price of financial assets. Second, they alter the timings of consumption over time. For the young, perhaps the second effect alone is significant. For the relatively old, we expect the first effect will be relatively important.

According to the author, a 1% unanticipated, but permanent, rise in real interest rates reduces consumption expenditure by about 7.6% with the 2000 population structure, but about 9% with the 2050 population structure, given the same pension system. That is to say, with more old people, the resulting decline in aggregate consumption is greater when interest rates rise. If a less generous state pension system is introduced, it will almost certainly mean higher private saving and a greater stock of assets, particularly those assets held by the relatively elderly, hence such a change will strengthen the first effect.

The third model tries to capture the effects of uncertainty in real returns, labour income, mortality, and imperfect financial markets on portfolio allocation and aggregate savings, using a life-cycle-saving, overlapping-generation model. To this end, the author assumes an exogenous rates of return on safe assets (2% of real expected return) and risky assets (expected return of 6% but a standard deviation of 17.5%) and he asks how savings and portfolio allocation evolve as demographics change and as pension arrangements vary. Individuals cannot borrow against labour income, and face the potential of being able to buy annuities, but at varying degrees of actuarial fairness. Annuities markets are assumed to be either actuarially fair, comprising annuities with a money's worth ratio of around 90%, as in the U.S. or U.K., or completely inefficient. Life expectancy is assumed to be stabilized at levels predicted for Japan in the year 2050, which I believe are 80.6 for males, and 86.7 for females. Major results of simulations are summarized as follows.

First, the overall stock of wealth is sensitive to pension arrangements. With state, unfunded pensions worth 20% of average wages, the stock of aggregate wealth is around double what it would be with state pensions worth 50% of average earnings.

Second, the fraction of population that is liquidity-constrained is substantial amongst the relatively young and amongst the very old. But the less generous state pensions are, the smaller will be the proportion of the elderly population that are credit-constrained, since a low state pension means a larger stock of private financial assets.

Third, the proportion of overall wealth that is invested in risky assets depends very sensitively on pension arrangements. The lower the state pension (a safe asset), the more people are exposed to rate-of-return uncertainty and the greater is the share of their safe assets. However, with lower state pensions, people hold more financial assets, and the overall level of investment in risky assets is generally higher. Note that portfolio allocation is likely to be highly sensitive to pension arrangements, but depends only to some extent upon how efficient annuity markets are.

3 Questions and comments

Let me comment on the three models introduced by Professor Miles in turn. The policy implication of the first model is that not only the real interest rate, but also nominal interest rates, will, on average, be lower in the future than they have been in the recent past. The author relates his results to the Taylor rule (Taylor (1993)) with a variable intercept term. I think this is a reasonable extension (See for example, Taylor (1995) and Hall (2000)). I have two comments on this model. My first comment is theoretical, and my second comment is based on Japanese experience.

First, theoretically, it would be interesting to introduce wealth effects of money into the first model. Ireland (2001) shows that the wealth effect of money works in a cash-in-advance model with population growth. His intuition is that while a low nominal interest rate makes the opportunity cost of holding money irrelevant, part of the future tax burden of today's monetary finance for the current generation will be transmitted to future generations if the population is growing. He therefore argues that monetary policy works, even under the liquidity trap, if there is population growth. One may also wonder if the issue of capital tax is relevant to this line of research.

Second, one may think that the results also suggest that the probability that a central bank faces a zero bound of nominal interest-rate increases under conditions of a large negative demand shock, given a low inflation rate and a low real interest rate (see, for example, Reifschneider et al. (2000)). Japanese experience of the zero-interest-rate policy from 1999 to 2000, in retrospect, suggests that depending on the nature of the shock, desirable monetary policy responses should be different, and that the zero-interest-rate policy is just one option among them. For example, suppose a central bank wishes to make large-scale outright purchases of long-term government bonds under the zero-interest-rate policy. In this case, monetary policy must be helped by a fiscal commitment that promises to protect the central bank's future capital losses from such an operation (See Fujiki, Okina and Shiratsuka (2001) and Goodfriend (2000, 2001)). To make such a commitment in a credible manner, we need a detailed study on the supply side of the economy, including the demographic changes and pension systems, which are assumed to be constant in those papers.

The second model shows two reasons why monetary policy might be more powerful in a greyer world. First, there would be more old people whose consumption might be more sensitive to interest-rate changes. Second, the ageing of the population may generate changes in pension arrangements which themselves increase the stock of financial assets. I agree with the points made by the author, and the main message obtained from this exercise is broadly consistent with the simulation of Deaton et al. (2000), which focuses on the effects of the introduction of social security on cross-sectional income distribution and wealth distribution. I have one theoretical comment. It would be nice if we could have some benchmark variables for the sake of monetary policy decision-making, and make the changes in the nominal interest rate as a function of those variables. How sensitive are aggregate utility measures, income distribution measures, or the distribution of consumption across generations to changes in the target variables and policy-reaction functions?

The third model is useful in that it enables us to understand the sensitivity of the level of saving and the allocation of that saving between risky assets and safe assets, to changes in demographics, to changes in the generosity of unfunded state pensions, and to the degree of efficiency of private annuity contracts. I will touch on two points of this model, based on current Japanese issues, but I hope these contain some general points of interest to those of you here.

First, it would be nice if Professor Miles could offer an intuitive explanation of the results shown in tables 3-5, namely, that the efficiency of annuity market is not so important compared with the generosity of the public pension scheme. To what extent should we take these results seriously in terms of economic policy-making? How large is the insurance value of annuity contracts in these specific examples, and what is the effect of annuity contracts on the lifetime utility of agents? I am moved to raise this question because Japanese corporate pension schemes partly provide pensions on behalf of the public pension system: a division of labour between the public pension system and the corporate pension scheme is a relevant concern to Japanese policy makers.

Second, Professor Miles shows that the proportion of risky assets to total assets and the total volume of risky assets in a greyer world should be distinguished, and that both of them are sensitive to the generosity of public pensions. Some economists say that Japanese citizens must have more risky assets, such as stocks, instead of deposits (see evidence, Figure 1), but economists are not careful about this distinction. I agree that the portfolio selection of Japanese citizens, or of the citizens of any other countries, should consider possible changes in public and corporate pension schemes, and changes in employment practices developed along with those schemes.

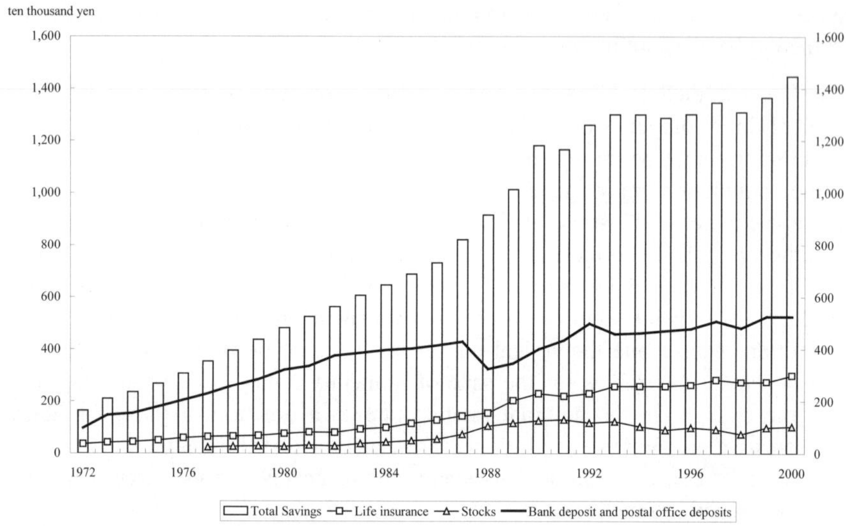

Figure 1: Japanese average household savings
Source: *Public Opinion Survey on Household Savings and Consumption 2000.*

4 Extensions for future study

In the final section, Professor Miles states that one implication of ageing may be a greater concern amongst populations in Europe about asset price volatility. The ECB may be required to attach much more weight to asset price volatility than it now does. I have two responses to that remark.

First, I agree that monetary policy should take into account many issues related to income, wealth or consumption distribution in a greyer world, and perhaps even now. For example, some argue that the Japanese zero-interest-rate policy has a negative effect on economic agents who largely depend on interest income, such as old age pensioners (see evidence in Figure 2).

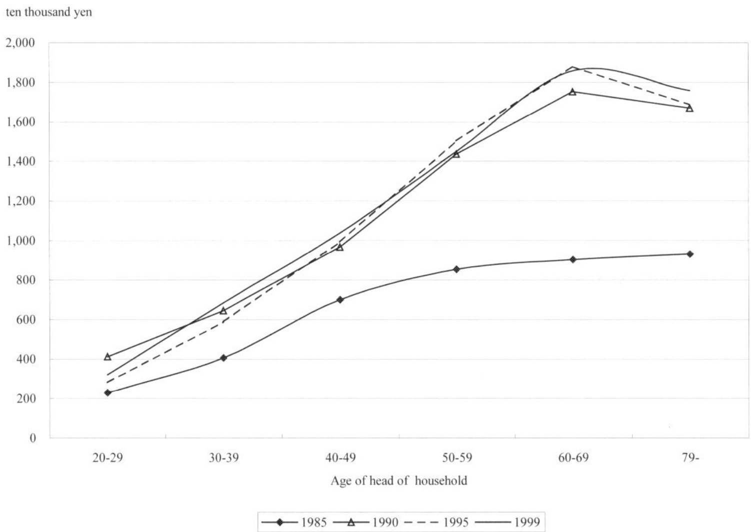

Figure 2: Japanese average household savings, by the age of the head of household
Source: Public Opinion Survey on Household Savings and Consumption 2000.

Regarding this point, BOJ deputy governor Yamaguchi said, „the elderly are the most affected, as low interest rates have continued for a long period, and we sympathize with their situation. However, let me emphasize that raising interest rates would not solve the problem. Households can earn interest income from their deposits because banks obtain interest income on their loans. In other words, the source of interest income received is economic growth, and interest income will not be generated if the economy does not grow (Yamaguchi (2000)).“ That is to say, given their success in controlling inflation, central bankers in the ageing society, or even now, must think seriously about the zero bound of the nominal interest rate. I do not have any clear theoretical answers to the question of whether the optimal monetary policy rule that takes account of demographic change should be defined as a feedback rule from asset prices, some distribution measure of income, consumption, or wealth. Furthermore, overcoming the measurement issues regarding asset prices, income or wealth across generations seems to be a formidable task. But we clearly need to develop a better understanding of the supply side of the economy before we go into a greyer world.

Second, I am wondering if the message contained in this paper could be stated in the opposite way: „fiscal policy should be different in a greyer world under a low nominal interest rate“. Monetary policy must be helped by fiscal policy in the presence of the risks of deflation and zero bound of nominal interest rate (Fujiki, Okina and Shiratsuka (2001) and Goodfriend (2000, 2001)). One may well wonder if the zero bound of nominal interest rates might also be a relevant concern to pension funds, the government bond market, or, even the labour market. No matter how monetary policy tries its best, fiscal discretion might distort monetary commitment (Dixit and Lambertini (2000)). This statement might be too abstract for you, so let me give you some examples. Suppose a public pay-as-you-go pension system invests only in government bonds. How could we protect this conservative investment policy from political pressure that tries to make public pension funds invest in stocks in a certain industry? Another example of the potential benefits of fiscal discipline is that, by maintaining a deep and liquid financial market in government bonds, the government could encourage the younger generation to invest in government bonds, and thus support the public pension system indirectly. In this way, demographic changes are a common challenge to both monetary and fiscal policy. Thank you.

References

Deaton, A., P.-O. Gourinchas and C. Paxson. 2000. *Social Security and Inequality over the Life Cycle*, National Bureau of Economic Research, Working Paper No. 7570.

Dixit, A. and L. Lambertini. 2000. *Fiscal Discretion Destroys Monetary Commitment.* (*http://econweb.sscnet.ucla.edu/lambertini/papers/fiscdisc.pdf*).

R. E. Hall. 2000. *Monetary Policy with Changing Financial and Labour-Market Fundamentals.* (*http://www.stanford.edu/~rehall/ Taylor%20rule%205-16-00%20cp.pdf*).

Fujiki, H., K. Okina and S. Shiratsuka. 2001. Monetary Policy under Zero Interest Rate - Viewpoints of Central Bank Economists - *Monetary and Economic Studies*, 19 (1), Institute for Monetary and Economic Studies, Bank of Japan, pp. 89-130. (*http://www.imes.boj.or.jp/english/publication/mes/2001/ me19-s1-8.pdf*).

Fujiki, H., S. K. Nakada and T. Tachibanaki. 2001. Structural Issues in the Japanese Labour Market: An Era of Variety, Equity, and Efficiency or an Era of Bipolarization? *Monetary and Economic Studies*, 19, S-1 , Bank of Japan, pp. 177-207. (*http://www.imes.boj.or.jp/english/publication/mes/2001/me19-s1-8.pdf*).

Goodfriend, M. 2000. Overcoming the Zero Bound on Interest Rate Policy. *Journal of Money, Credit, and Banking*, 32 (4), S, pp.1008-1035.

Goodfriend, M. 2001. Financial Stability, Deflation, and Monetary Policy. *Monetary and Economic Studies*, 19, S-1 , Bank of Japan, pp.143-167. (*http://www.imes.boj.or.jp/english/publication/mes/2001/me19-s1-7.pdf*).

Ireland, P. N. 2001. *The Real Balance Effect.* National Bureau of Economic Research, Working Paper No. 8136.

Reifschneider, D. and J. C. Williams. 2000. Three Lessons for Monetary Policy in a Low-Inflation Era, *Journal of Money, Credit and Banking*, 32 (4), pp. 936-966.

Taylor, J. B. 1995. Monetary Policy Implications of Greater Fiscal Discipline, in *Budget Deficits and Debt: Issues and Options*, A Symposium Sponsored by The Federal Reserve Bank of Kansas City.

Taylor, J. 1993. Discretion versus Policy Rules in Practice, *Carnegie-Rochester Conference Series on Public Policy* 39, pp. 195-214.

Walsh, C. E. 1988. *Monetary Theory and Policy*, Cambridge, MASS: MIT Press.

Yamaguchi, Y. Thinking behind Current Monetary Policy, Remarks at the Japan National Press Club on August 4, 2000. (*http://www.boj.or.jp/en/press/koen058.htm*).

Yashiro, N. 1999. *Shoshi Koureika no Keizaigaku* (Economics of The Low Fertility and Ageing Economy), Tokyo, Touyoukeizaishinposha (in Japanese).

Discussion

Benoît Mojon

David Miles' paper provides a useful description of some of the major challenges that ageing may add to the already difficult task of conducting monetary policy. In particular, central bankers should expect that ageing will change the equilibrium real interest rate, the elasticity of aggregate consumption with respect to the interest rate, the allocation of wealth among financial instruments and the preference of society with respect to the trade-off between price stability and output stability.

Miles arguments are illustrated within an overlapping-generation (OLG) framework, previously calibrated to European demography in Miles (1999). The major corner-stone of this model is its complex demographic structure, based on the past demographic profile and the United Nations projection in Europe for the next fifty years. Each cohort is optimising a utility function over consumption and leisure in a forward-looking manner over its lifetime. Other heroic assumptions of the model include the calibration of the utility function and of the production function[1]. The model provides a useful simulation tool, where the impact for the economy of ageing and of different pension systems can be thought of. However, the arbitrariness of the model itself and, within this class of model, of some of its calibrating assumptions call for an evaluation of its results outside the model framework.

Hence this comment proposes additional evidence or arguments, in order to complete our perspective on monetary policy in a "Greyer world". It focuses mainly on the effects of ageing on the transmission mechanism. It then discusses briefly the decline in the equilibrium real interest rate and the change in the preference of society for price stability.

1 Changing transmission mechanism?

The main aspect of the impact of ageing on monetary policy addressed by Miles is the likely evolution of the transmission mechanism. His two main arguments seem plausible. First, the scale and composition of wealth evolves over the life cycle,

1 Why, for instance is the depreciation rate negative?

and more dramatically if the importance of the PAYG pension systems declines. The permanent income model of consumption then delivers that, provided that financial wealth is itself interest-elastic, the aggregate consumption response to changes in the interest rate will increase with the proportion of old wealth-holders in the population. Second, the aggregate inter-temporal allocation of resources should be more interest-rate elastic as the proportion of (old) households, whose accumulated wealth protect from being liquidity-constrained, increases.

But the method used by Miles to quantify the changes in the transmission mechanism is not convincing. He evaluates the effect of a 1 percent permanent change in the real interest rate with a simplified version of his calibrated OLG model. The interest-rate, which is now exogenous, can be changed by the central bank and households invest all their wealth in bonds paying a fixed coupon every period. The wealth loss experienced by holders of bonds following a rise in the interest rate is maximised, and so is the reaction of aggregate consumption. This result should be taken with caution for at least two reasons.

First, the diversification of wealth would allow households not to change their consumption following temporary changes in the real interest rate triggered by monetary policy. Second, we do not know which one of the two determinants of the elasticity of consumption with respect to the interest rate, i.e. the substitution effect or the income effect, dominates.

1.1 Composition of wealth and monetary policy

Back in the early 1990-'s, French and Italian households saw their disposable income increase following rises in the short-term interest rate. This was reflecting the widespread investment into money mutual funds and in Treasury bills. This episode illustrates why the Miles' measure of the impact of interest rate on consumption can be misleading.

The largest vector of households' wealth in the OECD countries is housing. As for financial wealth, households hold not only stocks and bonds, but also bank accounts and short-term securities. For instance, Mojon (2000) reports that the latter two financial instruments are nearly as large as the former two in the portfolios of German, French, Italian and Spanish households. Although it is likely that asset prices (the prices of houses, of bonds or of stocks) would decrease following a rise in the interest rate, it is not clear whether households who save with a very long horizon would react to the cyclical pattern of the real interest rate. Recent empirical evidence by Boone, Giorno and Richardson (1998) shows that the propensity to consume financial wealth is very small, and hardly significant statistically, in large continental European countries. It ranges from 0.018 in Germany to 0.008 in Italy. In addition, the income effects associated with interest flows from bank deposits and short-term securities should offset the wealth effects of changes in asset prices. At the end of the day, the total effects of changes in the interest-rate on aggregate consumption will depend on the ability of households to compose their wealth in order to hedge against the interest-rate risk. If Miles could take the variety of financial instruments which makes this hedging feasible into account, the effects of the interest rate on aggregate consumption which he obtains, would be greatly reduced.

1.2 Real interest rate, inter-temporal allocation of consumption and ageing

This sections proposes to test Miles' conjecture that ageing alter the response of consumption to the real interest rate. The negative response of aggregate consumption to the interest rate obtained by Miles depends largely on the value chosen in the somewhat arbitrary calibration of the model, e.g. the inter-temporal elasticity of consumption. Quoting Blanchard and Fisher (1989): "Whether an increase in the rate of return on saving would in fact increase saving, is the subject of much controversy".

Empirical research on the determinants of the savings rate has found that the real interest rate had a positive impact on the savings rate. Masson, Bayoumi and Samiei (1998) obtain that, in a sample of 21 OECD countries for the period 1971-1993, the real-interest-rate impact on private savings is 0.21 (with a T-statistic of 4.4). They also show evidence that "demography" has an impact on the savings rate. The dependency ratio has a negative impact on savings. However, these regressions do not tell us whether an older society, with a higher dependency ratio would be characterised by a higher elasticity of consumption with respect to the real interest rate. The simplest way to test Miles' conjecture is to estimate the second derivative of the savings rate with respect to the real interest rate and the dependency ratio. To do so, I estimate the following equation, where S is the savings rate, r the real interest rate, D the dependency ratio, X is a vector-of-control variable and i is an index for countries.

$$S_{it} = k_i + a_i S_{it-1} + b_i r_{it} + c_i D_{it} + f_i r_{it} \times D_{it} + G_i X_{it}$$

The coefficient f is an estimate of the second derivative we are interested in. Table 1 reports the results for a sample of ten countries of the euro area (Austria, Belgium, Finland, France, Germany, Greece, Ireland, Italy, the Netherlands and Spain), for the period 1980-1998. The variables are defined as follows: the savings rate is that of households', the real interest rate is the government bond yield, minus inflation in that year, the dependency ratio is the proportion of the population below 20 and over 60 divided by the population between 20 and 60 and the control variable is just inflation, which had a very significant impact on the savings rate for European countries in the sample period[2]. Because the time

[2] Demographic data come from Eurostat and macroeconomic data from the OECD.

dimension is large with respect to the cross section, I allow for different slopes across countries, and remove them in a general-to-specific sequence, if they are not significant[3].

Table 1: The impact of ageing on the response of European households' savings rate to the real interest rate

	Regression 1		Regression 2	
Variables	Coefficient	P-value	Coefficient	P-value
S(t-1)	0.407	[.000]	0.467	[.000]
S(t-1) × D_ES	-0.472	[.018]	-0.580	[.003]
S(t-1) × D_FR	0.220	[.077]	0.165	[.194]
r(t)	0.295	[.000]	1.172	[.047]
r(t) × D_AT	0.568	[.086]	-6.939	[.149]
r(t) × D_BE	-0.485	[.105]	-0.426	[.161]
r(t) × D_IT	0.306	[.005]	0.271	[.028]
Inflation(t)	0.421	[.000]	0.378	[.000]
D(t)	-0.059	[.045]		
D(t) × D_AT			-0.386	[.069]
D(t) × D_BE	-0.892	[.002]	-0.830	[.005]
D(t) × D_ES	-0.348	[.000]	-0.348	[.000]
D(t) × D_FI	1.134	[.049]		
D(t) × D_FR	-0.555	[.001]	-0.495	[.003]
D(t) × D_IR	0.075	[.134]	0.045	[.286]
D(t) × D_IT	0.161	[.006]	0.150	[.017]
D(t) × r(t)			-0.011	[.097]
D(t)× r(t)× D_AT			0.088	[.117]
D(t)× r(t)× D_FI			-0.003	[.255]
D(t)× r(t)× D_IR			0.004	[.069]
	Ad. R-sqr = .935061		Ad. R-sqr = .932286	
	D.-W. = 1.88811		D.-W. = 1.86087	

Sample 1980-1998, AT, BE, DE, ES, FI, FR, GR, IR, IT, NL

[3] In view of the large number of explanatory variables, only coefficients of which the P-value was superior to 0.30 were constrained to zero.

Regression 1 corresponds to the benchmark equation of the savings rate à la Masson, Bayoumi and Samiei (1998), for this sample. The savings rate responds positively to the real interest rate in all countries except Belgium. And in Austria and in Italy this response is higher than the average of the countries. Surprisingly, the impact of inflation on savings is of the same magnitude in all countries. On the other hand, the effect of the dependency ratio varies across countries. It is positive in Italy, in Finland and in Ireland.

The inclusion of the interaction term between the real interest rate and the dependency ratio changes the set of significant coefficients to some extent. In particular, the dependency ratio remains negative and significant only in Austria, Belgium, Spain and France. The interaction of the real interest rate and the dependency ratio appears negative (except for Austria and Ireland), which invalidates Miles' conjecture. However, it is significant only at the 10 % level and it is fairly small. Overall, while these results should be taken as preliminary, it appears at least that an increase in the dependency ratio, such as has been experienced in Europe over the last two decades, has not been associated with a larger impact of the real interest rate on savings rates.

2 Negligible decline in the equilibrium real interest rate

In this section, I explain why the decline in the equilibrium real interest rate stressed by Miles is negligible. A priori, that decline should matter to a central bank that set the short-term nominal interest rate as the sum of its evaluation of the real interest rate and a reaction to the deviation of inflation and output from the target, i.e. following a Taylor rule. The argument is interesting and seems reasonable. However, the result of the model simulation seems hardly relevant from the point of view of central bankers. The maximum decline in the equilibrium real interest rate, which is obtained in the scenario where European PAYG pensions phase out from 2020 onwards, amounts to a fall of 30 basis points from 2000 to 2010, and another 35 basis points from 2010 to 2020. By way of comparison, monetary policy actions are often implemented through sequences of 25-basis-points changes in the interest rate. A decline of three or four basis points per annum in an unobserved component of the nominal interest rate is very unlikely to change the conduct of monetary policy.

3 Changing preferences

Miles also argues in the paper that a "Greyer world" has a greater preference for price stability. This is because, the larger the asset one has accumulated, the larger the risk of capital loss if inflation rises. This would simplify the life of central bankers since their natural taste for price stability won't be challenged by pressure from society to stabilise output. However, anticipating that any policy-related trade-off will be investigated by economists, it is worth spelling out why ageing will not unilaterally bend society towards preferring inflation stabilisation.

The first motivation of a "greyer society" not to disregard output stabilisation completely is to preserve high participation rates. The paper of Miles, and most of the demography-based OLG models, tend to assume the full employment of the population of working age. This is actually far from reality. Europe is currently characterised by participation rates that are significantly below those of the US and Japan, especially for people over 55. The literature on the financing of pensions in Europe tends to consider that an increase in the participation rate is one of the easiest ways of limiting the burden of pensions on society. As labour is becoming a scarce resource, it is even more crucial that as many people as possible actually work, either to increase the return on capital or to pay pensions within PAYG systems.

The second reason why the preferences of society for inflation stabilisation may not go as expected by Miles is actually much more perverse. For instance, the 1996 reform of the French PAYG system led to a gradually reduced burden of pensions, by indexing them to prices. In this way, only wage-earners may benefit from productivity gains. However, people drawing pensions indexed to prices may get a larger share of output if prices rise more than wages. Paradoxically, we may expect political pressure in favour of inflation, because people drawing pensions indexed to prices become a key part of the electorate!

References

Blanchard, O.-J. and S. Fisher. 1989. Lectures on Macroeconomics, MIT Press.

Boone L., C. Giorno and P. Richardson. 1998. *Stock Market Fluctuation and Consumption Behaviour: Some Recent Evidence,* Economics Department Working Papers 208, December, OECD.

Masson P., T. Bayoumi and H. Samiei. 1998. International Evidence on the Determinants of Private Savings, *The World Bank Economic Review,* 12, 3, pp. 483-501.

Miles, D. 1999. *Modelling the Impact of Demographic Change Upon the Economy,* The Economic Journal, 109, 452, pp. 1-37.

Mojon, B. 2000. *Financial Structure and the Interest Rate Channel of ECB Monetary Policy,* ECB Working Paper No. 40.

Aging as a challenge for economic policy Panel Discussion

Claus F. Hofmann

1 Objectives of old-age provision

The objective of the German system of old-age provision is to secure an appropriate standard of living for older people which is geared to the income they used to earn. The statutory pension insurance is the most important pillar of old-age provision. It combines the principle of fair contributions with the principles of solidarity and social equity in a way that has proved its worth. The demographic development and ageing of society show that solidarity also includes fair burden-sharing between the generations. Economic growth, a high level of employment and sound public finances constitute important general conditions for safe old-age pensions.

2 The challenge

2.1 Demographic change

In Germany, just as in other industrial states, the population is ageing: the share of older people in the population is rising, while the share of younger people is declining.

Out of 100 people in Germany, currently about 21 are children and young people, 63 are aged 20 to 65, and 16 are 65 and older. According to the most probable variant of the latest model calculations of the federal government, the share of people aged 65 and older will rise from today's 16% to 30% by the year 2050 (assuming continuously low birth rates, distincly increasing life expectancy and an average net immigration of 200,000 foreigners per year). The elderly dependency ratio (established as the numerical ratio of persons aged 60 and older to persons aged 20-60) came to 37% in 1970, has increased to 40% today, after some fluctuations, will rise to 70% by the year 2030 given the assumptions explained above, and will climb to as much as 76% by the year 2050. Even if we succeeded in increasing the effective pension age from the current just under 60 years of age to 65 years, the elderly dependency ratio, modified accordingly (numerical ratio of persons aged 65 and older to persons aged 20 to 65), would still increase from 25% today to 47% by 2030 and 53% by 2050.

The reasons for the ageing of the population are the continuous increase in life expectancy and permanently low birth rates. For more than 25 years, Germany's population has been reproducing at a rate of only two-thirds of that of the previous generation. Women and men have kept postponing starting a family. In 1998, women in Germany were 28 years old on average when getting married for the first time (1991: 26 years). In 1998, women had their first child at an average age of 29 years (1991: 27 years). Furthermore, the number of people not having children has increased considerably in recent years. About 30% of the 1965 cohort of women in West Germany will remain without children. Among women born in 1950, only 15% did not have children. In East Germany, where it used to be relatively unusual not to have children, this figure has soared since reunification, so that about 25% of East German women of the 1965 cohort will not have children.

While birth rates have been declining, average life expectancy has risen at the same time. In the past 100 years, average life expectancy in Germany has doubled. In Western Germany, it is currently 74 years for men and over 80 years for women (mortality chart of 1996/98). It seems likely that life expectancy will continue to rise. According to the federal government's model calculations, a further increase to 78.1 (men) and 84.5 years (women) by the year 2050 is to be expected. In the pension projections, it is assumed that the life expectancy of 65-year-old men/women will rise by 17.0 and 21.6 years, respectively. That would also correspond to the trend of increasing life expectancy Germany has experienced in the recent past.

Demographic ageing constitutes a central social challenge, for which society, the business world and politics, employment systems and social security schemes have to be prepared.

2.2 Employment pattern

In the next 20 years, it can be assumed that the employment participation of certain groups will increase due to institutional changes, changes in the attitude towards, employment (in particular, among women) and expected bottlenecks in the labour market. It is presumed that, above all, there will be increasing employment participation of women in Western Germany, of men and women under 30 years of age, and of 55 to under 65 years. The demands in terms of the flexibility of the employees will continue to grow in the future.

2.3 Key problems

In particular, among young people, there is widespread uncertainty due to the developments described above, about whether they will receive sufficient pension benefits in old age from the statutory pension-insurance scheme in spite of the high contributions they are paying in during their working life. Doubts are gaining ground as to whether future contributors will be prepared to accept the financial burden on their incomes caused by the contribution payments to the statutory pension-insurance scheme that will be required by them. In order to restore certainty and trust in statutory old-age provision, contributors of today and tomorrow need a clear indication that the financial burden on their incomes will not exceed a certain contribution rate. This requires a forward-looking reform of old-age provision that is viable over the long term. At the same time and as a result of the globalisation of markets, the need for improving the conditions for more growth and employment and for safeguarding Germany as an industrial location by curbing non-wage costs has intensified. Thus, through a viable and sustainable stabilisation of the contribution rate, not only has fair burden-sharing between the generations to be ensured, but also a contribution has to be made to limiting non-wage costs.

2.4 Scenarios

In Germany, an annual pension-insurance report describes past developments and presents the prospects for the future through projections over a period of 15 years. In addition, the federal government elaborates model calculations for the time horizon up to the year 2030. Substantial measures have been taken in the past 25 years. Without further reforms, however, the long-term development of the contribution rate in the pension-insurance scheme would still be unacceptable:

Year	Contribution rate in %	Net level of pensions in %
2001	19.1	69
2020	20.6	69.3
2030	23.6	69.4

3. How can the challenge be tackled?

3.1 Employment rate

According to the conclusions of the Lisbon summit, the federal government striving to increase the employment rate to an overall 70%, and to more than 60% for women by the year 2010. Currently, the employment rate in Germany, including minor employment, amounts to just under 67%. According to the projections of the pension model, the employment rate will rise to about 73% by 2010. The employment rate of women aged 15 to under 65 came to about 57% in April 1999 (microcensus figures). A rate distincly exceeding 60% by 2010 can be expected in this case as well.

3.2 Employment policy

Labour-market policy plays a substantial part in mobilising society's full employment potential. A decisive factor in this context is to increasing the employment rate of older people. For many years, early retirement has weighed heavily on the pension-insurance scheme. In order to correct this, the 1996 Act on Promoting a Smooth Transition to Retirement [*Gesetz zur Förderung eines gleitenden Übergangs in den Ruhestand*] constituted a significant step towards ensuring that companies could benefit from the know-how, expertise and experience of older employees for as long as possible. At the same time, the obligation to refill vacancies opened up employment opportunities to unemployed persons and apprentices. Improvements and simplifications of procedures were built into the Act on the Further Development of Old-Age Part-Time Work [*Gesetz zur Fortentwicklung der Altersteilzeit*], which was adopted in 1999 and is based on a decision by the Alliance for Jobs, Training and Competitiveness. Those improvements facilitate an extension of the group of people eligible for schemes providing for a smooth transition to retirement and have created additional incentives for the recruitment of new employees. On the other hand, it is currently out of the question to raise the retirement age to more than 65 years, given the fact that the labour-market situation is currently, and will remain, strained for years to come.

In order to increase the employment participation of women, obstacles (such as problems in reconciling family and working life, or a better recognition of the special situation of women in pension legislation) have to be systematically removed. Moreover, a correction of the gender wage gap constitutes an important prerequisite for giving women incentives to work.

In the Alliance for Jobs, Training and Competitiveness, a decision was taken on 9 January 2001 on improving the employment prospects of older employees.

Among other things, the decision aims at maintaining the employability of older people by adjusting working conditions, work organisation and the social policy framework, initiating a qualification campaign that will benefit older employees. In this context, further training offers are to be developed for this group of people, in particular, in-company training for older employees is to be extended. In addition, active labour-market-policy instruments are to be designed in such a way that they increase the employment of older workers, and duly recognise the difficulties unemployed persons still encounter in their integration into working life.

3.3 Reforming the Pension System

The solidarity-based pension system has proved its worth and remains the foundation of any old-age provision scheme in Germany that is pay-as-you-go funded and built on solidarity. The top priority is to reduce the contribution rate, and keep it stable on a long-term basis. At the same time, a sufficient pension-replacement ratio is to be ensured. The centrepiece of the measures is strengthening supplementary, (that is private and occupational) old-age provision. An adequate standard of living will be ensured and guaranteed even more in the future if the pensions of the statutory pension insurance scheme are regularly topped up by benefits from private and/or occupational old-age provision schemes. This requires substantial government subsidies, in particular for earners of middle-sized and low incomes. Strengthening the capital-funded pillars of old-age provision combines the advantages of pay-as-you-go systems with those of capital-funded schemes, and potential future risks may thus be minimized.

4. Implemented and current reforms

4.1 Implemented reforms

Reform measures in the area of pension-insurance provision have been taken over the past 25 years, more or less, in order to meet the challenges posed by an ageing population. They can be described as follows:

- Through a range of individual measures, the level of benefits was limited either across-the-board or through cuts in individual benefit components, in order to avoid, or at least contain, further increases in the pension-insurance-contribution rate. Retirement age was raised progressively from 60 and 63 years of age (women and men, respectively) to a uniform 65 years. Drawing pensions early is possible only if deductions are accepted (0.3% per month of early pension receipt). Another recent measure was the adjustment of pensions according to the rise in prices in the year 2000.

- Since 1999, the group of insured persons has increased significantly. Through new provisions on minor employment, about 4 million employment relationships were newly included in pension insurance. Self-employed persons whose status resembled that of permanent employees were also included in the statutory pension-insurance scheme.
- In 1986, provisions were introduced according to which child-raising periods are recognised as part of acquired rights, which since then have been extended both in terms of the period of time that is being credited as well as in terms of the amount of the credit.
- Furthermore, the government subsidy towards pension-insurance expenditure was increased. This measure was introduced in the context of the 1992 Pension Reform Act and has been strengthened considerably since 1998. It was ensured that benefits that were not covered by contributions were funded out of taxes. The transfer of one percentage point of the yield of turnover tax and of the revenue generated by the ecological tax reform to the pension-insurance scheme should be mentioned here. The marked expansion of the tax-funded share also has to be seen against the backdrop of the financial burden the pension-insurance system has had to cope with since German reunification, though its significance will decline over a longer-time horizon.

By means of these measures, the contribution rate has been lowered since 1998 from 20.3% to today's 19.1%.

4.2 Current reforms: core elements of the new reform in Germany

The German pension reform is currently subject to legislative procedures, and requires in part the approval of the Bundesrat (the chamber of parliament in which the Land governments are represented). The main points are the following:

- **Modified pension adjustment**

  With a return to wage-oriented pension adjustment as from 2001, it is ensured that pensioners participate in the growth of the economy, which is reflected in wage movements. Changes in taxes and charges that do not be concern old-age provision will not be included in the adjustment formula in future. This way, tax-legislation changes will no longer affect the adjustment of pensions. Since an adequate standard of living in old age can only be ensured over the long term basis by supplementary old-age provision, it is reasonable that spending on private old-age provision will be taken into account in the adjustment formula from 2003 onwards.

From 2011 on, demographic changes will be considered much more strongly in the pension-adjustment formula. The formula, modified along those lines, will ensure that both existing and future pensioners will be helping to achieve the necessary economies through a somewhat reduced growth rate of the pension level. The pension level will be between 67% and 68% in the year 2030. The contribution rate will remain under 20% up to the year 2020, and will not exceed 22% by the year 2030.

If the contribution rate exceeds the indicated values not only temporarily but permanently due to developments that cannot be predicted today, the federal government is obliged to submit appropriate proposals to the legislative bodies. The same applies if the net level of pensions remains below 67% according to forecasts.

- **Modernising the system of old-age security by building up capital-funded old-age provision**

  Building up supplementary capital-funded old-age provision on a large scale will put old-age security on a broader financial basis, permitting pensioners to maintain, in old age, the standard of living they have reached during their working lives. In order to enable employees with low and middle incomes to meet the cost of supplementary capital-funded old-age provision, the government will grants effective subsidies for the purpose of building up private old-age provision in the form of tax allowances and tax relief. Supplementary private provision is intended to complement the statutory pension scheme. The subsidies are designed in such a way that especially employees earning low incomes will enjoy the maximum rate of subsidy, thus enabling that group of people to participate in private provision schemes. Private pension savings will be phased in, in four steps, from 2002 to 2008. From 2008 on, persons saving an overall additional 4% of their gross income per year, in terms of both their own contribution and the government subsidy, will receive the maximum rate of subsidy. The maximum allowance equals an annual DM 300 for singles, DM 600 for married couples and DM 360 per child. If the amount of tax saved, when the option of tax relief is taken exceeds the allowance, the difference is credited to the taxpayer. The government will make about DM 20 billion available for this purpose.

- **Extension of child-related benefits to improve individual old-age security for women, and pension-splitting for married couples**

  The pension reform substantially strengthens individual security, especially of women who have raised children. The main provisions include:

  - The child-related upgrading of contribution periods for calculating pensions
  - Additional benefits when raising several children

- **Strengthening occupational old-age provision**

  - Occupational old-age provision will be strengthened by improving its underlying conditions.

- **Preventing poverty due to a false sense of shame**

  Often enough, especially older people do not claim social security benefits to which they are entitled because they are worried that their children may be held responsible for providing for them. This is major cause for poverty due to a false sense of shame, which the federal government wants to prevent. Therefore, the introduction of a special Act on Need-Oriented Basic Protection in Old Age and in the Event of Reduced Earning Capacity [*Gesetz über eine bedarfsorientierte Grundsicherung im Alter und bei Erwerbsminderung - GSiG*] is envisaged as part of the pension reform, according to which persons aged 65 and over who are considered to be needy within the meaning of the legislation on social assistance, as well as adults with a permanent full reduction of earning capacity for medical reasons, will receive benefits for the purpose of safeguarding their living expenses. Their maintenance claim on their children and parents will not be considered and non-recurring benefits will be consolidated into lump-sums.

4.3 Effects of the measures on old-age provision

The measures will have the following effects on old-age provision and the financing thereof:

1. Legislation in force

(excluding the demographic factor, including the Act on the reform of pensions on account of reduced earning capacity)

Year	2001	2002	2003	2004	2005	2010	2020	2030
Contribution rate	19.1	19.2	19.1	19.2	19.1	19.5	20.6	23.6
Pension level	69.0	69.7	69.7	70.2	68.1	69.5	69.3	69.4
2. Overall effect of the reform								
Contribution rate	19.1	19.1	18.8	18.9	18.7	18.5	19.6	22.0
Impact on the contribution rate	0.0	-0.1	-0.3	-0.3	-0.4	-1.0	-1.0	-1.6
Pension level	69.1	70.0	69.3	70.6	68.3	69.0	69.1	67.9

Platon Tinios

1 A view from the South

In early 2001, Greece is undergoing the political economy shock of an attempt at a thoroughgoing reform of its pension system. This reform can easily be characterized as overdue, given that one can find authoritative statements referring to the "extreme urgency" (*sic*) of such a reform which go as far back as 1958.[1] Greece, like other countries, can be characterized as attempting a *"Reform by Instalments"*.

Quite apart from the precise features of this reform, an interesting question is why is it taking so long? Why is the instalment plan so protracted? In economics we learn that, even when an economic agent insists on behaving irrationally, the *market as a whole* will still behave rationally. The budget constraint will ensure that irrational behaviour will, in a sufficient number of cases, be infeasible. Similarly, the budget constraint ensures that irrationality cannot continue indefinitely, as, sooner or later, it will run up against the budget constraint and will have to be revised as infeasible.

The inability to put into practice a reform package that was first outlined more than one generation ago, quite apart from whatever reflections there may be on political economy, signals the failure of this natural selection "rationalizing" mechanism. The reason must be sought in the existence of soft and fungible budget constraints in the Greek pension system.

The Greek economy is in many ways characteristic of the economies of the European 'South'. In demographic terms, it is characterized by low fertility, which, combined with the echo effects of past migration history, leads to the dependency ratio rising precipitously in the early decades of the century. The labour market has large SME and self-employed sectors, but is generally rigid, especially in the larger public-sector firms.

The social security system is financed on PAYG lines, but on a *sectoral* basis, by means of highly-fragmented, occupationally-based pension funds. These funds allow the formation of what has been called the "Mediterranean Welfare State"[2] - i.e. the coexistence of islands of privilege in a sea of generally insufficient

---

1 See Tinios, 2001, for an outline of the drawn-out instalments of this pension reform. Börsh-Supan and Tinios, 2000, provide an analytical overview.

2 See Ferrera, 1996.

provision. A sectorally-differentiated pensions system requires very stringent rules of operation and transparency in order to function.[3] In the PAYG context such a system is lacking. If that is combined with the possibility of shifting part of the pension burden of an occupational group onto other groups or onto consumers at large, the budget constraint becomes so fungible as to disappear completely. Thus, the ability common to PAYG pension financing to shift the burden among generations, is compounded by *horizontal* shifting between population groups at a point in time. The operation of this system can be proved to lead to secular tendencies towards increasing pension expenditure. Indeed, it may serve as an explanation of why the existence of a pension-financing problem long predates the emergence in Greece of a demographic dependency problem.

In such a pension system, the last remaining operative budget constraint, the "last line of defence" in a manner of speaking, is the macroeconomic budget constraint of the Government as a whole. It was the pressure of extreme cash problems in the early '90s, that led to the first meaningful phase of pension reform in Greece, that of 1990-2.

How will this system evolve over time? In particular, what is the likely effect of EMU membership on the mechanics of pension evolution and on the political economy of pension reform?

Operating within EMU, and, in particular, the Stability and Growth Pact, could be seen as serving to provide the 'missing budget constraint'. Such a budget constraint is necessarily general in nature, given that it constrains the *totality* of the General Government deficit. However, over time, and given the rising trends of pension expenditure, the concerns about sustainability are likely to play an increasingly important role. The SGP will limit the possibility of shifting the pension burden over time; meeting the SGP limit is bound to raise the question of the relative merits of different kinds of expenditure.

In this way, operating within EMU will gradually alter the parameters of the political economy of pension reform. This will come about, both as a result of the operation of the SGP limit, but also, in a more diffuse way, through operating in an environment of greater *overall* transparency. The changed objective situation vis-á-vis the political economy of pension reform will be felt gradually, though it will have an important cumulative effect over the coming years.

Pension reform is a long-term process which can stretch over long periods of time. In this context, the changed macroeconomic and policy environment in EMU is likely to make a significant difference, in the direction of increasing the possibility of meaningful reform.

---

3 Such rules could be provided by means of devices such as the International Accounting Standards (IAS-19), which are adopted by many internationally quoted companies. However, no equivalent disclosure and transparency obligations exist for public pension providers.

References

Börsch-Supan, A. and P. Tinios. 2001. The Greek Pension System: Strategic Framework for Reform, in R. C. Bryant, N. C. Garganas and G. S. Tavlas, eds. *Greece's Economic Performance and Prospects*. Bank of Greece-Brookings Institution.

Ferrera, M. 1996. The Southern Model' of Welfare in Social Europe. *Journal of European Social Policy*, 6, pp. 1.

Tinios, P. 2001. *Economy, Society, Pensions: Hidden Treasure?,* Papazisis, Athens.

Martin Werding

1 The process of ageing: a European perspective

An obvious question to start with is why there should be a European perspective on the problems associated with large-scale demographic ageing. The important thing to notice in this context is that a European perspective on these problems - presuming there is one - arises not only because the relevant attitudes differ across continents, but also because problems are different.

Figure 1 is meant to illustrate this based on (projected) old-age dependency ratios for the U.S., the EU-15, and Japan for the period from 1950 to 2050. Figure 2 compares the future development of old-age dependency for a number of EU countries, showing some degree of diversity. On average, the situation is expected to be much more serious in Europe than in the U.S. Projections for some European countries - in particular, Italy and Spain - are close to the extreme case of Japan. In spite of low fertility, Germany is in the range of the EU average[1], the main reason being that it has received a substantial amount of net immigration in the past and is expected to do so in the future.[2]

2 Public pensions

It is easy to understand that the process of ageing will have a strong impact on national pension systems.[3] At the same time, problems of financing for future pensions will be more pressing where pension systems are more ambitious. This is where attitudes differ a lot - regarding the role of the state for old-age provision in general and the share in retirement income to be covered by public pensions on an

---

1 This still holds if one calculates an EU-14 average, excluding Germany.

2 With zero net migration, the old-age dependency ratio projected for Germany could easily go up to about 55 % by the year of 2050 (54,7 % according to *Statistisches Bundesamt* 2000, scenario 0).

3 Similar things can be said regarding the provision of nursing care and general health care. In the latter case, however, cost dynamics may be much more driven by technological progress than by changes in age structure alone. (For evidence related to the case of Germany, *cf.* Breyer and Ulrich 2000.) In order to do full justice to at least one very prominent issue, I confine my attention to the systems providing old-age income.

individual level. In fact, the effects of ageing may be more dramatic in (continental) Europe than even in Japan. As a consequence, adjustments are needed more urgently in a European context and, at the same time, they may be harder to accomplish than elsewhere.[4]

Structural reforms of public pension systems that are currently being considered around the world - like introducing actuarial reductions in benefits in the case of early retirement, or actuarial responses to increasing life-expectancy - are clearly important. But they will not do the job of bringing European public pension schemes back into balance over the next 30 to 50 years. In fact, Germany and other European countries could do without using these options, if they succeed in driving down the overall level of public pensions by an appropriate margin.

Nonetheless, there is an important lesson to be learned from countries outside Europe, such as the U.S., Australia or New-Zealand. Actuarial solutions offer an important benchmark for what should be relevant also in the context of public old-age provision. Cushioned by generous public pension schemes, many Europeans have lost a natural sense of the simple fact that protection requires some sort of provision - against the individual risks of disability and longevity, for their own old-age consumption and for the case of survivors. The lesson is that someone has to pay for future pension claims, while liabilities passed on to future contributors cannot go unlimited. In this sense, the recent debate on pension reform in Germany may have achieved a lot in terms of strengthening this kind of sensitivity, even if the decisions that have finally been taken fall short of what is actually needed.

Today, the ratio of public pensions to German GDP is roughly 12 %. In spite of the latest reform, it can be expected to go up by more than 40 per cent during the next 35 years.[5] In order to render the German pension system more sustainable, introducing "notional individual accounts" - a key element of the pension reforms enacted in Italy (1995), Sweden (1998) and many countries in Eastern Europe - is not an option. The reason is that they are already built in the traditional form of social old-age insurance in Germany. As a consequence, other means are needed to curb the level of pension spending.

---

4 It should be mentioned that, even inside the EU, the figures presented here can be misleading with respect to where public pension schemes will be affected most seriously by population ageing. Italy, for example, adopted a remarkable reform package in 1995, rendering the situation much more sustainable than in Germany or France over the long term.

5 *Cf.* Werding and Blau (2001). Note that in terms of the overall level of future pension expenditure, the current reform is not very different from the 1999 reform act which has been assumed for these projections.

Europe is, in itself, rich in examples of individual mechanisms of how this can be achieved. Given similar problems, national compromises regarding effective reforms to be implemented differ a lot. As a consequence, what is regarded being impossible in one country may be in practice in another, thus yielding a large set of illustrative settings to be explored by international comparisons. For the case of Germany, for instance, switching from wage indexation of pensions to a French or U.S. style CPI indexation might be an option. The Ifo Institute has even thought it defensible to fix contribution rates at their current, high levels - around 20 % - and then, like the Austrians, see what happens to pension benefits.[6] In any case, those who are facing lower public pensions should be expected to engage more heavily in precautionary saving.

Regarding private provision, the road which has been taken by the new German reform and also, at an earlier stage and more distinctly, by the 1998 reform enacted in Sweden is a promising one. The cut in pension levels envisaged so far is a small one. But, for the individuals affected, it is meant to be off-set by an increase in precautionary saving. Hence, it represents a partial shift from unfunded public pensions towards more funded elements within the overall portfolio generating retirement income in Germany. As a consequence, new questions arise that neither politicians nor individuals are fully prepared to answer - for instance: what kind of assets are suitable for true "life-cycle" saving? How much regulation is needed if we move part of the old-age provision to financial markets? Should some level of private provision be mandatory? These questions have been dealt with in many other countries. Assessing international experience may be very helpful in order likewise to answer those questions in Germany and a number of other European countries.[7]

It should be noted that the partial transition from pay-as-you-go institutions to capital-reserve systems is not a matter of the higher returns earned in the latter. It has been argued before that there is no efficiency gain involved in converting unfunded pension systems into funded schemes (see Sinn, 2000 for a survey). Even the *Bundesbank* (1999) has engaged in spreading such findings around. The true reason why funding is useful is that many advanced societies are running short of human capital which pay-as-you-go schemes effectively are based on. Therefore, the generations that are currently active have to invest more in real capital and

---

6 *Cf.* Sinn and Werding (2000). While pure cost-of-living adjustments of pensions after award would reduce the benefit level from a current 70 % (of average wages) to less than 60 %, constant contribution rates would require reductions to around 50 % by the year 2050.

7 Regarding the first question, perhaps, the answer can not be entirely left to the market. Regarding the second question, however, there may be a tendency towards over-regulation in the current reform act. Regarding the third, the answer definitely should be yes if public pensions are to be cut down more drastically than they have been now.

financial markets. Ultimately, an important justification of why individuals should be expected to save more for their retirement is that they can take the resources they did not spend on children, thus causing the decline in fertility and a large part of the ageing problem, and place them on the capital market for other forms of old-age provision.[8]

3 Labour markets

Demographic ageing may lead to problems that are relevant to economic policy not only with respect to public pensions. So far, the discussion about the potential consequences of ageing has gone through several stages. Difficulties in financing future pensions are just the most visible impact, which became apparent back in the early 1980s. Since the 1990s, the effects of ageing populations on labour markets and potential shortages in labour supply have moved to the centre-stage of policy debates. Most recently, the scope of discussions has widened again, now also encompassing the broader issue of how to sustain economic growth and technological progress in ageing societies. Regarding the consequences of ageing on labour markets and aggregate growth, however, establishing hard conclusions is clearly a challenge.

Quite palpably, the increases in old-age dependency which are projected to take place in many industrialised countries during the next few decades will go along with a decline in the working-age population, while total population starts declining at a later stage in some, but not all, of the countries affected. The demographic impact on labour supply may be mitigated by higher participation rates, especially among women and older workers. But, under realistic assumptions, these tendencies will not off-set the trend in the demographic fundamentals.

---

8 The latter idea can have important implications for the way in which individual pension claims are determined. Just cutting the overall level of benefits defines liabilities for age-cohorts, not for individuals. By the same line of argumentation, therefore, unfunded pensions ultimately could be made contingent on the amount of child expenditure each individual has taken responsibility for - not, or not only, on the contributions she has made to financing pensions paid to her parents' generation. This is what the German *Bundesverfassungsgericht* has pointed out in a long series of decisions. In an international perspective, the Constitutional Court is not doing full justice to the German pension system, since it is one among the very few where child-rearing activities are reflected in terms of higher benefits to some extent at least. On the other hand, the system is still far away from fully acknowledging this point (e.g. Sinn, 1997; Werding, 1998; Sinn and Werding, 2000).

Realistic predictions regarding future labour demand are much harder to obtain. In order to assess potential changes in the labour-market situation, one would have to predict both the level and the curvature of periodic labour-demand schedules over the entire projection horizon, thus showing how employment will effectively depend on wages and institutional features, such as wage-bargaining procedures or employment-protection legislation, and taking into account the future patterns of real investment and international trade. Instead, long-term projections of labour demand are mostly confined to yielding point estimates of the level and structure of employment at some future period in time, given a large set of assumptions which need not become reality.

In any case, it is far from clear that labour markets in Europe will switch from current unemployment to excess demand at some point in the future, just because the labour force starts declining for demographic reasons. There are many causes, both exogenous and endogenous ones, why labour demand could go down as well - for instance, because demand for non-tradeables decreases with lower population or because the countries affected are losing competitiveness at an international level. In the absence of any structural reforms on labour (and goods) markets, additional shortages could be hypothesized to arise in the labour-market segments for some professions and qualifications.[9] Mismatches may become more pronounced than they are today. But the situation need not be fundamentally different.

4 Productivity and aggregate growth

If long-term projections for European labour markets are difficult to make, predictions regarding the impact of ageing on overall economic activity are truly impossible. In contrast to the caution regarding future labour-market outcomes, one aspect should be stressed here. Contrary to a wide-spread belief, there is no clear evidence that older members of the work-force are less productive, or less creative, than younger people. Research in this area is limited by the fact that a proper concept of how to measure individual productivity is lacking. Broadly-based displacement of older workers, such as has been observed in many European countries during the last decade, need not be taken as indicating human resource management which is actively selecting among different age-groups.[10] Rather, it may reflect one socially acceptable way of reducing firm-level

---

9 *Cf.* Munz and Ochel, 2001.

10 There is no evidence that decisions to hire or fire become more selective with respect to age in labour market segments for highly-skilled workers, or in sectors which are very research-intensive. (For the entire context, *cf.* Hofmann and Meier, 2001, ch. 5.)

employment as a whole. In addition, it may have been encouraged by the third-party payments involved in generous early retirement programmes.

On the other hand, it is highly likely that the ways in which elderly people are productive and the ways in which they create innovations are different from those of younger workers. Therefore, new management techniques may be required to exploit their capacities, and the organisation of production processes may have to adapt. So far, too little is known about these aspects. One of the most important steps that have to be taken in ageing societies is to address these problems in good time and, if necessary, to respond to them quickly. Unfortunately, the latter may not be an area of competitive advantage for many European countries.

5 Conclusions

As indicated before, the direct impact of demographic ageing on existing pension schemes should be tackled directly by means of pension reforms. Among other things, this is true because there should be closer checks on non-wage labour costs before we enter a period of time when the flexibility of labour markets can be expected to be of paramount importance. For the indirect, and less obvious, effects of ageing on labour markets and on economic growth, there is no recipe that is similarly clear. In addition, the role of policies may be rather limited here. Perhaps, the most promising path is to liberalise markets in order to ease adaptations of any kind.

This can best be illustrated by an example of what should *not* be done. Currently, Germany is considering the options for installing an explicit immigration policy. Migration is generally expected to play some role in solving the problems of demographic ageing.[11] Nevertheless, there is wide agreement that labour mobility should be restricted until around 2010 because of current unemployment, while it should be encouraged in the following years, *i.e.* when the

---

11 First of all, this is true of aggregate growth. It is a standard result of migration theory that, given flexible labour markets, there are gains from migration for both the sending and the receiving country. Regarding labour markets in the host country, this may go along with higher pressure on wages, at least in some labour market segments. On the other hand, the burden imposed on domestic workers can in theory be offset out of the "immigration surplus" falling on the same country. Permanent immigration will also have a beneficial effect on unfunded pension systems that are "intra-generationally fair" (Sinn 1997), while it would be detrimental to the pension scheme in the immigrants' home country. Things are different with temporary migration, or in the case of "welfare migration" that is induced by redistributory welfare programmes.

demographic decline in the labour force gains momentum. Judging by our prior considerations, both these approaches appear to be inappropriate.

The true lesson is that, without much delay, labour markets have to be opened up to competitive forces through decentralised wage-bargaining, deregulation, and revisions of current welfare programmes. In addition, they should be made accessible (and attractive) to international migration as soon as possible, thus taking the full challenge of globalisation.[12] Otherwise, the combination of higher direct investment abroad and expanded international trade may have much the same effects on domestic wages as in the case of more competition on labour markets, while the long-term level of economic activity would be much lower than in the alternative scenario.

References

Breyer, F. and V. Ulrich. 2000. Gesundheitsausgaben, Alter und medizinischer Fortschritt: Eine Regressionsanalyse, *Jahrbücher für Nationalökonomie und Statistik,* 220, pp. 1-17.

Deutsche Bundesbank. 1999. Möglichkeiten und Grenzen einer verstärkten Kapitaldeckung der gesetzlichen Alterssicherung in Deutschland. *Monatsberichte* 12/1999, pp. 15-31.

Eurostat. 2000. *Revised Long-Term National Population Scenarios for the European Union,* final report prepared by Statistics Netherlands, Luxembourg and Brussels.

Hofmann, H. and V. Meier. 2001. *Beschäftigungseffekte und demographischer Wandel: Eine Literaturstudie*, ifo Institut: München.

Munz, S. and W. Ochel. 2001. *Fachkräftebedarf bei hoher Arbeitslosigkeit*, ifo Institut: München.

OECD. 1988. *Ageing Populations: The Social Policy Implications*, OECD: Paris.

Sinn, H.-W. 1997. The Value of Children and Immigrants in a Pay-as-you-go Pension System: A Proposal for a Partial Transition to a Funded System, *NBER Working Paper* No. 6229.

Sinn, H.-W. 2000. Why a funded system is useful and why it is not useful, *International Tax and Public Finance,* 7, pp. 389-410.

Sinn, H.-W. and M. Werding. 2000. Rentenniveausenkung und Teilkapitaldeckung: Ifo Empfehlungen zur Konsolidierung des Umlageverfahrens, *ifo Schnelldienst* 53 (18/2000), pp. 12-25.

12 In fact, access to labour markets in Western Europe *will* be given to EU nationals from a number of Central and Eastern European countries following EU enlargement. Given the principle of free mobility for the EU labour force, mechanisms will have to be devised which make sure a meaningful self-selection among potential migrants.

Statistical Office of Japan. 2001. *Statistical Yearbook of Japan* 2001.

Statistisches Bundesamt. 2000. *Bevölkerungsentwicklung Deutschlands bis zum Jahr 2050. Ergebnisse der 9. koordinierten Bevölkerungsvorausberechnung*, Statistisches Bundesamt: Wiesbaden.

U.S. Bureau of Census–online, http://www.census.gov.

Werding, M. 1998. *Zur Rekonstruktion des Generationenvertrages*, Mohr–Siebeck: Tübingen.

Werding, M. and H. Blau. 2001. *Auswirkungen des demographischen Wandels auf die öffentlichen Finanzen: Modellrechnungen für die staatlichen Alterssicherungssysteme*, ifo Institut: München.

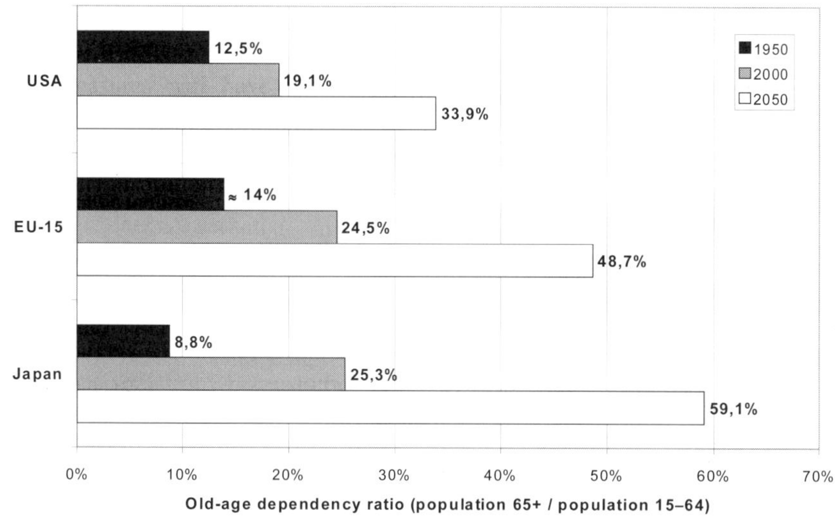

Figure 1: Old-age dependency around the world (1950–2050)

Sources 1950: OECD 1988. 2000–2050: U.S. Bureau of Census-online, Eurostat 2000, Statistical Yearbook of Japan 2001.

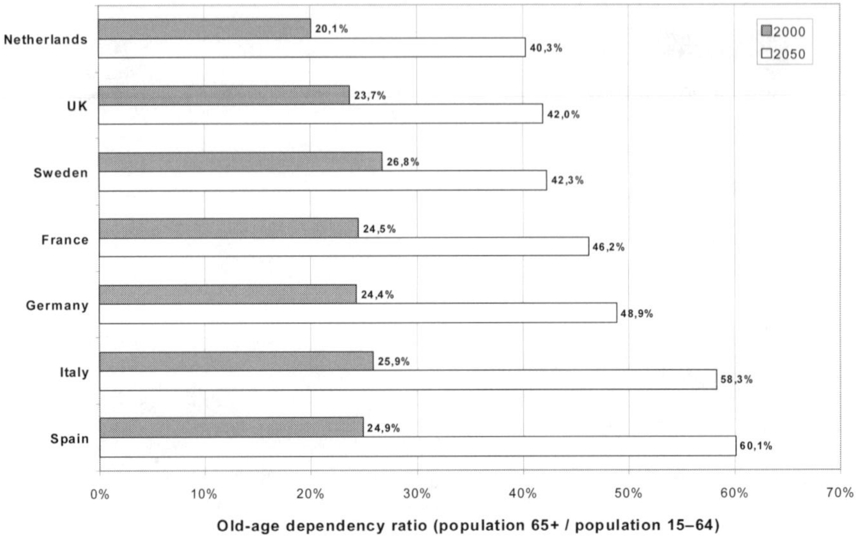

Figure 2: Old-age dependency in the European Union (2000–2050)

Source Eurostat 2000.

David A. Wise

There seem to me to be two overriding problems faced by ageing societies. One is the financing of social security programs. The other is paying for health care. The first problem has arisen because, under pay-as-you-go social systems, governments have over-promised. That is, they have made promises that cannot be kept. It is clear now that financial commitments made in the past are not sustainable.

The second problem faced by ageing societies is health care. In my view, this issue arises primarily because of advances in medical technology. (More older persons means more care as well, but the bulk of cost increases come from greater expenditure over time on persons of the same age.) Better medical technology, of course, is not bad. It will improve health care in the future. But more advanced technologies also cost more. Like computers, that do more per dollar of cost, medical care will also do more per dollar of expenditure. Even persons spending their own money would likely be willing to allocate more to health care because they get more for the money. But only a few instances are individuals paying their own money for a given episode of health care. Care is paid for through public or private insurance. Even countries that limit health-care expenditures through restrictions on supply, will, I believe, face increasing pressure to provide more care at higher cost.

It is important to keep in mind that these are problems for the current young, not the current old. Both problems combined mean that the young in all countries will have to save more; they will have to consume less than would otherwise be possible. In my view, in adopting policies to address these problems, the guiding principle should be this: while repairing the financial imbalances caused by social security over-promises, and while preparing for the greater health-care expenditures that better medical technology will require, what policies minimize the reduction in the consumption of the young?

Now let me turn to some comments on social security reform. In part, the over-promises were made because governments did not adequately anticipate demographic change, the ageing of societies. But perhaps just as important, social security provisions themselves often provide enormous incentives to leave the labour force early; work is penalized. These effects are detailed in Gruber and Wise (1999). Because of ageing, a larger fraction of the population is receiving social security benefits, compared to the fraction of persons in the labour force and paying for the benefits. Earlier retirement exacerbates the problem caused by ageing populations, taking an even greater share of the population out of the work-force.

What of reform? Again, I emphasize my view that the goal should be to bring social security programs into financial balance with the least effect on the consumption of the young. There are three categories of reform. (1) In my view, the first should be to remove the program provisions that penalize work at older ages. Removing the incentives to retire early should be a feature of any reform that maintains a defined-benefit component. (These incentive effects are automatically eliminated to the extent that defined-contribution systems are adopted.) (2) Incremental reforms can be adopted, while maintaining the unfunded pay-as-you-go system. Perhaps the most important of these is increasing the benefit-eligibility ages, especially the age at which benefits are first available. (In some countries, this is the social security early retirement age. In other countries, it is the eligibility age for "disability"-benefits, or for special unemployment benefits, that effectively provide early retirement benefits before the social security early retirement age.) (3) Fundamental reform is the third category. Perhaps the most commonly discussed fundamental reform is funding, typically through the adoption of personal social security accounts.

In many countries, it is likely that reform will involve all three categories. I will not try to evaluate the relative merits of each here. But I believe it is important to approach reform with a balanced understanding of risk. We often hear concerns about the market risk of personal accounts. This risk is often emphasized, with the implicit assumption that current pay-as-you-go systems present no risk. In fact both risks are important. The increasing recognition that current systems are not sustainable and the retrenchment evident in many countries highlights, the political risk of over-promise, or perhaps the "risk" of lower death rates and increasing life expectancy. Young persons who might have counted on the same benefits as current retirees have now sustained an important negative shock.

Now let me make a few comments about the private sector. In some countries, most retirement benefits are provided through the public sector. In others, the private sector is equally important. When I say that the young should plan to save more, I have in mind both public and private saving. And I believe that the need for increased private saving will likely be important in all countries, even those where retirement income now comes largely through the public sector. First, let me emphasize that personal saving, as measured by National Income and Product Accounts, may well present an inadequate and even misleading understanding of saving for retirement. A detailed exposition of this issue is presented in Poterba, Venti, and Wise (2001).

How can private saving be encouraged? The first way to encourage private saving is to avoid over-promises with respect to social security benefits. Persons who regard these promises as reliable will save less on their own. And, persons must not be discouraged from working and earning more at older ages, a partial substitute for saving more. Then I think we must turn aside from conventional theories of saving that are based on rates of return and life-cycle conceptions of saving. Conventional theories leave most of saving behaviour unexplained. In my view, more recent "behavioural" theories of saving are likely to help much more. In particular, I believe that institutional arrangements for saving are extremely important.

Here are some examples. The Individual Retirement Account program was very important in the United States between 1982 and 1987. Under this programme, contributions to the IRA are not taxed when made; tax is paid when funds are withdrawn from the account. Between 1982 and 1987, perhaps 16 percent of families contributed to the accounts each year. The 401(k) plan in the United States is economically almost the same as the IRA. Contributions are not taxed. But this program is run through employers. Contributions are by payroll deduction. Under the payroll deduction arrangement, the contribution rate is about 80 percent, compared to 16 percent under the IRA program. Recent evidence shows that under 401(k) plans with a default automatic contribution the contribution rate is over 95 percent. That is, if an employee has to choose <u>not</u> to contribute, almost all do contribute. The Registered Retirement Saving Program in Canada has been in existence for several decades. The program is economically equivalent to the IRA program in the United States. But contributions can be made by payroll deduction. In addition, in Canada, each tax-filer now receives a letter each year from the Minister of National Revenue advising the recipient of the allowable RRSP contribution for the year. Although this practice was adopted because of the complex Canadian contribution limits, it is evident that the practice may also serve to promote saving under the program. About 50 percent of eligible <u>persons</u> now contribute to the RRSP, as compared to 16 percent of <u>families</u> who participated in the IRA program.

Next let me emphasize that, in discussions of the reform of public programs, it seems to me important to monitor what is happening in the private sector. In many countries, the social security system is the principle source of retirement benefits. In other countries, the social security system is only one of the important sources of retirement support. In the United Kingdom and the Netherlands, for example, employer-provided pension plans are a key source of benefits. The same is true in the United States. In these countries, some retirees receive defined benefits in the form of annuities from employers as well as from the social security system. Indeed, in some countries - like the United States - the employer-provided benefits are often integrated with social security benefits. The early-retirement-incentive effects of these private defined-benefit plans are very similar to those present in social security systems. Indeed, social security and employer-provided defined-benefit systems may interact in important ways.

The most dramatic change in retirement saving in the United States is the conversion to individual retirement-saving plans (Individual Retirement Accounts, 401(k) plans, and other), which are essentially the same as a voluntary version of personal social security accounts. Individuals must decide how much to contribute to the accounts, how to invest the contributions, and how to withdraw funds after retirement. In 1980, almost 92 percent of pension-plan contributions were to traditional employer-provided plans, and about 64 percent of these contributions were to conventional defined-benefit plans. Today, over 60 percent of contributions are to personal retirement accounts, including 401(k), IRA, and other plans. Including employer-provided (non-401(k)) defined-contribution plans, about 85 percent of contributions are to plans controlled in large measure by individuals. It is perhaps not an exaggeration to say that the personal control of retirement saving is progressing more quickly than any resolution of the debate about Social Security personal accounts. Universal 401(k) coverage would, indeed, look much like a social security system with personal accounts. As I mentioned above, these personal accounts have none of the early-retirement incentive inherent in defined-benefits (public or private) systems.

To close, I have two short comments. First, it is important to ask how countries so consistently made social security promises that can't be kept. A detailed answer is not possible here, and would have many parts. But an important part must be the way that public accounting is done. In many instances, it is easier to make promises to commit future revenues that are not adequately and transparently accounted for in current accounting practices. Second, and perhaps more closely related to central bank policies, in some countries an increasingly large fraction of households participate in equity markets. Central bank policies may be swayed to accommodate a larger fraction of the public in a way that might not occur when only a small fraction of persons have equity investments. Whether such a change would be good or bad is unclear.

References

Gruber, J. and D.A. Wise. 1999. *Social Security and Retirement Around the World,* University of Chicago Press.

Poterba, J. M., S. F. Venti and D. A. Wise. 2001. The Transition to Personal Accounts and Increasing Retirement Wealth: Macro and Micro Evidence.

How pension reforms are changing financial systems
Panel discussion

Joseph Bisignano

It is with some hesitation that I embark on a discussion of how pension reforms are changing financial systems, particularly in a short commentary. The first reason is that pension reform is an extremely complex issue. As Orszag and Stiglitz (1999) have noted, while pension reforms are needed in many countries, it is unwise to think that the same set of reforms is optimal for all countries. Hence, a serious analysis would have to look at the exact set of reforms under consideration, in particular, the issues of privatisation, prefunding, diversification, and the choice between defined-benefit versus defined-contribution programs. Secondly, the structure of financial systems in the developed countries, while converging to some degree, are both evolving rapidly and yet still quite different. Pension reforms may have little impact on some financial structures and enormous consequences for others. And thirdly, among experts on pension reform there is considerable debate over the effect of pension reform on private saving and financial development. The darling of all pension restructuring appears to be the Chilean 1981 reform program. Although most researchers have found that the impact of the reforms has been to raise private saving, Gray and Weig (1999, p. 21) suggest that even here there is evidence that the impact may have been negative.[1] There is still a lot of uncertainty over the impact of pension reforms on private saving, and the distortions eliminated or created by moving from public to private programs. In short, what I say here should be read as controlled conjectures rather than strongly held opinions.

When we talk loosely of pension reform, I assume most of us have in mind a move from what the World Bank (1994, Chapter 7) describes as Pillar 1 (a public pillar: a defined-benefits, pay-as-you-go, public pension system) to Pillar 2 (a private pillar: a mandatory, defined-contribution, fully-funded system, privately managed but publicly regulated,) combined with some aspects of Pillar 3 (a voluntary pillar: occupational or personal saving plans). Many of us may also have in mind several assumed benefits of Pillar 2 over Pillar 1 systems, and the

---

1 Holzmann (1997) reports that the impact of the Chilean pension reform in 1981 - from an unfunded to a funded system - was negative until 1989 and then small after that. The World Bank (1994) study concludes that funded pension plans have the potential to increase private saving, but not pay-as-you-go plans.

problems with the latter. These include, using Orszag and Stiglitz's macroeconomic classification: "individual accounts raise national saving; rates of return are higher under individual accounts; declining rates of return on pay-as-you-go systems reflect fundamental problems; and investment of public funds in equities has no macroeconomic effects". Orszag and Stiglitz provide arguments why all four of these might best be called "myths". Indeed, they argue that a Pillar 2 private managed-fund approach may not be best for many countries.

I mention the difficulties of reaching broad conclusions on the macroeconomic impacts of Pillar 2 as a way of cautioning us from thinking that we can be very sure what specific impacts pension-fund reform may have on financial systems. At the same time, the careful study by Holzmann (1997) on the impact of the 1981 Chilean pension reform suggests that the consequences for the financial system may be substantial. He found that the development of funded pension schemes in Chile, together with supporting legislation in related areas, appeared to contribute considerably to the development of financial markets. By 1994 pension-fund assets amounted to about 40 percent of outstanding financial instruments and roughly 41 percent of GDP. Qualitatively, funded pension funds contributed to the development of a wide variety of new financial instruments, a diversified investment menu and in general a much more sophisticated financial system. What should be noted in his conclusion on the role of pension reform in contributing to economic growth in Chile is the importance of a comprehensive micro- and macroeconomic programme aided by a tight fiscal policy. Pension-fund reform may indeed make a major contribution to financial development and growth if combined with needed institutional structural reform in other areas.

Financial innovation and development does not necessarily require major institutional reform. Private fully funded pension funds, which have been around for some time can make a major contribution to economic growth. In recent years we have seen the role U.S. venture-capital funds have played in the financing of young firms in the technology sector. As much as two-thirds of the new investment of venture-capital funds has gone to technology-based firms. What may be found surprising is the evidence of Black and Gilson (1998) that, for the ten-year period up to 1995, about 40 percent of the capital raised by venture-capital funds came from pension funds. Pension funds have contributed greatly to the development of the "new economy" and the increase in U.S. productivity in the second half of the 1990s. At the same time, they might also be conjectured to have contributed to the bubble in technology stocks during this period.

This example is not meant merely to serve as an illustration of one of the possible contributions to the economic development of a financial system with a thriving private pension-fund scheme. Rather, it is to show that the contribution of pension funds to the financing of young technology firms via venture-capital funds depends on another important link in the financial system, a vibrant equity market. Black and Gilson emphasise that what is vital in the U.S. venture-capital market, assisted by pension funds, is the ability of venture-capital firms to exit from a successful investment in a start-up firm with the use of an IPO, via the equity market. Venture-capital investors provide not only capital but also monitoring and advisory services. Their ability to exit via an IPO allows them to recycle their capital and expertise to other start-ups. This would not be possible, they argue, if the financial system were bank-centred. The close link between the private and the public equity market, as in the U.S., permits both financial capital and venture-capital investor-monitoring skills to be used efficiently. It is the flexibility to reemploy capital which equity markets provide venture-capital firms, with healthy funding by pension funds, which was the central element in the financing of the U.S. technology boom during the 1990s. Pension funds were an important link in the financial food chain whose contribution would arguably been much less had the U.S. had a bank-centred financial system. This is an example of the difficulty of speculating on the impact of pension-fund reform on financial systems. We need to know the initial conditions of the financial system and the direction of recent changes.

Looking at the financial landscape broadly, the initial conditions in terms of pension assets in the major industrial countries are quite striking. Culhane (2001) of Goldman Sachs estimates that, of the $13 trillion in pension assets in 1999, the U.S., U.K., and Japan held 82 percent. Germany, France, Italy, and Spain together held only 2.2 percent of total world pension assets. Given that European governments with largely unfunded pension liabilities are facing enormous future contingent pension liabilities, Goldman Sachs expects continental Europe to have the strongest demand for pension assets, in particular equity. This growth will of course depend on pension legislation, which to date has been slow to arrive in France and Germany. With pension funds and "comparable facilities" equivalent to only 3.3 percent of GDP in Germany, according to figures provided us by Jürgen Pfister, compared with 85.6 percent in its forward-looking neighbour the Netherlands, it is not surprising that securities firms and insurance companies are expecting rapid growth in pension assets in the Federal Republic.

If private European pension funds are to build up high-yielding equity portfolios they will need a source of supply. At present, in some countries the domestic supply is quite limited. As Desario (2001) reports, there is very limited availability of shares in Italy because of the reluctance of Italian firms to share corporate control and to yield to disclosure standards for exchange listing. Yet the appetite for shares in Italy is considerable. Last year, Italian investment funds' holdings of foreign shares were three times their holdings of Italian shares, while they were less than their Italian share portfolio as recently as 1996. An equity culture is developing in Italy, but it is being largely satisfied with foreign assets. This helps explain the enormous net purchases of U.S. shares by Europeans in recent years, and the recent level of U.S. share prices.

The necessary development of an equity culture, together with the relaxation of pension-fund investment restrictions and greater portfolio diversification, will likely lead to major changes in financial structure in European finance. How it will come about will be interesting to observe. In a number of European countries, firms have typically been primarily financed with internal funds and debt, mostly bank debt, with equity playing a modest role. As we have seen in Germany, where only ten percent of the population are said to own shares, developing an equity culture can be a passionate and volatile affair. The new German equity culture was given a rude setback when the Nexam 50 Index on the Neuer Markt fell by 70 percent from its peak to early this year, compared with the 45 percent decline in the Nasdaq. Such volatility can quickly dampen one's appetite for equity, at least for a time. One consequence of the fall has been for the Deutsche Börse to require greater firm disclosure, illustrating a small but important aspect of the interrelationship between corporate financing, corporate governance and public disclosure in developing capital markets.

The potential impact of pension-fund reform on the growth of institutional investment and the risk involved in this transformation are nicely illustrated in the paper written for this conference by Phil Davis. One of the concerns often mentioned in conjunction with the growth of securities markets, cross-border investment, and the large and growing financial resources available to institutional investors is the danger of "herding" by institutional investors, as was recently seen during the Asian financial crisis and in the aftermath of the Russian crisis.[2] Given similar information, risk-control techniques, investment decisions and uncertainty, it is not surprising that portfolio managers have often been seen to take herd-like investment decisions. This remains a serious concern with regulatory authorities, central banks and international institutions.[3] The recent behaviour of the Neuer Markt may be one such illustration in Europe.

As Europe looks to encounter a significant transformation of its financial systems, in part induced by pension reform, it might ask whether it will be an encounter of the "third kind". Since the late 1980s, two countries with very different financial systems, but with similar underlying economic conditions, saw the rise of spectacular asset-price bubbles and equally spectacular collapses: Japan in the second half of the late 1980s, and the U.S. in the second half of the late 1990s, both saw the emergencies of bubbles. The recent financial transformation in the U.S. was the direct result not of deregulation but of innovation, the continued growth of institutional investment, and a rapacious appetite for equities by individual investors. Japan, in contrast, saw many of its problems result from the attempts by banks to recover ground lost when corporations sought to finance themselves in the capital markets and not by their traditional reliance on commercial and long-term credit banks. In both cases, inflation was low during a period of rapid private credit growth. The bubble in Japan was in both equities and real estate, but in the U.S. was restricted to the equity market. The equity-market bubble in both countries was also inflated by the strong growth in business fixed

---

2 For a nice review of the herding literature, see Bikhchandani and Sharma (2000).

3 See Eichengreen et. al. (1998) on market dynamics in this new financial environment.

investment and an assumed partly permanent increase in productivity. And in both cases, the combination of low and stable inflation and monetary ease arguably played a role in the formation of the bubble.

A question facing Europe is whether pension reform and the growth of institutional investment could potentially unleash a strong demand for longer-term financial assets, in particular equities, and contribute to an asset-price bubble. Will pension reform contribute to a greater availability of credit? It may pay to ask what unintentional fallout there might be from pension reform which might complicate the life of central banks and supervisory authorities. Another question that might be asked is whether the strong and singular focus on low inflation could detract attention from a boom in credit growth and the strong demand for financial assets by individuals and institutional investors that might result from pension reform.

The asset-price bubbles coupled with over-investment cycles which hit Japan in the 1980s and the U.S. in the 1990s should make us stop and ask whether low inflation alone will be sufficient to secure financial stability in Europe during a period of potentially substantial and rapid financial change. It should be recalled that during the 1930s Fredrick Hayek and Gottfried Haberler both cautioned that inflation was not an appropriate target for central banks and that the business cycle would not necessarily be dampened by the arrival of low and stable inflation. While low inflation was important for economic stability, they argued that it was primarily the excess growth of credit which induced changes in relative prices and fluctuations in business activity. As Japan learned, a period of prolonged monetary ease might result in a major credit cycle and asset-price bubble without any serious rise in the prices of goods and services. The impact on credit growth from structural change in the financial system may be unexpectedly large. Singular focus on price stability by the monetary authorities might detract attention from the damaging effects the build-up of an asset-price bubble can potentially have.[4] Europe might wish to recall the conclusion reached by Yutaka Yamaguchi (2000), Deputy Governor of the Bank of Japan, when looking back on the economic experience of Japan during the 1990s: "A key question for a central bank is whether swings in economic activity can be avoided only if price stability is maintained by appropriate monetary policy, ... by looking at the experience of Japan, my preliminary answer is, to my regret, No".

To conclude, if pension reform is to unleash significant change in the structure of European finance, and create a sizeable demand for longer term financial assets, European central banks may wish to ask whether their information variables or intermediate targets of monetary policy should be altered. Specifically, they may wish to be on the lookout for "excessive" credit growth and the formation of

4 See Bernard and Bisignano (2001) on the relationship between monetary policy and asset-price bubbles in Japan and the U.S.

incipient asset-price bubbles. As we have seen in both developed and developing countries, inappropriate monetary policy during a period of structural change in the financial system could result in big surprises. And the largest changes may result from disruptions caused by excesses in credit growth and unsustainable asset returns. Few countries have avoided difficulties from major changes in financial regimes, such as pension reform might produce in Europe.

References:

Bernard, H. and J. Bisignano. 2001. Bubbles, Crashes and Failed Paradigms in Central Banking. Paper prepared for the session *Asset Price Bubbles: Implications for Monetary and Regulatory Policies*. Western Economic Association Meetings, San Francisco, 5-7 July 2001.

Bikhchandani, S. and S. Sharma. 2000. *Herd Behaviour in Financial Markets: A Review*. IMF Working Paper No. 00/48, March.

Black, B. S. and R. J. Gilson. 1998. Venture Capital and the Structure of Capital Markets: Banks Versus Stock Markets. *Journal of Financial Economics*.

Desario, V. 2001. Financial Savings in Italy: Instruments, Intermediaries and Markets. Address at the conference on *The New Economy and its Impact on Finance and Insurance*. Università Luiss-Guido Carli, Rome, 10 January.

Gray, C. and D. Weig. 1999. *Pension System Issues and Their Relation to Economic Growth*. Harvard Institute for International Development, CAER II Discussion Paper No. 41, July.

Culhane, M. M. 2001. *Global Economic Commentary: Capital Market Implications of Global Aging*. Goldman Sachs, 28 February.

Eichengreen, B., D. Mathieson, B. Chadha, A. Jansen, L. Kodres and S. Sharma. *Hedge Funds and Financial Market Dynamics*. International Monetary Fund, Occasional Paper No. 166.

Holzmann, R. 1997. *Pension Reform, Financial Market Development, and Economic Growth: Preliminary Evidence from Chile*. International Monetary Fund Staff Papers, June.

Orszag, P. R. and J. E. Stiglitz. 1999. Rethinking Pension Reform: Ten Myths about Social Security Systems. Paper presented at the World Bank Conference on *New Ideas About Old Age Security*, 14-15 September.

World Bank. 1994. *Averting the Old Age Crisis*. Washington, D.C.

Yamaguchi, Y. 1999. Asset Price and Monetary Policy: Japan's Experience. Remarks made at the symposium sponsored by the Federal Reserve of Kansas City, Jackson Hole, Wyoming, 26-28 August.

Hans Blommestein

Eight implications of population ageing and pension fund growth for the financial sector

1 The Growth of Institutional Assets is a Push for the Modernization of the Capital-Market Infrastructure and Financial Innovations

The first implication of the ageing of populations in the OECD area and the associated growth of pension-fund and other institutional assets is part of the push for the modernization of the capital-market infrastructure.[1] Pension-fund growth is leading to an increased demand for professional-fund-management services. Professional portfolio managers, in turn, have an important influence on financial markets through investment and trading strategies. Pension funds and other financial institutions involved in the retirement-income industry are also pushing for changes in laws and regulations that would contribute to the modernization of the capital-market infrastructure. Pension funds and other institutional investors can also be expected to demand better corporate governance practices.

The process of financial innovation has been driven strongly by the growth of the institutional sector.[2] Sophisticated trading arrangements and investment techniques have been developed in response to the needs of pension funds and other institutional investors. The growing importance of institutional investors is also generating an increasing demand for risk-transfer techniques, that enable the investor to choose the desired combinations of return and risk. Such techniques include both securitisation, which enables the investor to transfer the credit risk as well as the market risk, and derivatives, whereby market or price risk is reallocated among participants. A more recent development is credit derivatives, which enable market participants to transfer credit risk separately. The demand for risk-transfer techniques has been strongly driven by the nature of the liabilities of the different types of institutional investors and regulatory requirements. For example, defined-benefit schemes and strict minimum-funding requirements have stimulated demand for hedging by pension funds. In order to minimize the costs of hedging, pension funds and life-insurance companies have an incentive to immunize their defined-benefit liabilities via an investment strategy of duration

---

1 See Blommestein (1998b); Walker and Lefort (2000).

2 Bodie (1990).

matching. The requirement of a fixed duration for investment instruments has stimulated innovations such as zero-coupon bonds, collateralized mortgage obligations, strips in government securities markets and guaranteed-income contracts; immunization strategies have also stimulated the development of markets for index options and futures. Fixed-duration instruments have also spurred the process of the securitization of mortgages in the case of collateralized mortgage obligations and of loans and private placements in the case of GICs (Guaranteed-Income Contracts).

2 The Efficiency of the Financial Market Infrastructure Influences the Investment Strategies of Pension Funds

The efficiency of the financial market infrastructure influences the ability of pension funds to implement asset-investment strategies in accordance with planned or desired risk-return profiles. This includes: the degree of indexing in managing investment portfolios (the more efficient markets are, the more a passive-indexing strategy seems warranted); the possibility to diversify - including the scope for cross-border investments; the ability to use risk-sharing and risk-management tools; and the use of sophisticated financial instruments and trading strategies. A well-functioning funded pension system requires a stable and efficient financial-market infrastructure, consisting of the legal framework, the financial accounting system, the regulatory and supervisory framework, clearing and settlement systems, and the micro-structure for trading securities. Most industrial countries have made considerable progress in the development of a solid regulatory and supervisory framework, although much still needs to be done. Several countries in the OECD area need to make further progress in establishing the proper legal and regulatory basis for dealing with take-overs, minority-shareholders protection, insider trading and institutional investor operations.

Weaknesses in the financial market infrastructure of emerging markets have a clear adverse impact on the investment strategies of pension funds and other institutional investors involved in the management of retirement assets. They need therefore to be addressed urgently. For example, the financial turmoil in emerging markets during the last couple of years demonstrates that lack of transparency and inadequate disclosure standards can prolong or exacerbate a confidence crisis.

3 Impact on Size and Structure of Financial Markets

Modernization and higher efficiency of the capital-market infrastructure, in turn, would promote the growth of securities markets because they become more attractive for investment by pension-funds. In other words, there is a virtuous cycle between an expanding pension-fund sector and the development of domestic capital markets. Figure 1a shows clearly that the size of outstanding institutional assets is having a positive impact on the size of capital markets.[3] The overwhelming importance of pension funds becomes clear by comparing figures 1b and 1a.[4] The long-term liabilities of pension funds makes them specifically suitable to promote the growth of capital markets. There is here a clear virtuous cycle at work. The growth of institutional assets (specifically, pension assets) increases the size of capital markets. An enlarged capital market would allow a more efficient allocation of risk. This in turn means that pension funds have a larger scope for implementing more sophisticated risk-sharing and risk-management tools.[5]

---

3 The data on assets are taken from OECD Institutional Investors Statistical Yearbook and the data on market capitalization from FIBV.

4 This conclusion is also supported by formal regression analyses. Details can be obtained on request from the author.

5 Risk-sharing tools are of crucial importance for the operation of pension systems and pension funds. See Bohn (2001).

Figure 1a. Market Capitalization and Total Financial Assets of Institutional Investors in 1988
(US$ Billion)

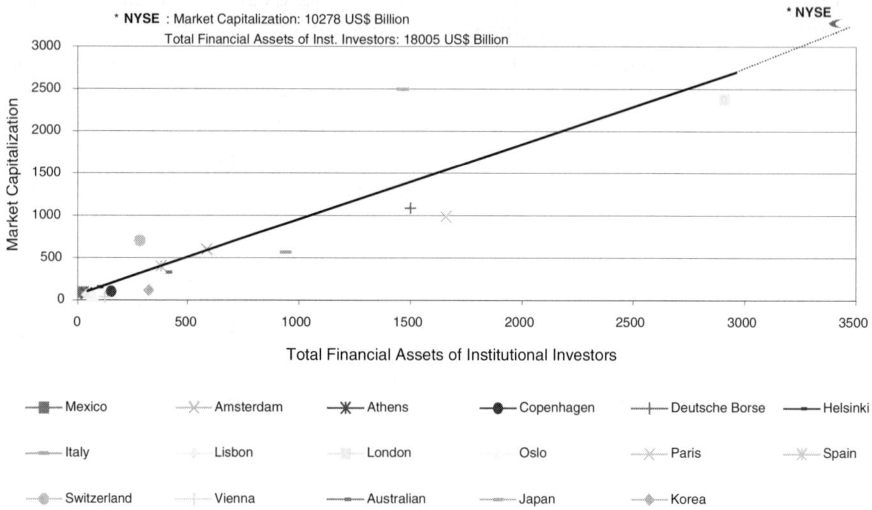

Figure 1b. Market Capitalization and Total Financial Pension Assets of Pension Funds in 1998
(US$ Billion)

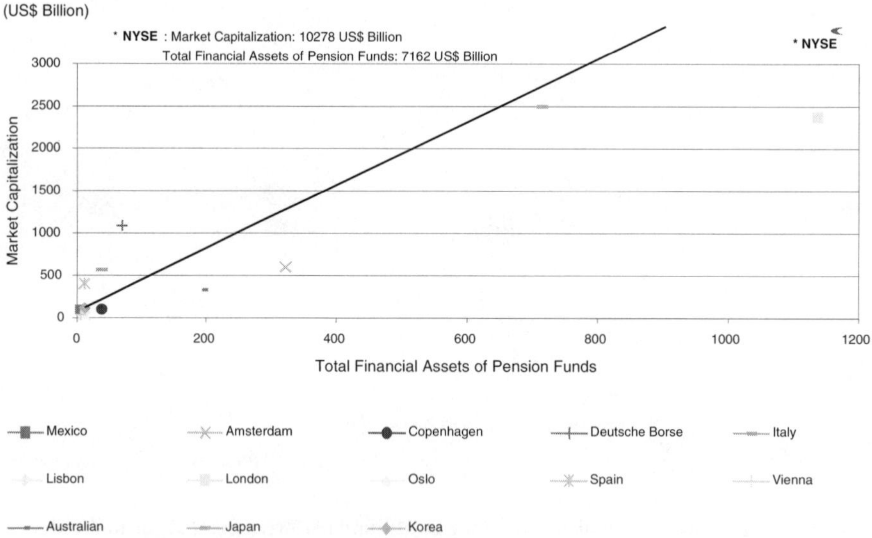

The growth of the pension-fund sector has not only had an impact on the *size* of capital markets but also on the *structure* of financial markets. The latter effect stems from the increased institutionalization of savings, which means that securities are increasingly being held by large, informed investors whose investment decisions are driven by relative asset returns. Institutional investors make rapid portfolio adjustments to changes in relative returns, which aids price efficiency and contributes to a more efficient allocation of funds and risks. This can only be done when the capital-market infrastructure is efficient. This increase in institutional savings also implies a shift in the composition of the portfolios of the household sector toward long-term assets and correspondingly higher returns. Countries with large funded pension schemes tend to have highly-developed securities markets, while equity markets are relatively underdeveloped in countries with small pension-fund sectors. The growth of a dynamic pension sector can be expected to contribute to a stronger role of capital-market intermediation. In particular, pension funds that are investing significant parts of their portfolios in equities would pressure for changes in laws and regulations of companies that usually can be found in "bank-dominated" financial systems.

4 Increased Supply of Risk Capital

As a consequence of the increased *demand* for long-term assets by institutional asset managers and households, the increased institutionalization of savings has, in turn, led to an increased supply of long-term funds, including risk capital. Given their focus on real returns, pension funds are particularly beneficial to the development of equity markets. Although pension funds could in principle also grow by investing in loans and real estate, their greatest comparative advantage is in capital-market operations. The resulting *increase in the supply of risk capital* has stimulated new businesses and job growth.[6] Increasingly, pension funds have been investing in blue chips and small caps. In addition, a part of their huge portfolios has been allocated to the private equity market, including venture-capital operations (such as high-tech projects). Reportedly, pension funds have also become significant investors in *hedge-fund* operations, by allocating part of their portfolio in a range of hedge-funds. In sum, the increase in demand for long-term assets on the one hand, and the increase in the supply of risk capital, on the other, have been important factors in changing the structure of capital markets.

---

6 Blommestein (1997).

5 Growth in Cross-Border Investments

The trend toward more investment in *foreign* assets, including emerging markets, can be expected to continue. The scale of flows from the 'ageing' mature industrial countries to the "younger" emerging markets and the broadening of market access constitute evidence that, in the 1990s, global financial market integration was rapidly increasing. With continued efforts to liberalize cross-border financial flows and to strengthen capital markets in developing countries, this trend is likely to persist as pension funds and other institutional investors continue to seek greater diversification of portfolios.

Liberalization of cross-border financial flows and the strengthening of the domestic financial infrastructure contribute to increasing the breadth and depth of financial markets across the world. These developments also facilitate greater diversification of pension-fund portfolios. Fund managers can thus improve the return-to-risk ratio of the portfolios, which would help ensure that there are sufficient funds to pay the benefits to retirees. These structural changes should also facilitate the flow of funds from savers to investors, leading to a more efficient allocation of resources and risks in the economy. As such, these developments would have a positive impact on growth and living standards.[7]

But the benefits of investments in emerging markets need to be carefully assessed against the risks… [8]**.**

Studies indicate that *international portfolio diversification* strongly enhances the power of portfolio diversification. Nevertheless, increased international diversifications, may not be as beneficial as they first appear. For example, over the last 15 years or so, the G-7 stock markets have given better returns than the emerging markets. Taking a 20-year period shows that, although investments in the US equity market over the 1975-1995 period would have given US pension funds both *higher* returns and *lower* risks than the emerging markets as a group. However, even then there could still be benefits from diversification. The asset allocation mix is dependent on the desired risk-return profile of pension funds in question. For example, in the 1975-1995 period US pension funds could have reduced their risk by giving up some return. By investing 30 per cent of their equity portfolio in emerging-market equities they would have minimized their risk. Naturally, the future might bring better news in terms of higher expected returns and/or lower risks, especially in the light of further improvements in the financial infrastructure in emerging securities markets, as well as a strengthening of the domestic institutional-investor base in emerging markets. A similar reasoning for three asset classes (US equity, international and emerging-market equities and a broader market composite) and a more recent period leads to the

---

7 See Börsch-Supan, Ludwig and Winter (2001).

8 Blommestein (1998).

conclusion that US pension funds could have minimized their risk by investing 20 per cent of their portfolios in international markets and 10 per cent in emerging markets (Figure 2).

Figure 2. Risk vs. Return: US stocks, International and Emerging Equity Markets: 1992-1997

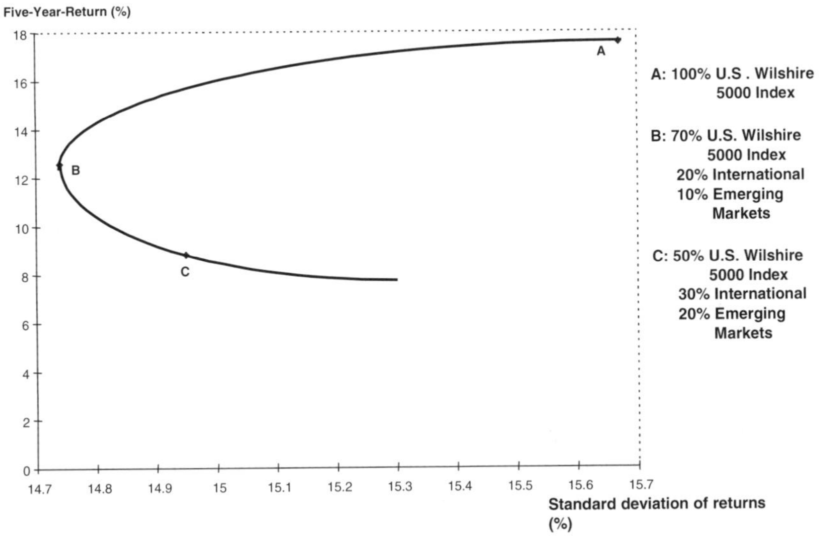

.....while the benefits of international diversification may be decreasing.

At the same time, analysts have pointed out that the benefits of international diversification may be decreasing. It is argued that increasing financial integration is leading to an increase in the correlation of returns, which reduces the potential for reducing risk through international diversification. For example, the increase in the integration of bond markets - especially in Western Europe - has reduced the potential for reducing risk on a bond portfolio through diversification. Moreover, the fact that an increasing amount of institutional money is managed using the concept of diversification is causing the benefits of diversification to get smaller. High correlation of returns between countries has in some cases led to a restructuring of portfolios by diversifying by sectors. Also, the correlations between OECD stock markets and equity markets in emerging countries have increased. Table 1 illustrates this point by showing the increase in correlations between US and emerging stock markets.

Table 1: The Benefits of Diversified Risk – Correlations Between US and Emerging Markets

	Correlation Between S&P 500 and:	
	1975-95	**1990-95**
All Emerging Markets	0.27	0.41
Latin America	0.24	0.38
East Asia	0.16	0.27
South Asia	0.24	0.23
Africa	0.07	0.03

Source: ICFA

Moreover, the fact that an increasing amount of institutional money is managed using diversification is causing the benefits of diversification to become smaller. The potential benefits of international diversification are also reduced by the fact that downside market-movements occur much more in parallel than upside ones. There are also studies that show that shocks in volatility are closely linked with rising correlations, in particular in the case of stock markets. Unfortunately, the fact that most assets seem to move uniformly during market-crash situations reduces the benefits of controlling downside risks using investment strategies based on diversified benchmarks. Although there is evidence that the risk-reducing benefits of international investments have become less powerful, studies show that they are still positive, even during sharp downside moves of securities markets.

High correlation of returns between countries has in some cases led to a restructuring of portfolios by diversifying across sectors. For example, studies found that diversifying across countries, but staying within a single industry, reduces volatility more than diversifying across industries in a single country, even though both portfolios carry the same average return.

6 Pressure on Financial Asset Prices?

There have been concerns that during the bubble period before the crash of 2000 the growing demand for high-quality private securities (equity and corporate bonds) associated with the growth of advance-funded pension systems and decreasing public-sector-borrowing requirements in many OECD jurisdictions had put *strong upward pressure on financial asset prices*. This latest boom in financial markets, in particular the stock markets in many OECD countries, may in part have been driven by population ageing and pension reform and the associated growth in pension-fund assets. As noted above, pension funds and other institutional investors also invested part of their funds in high-risk IPOs, such as new dot.coms.

Another channel that would influence changes in asset prices is the shift from government securities to private debt and equity instruments; for example, in the context of the privatization of social security systems. Critics have argued that allowing social security trust funds to invest in equities would primarily represent a reallocation of assets between those held in trust funds and those held - either directly or indirectly - by households. It could improve the financial position of the trust funds, because of equities' historically higher average returns, but for a given level of saving it would *not* increase the returns for the country as a whole. Investing a portion of the trust funds in equities would *raise the price and lower the return on equities, and lower the price and raise the return on government securities.* Higher yields on government debt would raise the interest costs and, all things being equal, the non-Social Security portion of the deficit. The analysis becomes even more complicated when one allows the possibility that the initial effects on rates of return could be moderated, as corporations restructured their finances to take advantage of cheaper equity financing, and as international buyers increased their purchases of now higher-yielding government securities.

In order to mitigate these possible price pressures, it has been proposed by some analysts to undertake pension reform (leading to an increase in demand for equity) and privatization (leading to an increase in supply) at the same time.[9] This would permit, at least over the medium-term, a somewhat more balanced development in private and public securities markets, although the impact of privatizations in the OECD area should not be exaggerated.

Some economists have argued that even without privatization, the supply of equity would increase. According to this view[10], as stock prices rise, companies realise they can raise more capital by issuing new shares. Once the increasing supply of stock catches up with heightened demand, stock prices would stabilize.

7 Impact on the Risk-Premium

In a somewhat *longer-term perspective*, population ageing and pension reform have an impact on the *risk-premium* (that is, the difference between the returns on stocks and the yield on bonds). Because asset preferences vary across age-groups, the ageing of the baby-boom generation could affect both the absolute and the

9 Merton (1992), Merton and Bodie (1995).

10 Abel (2000).

relative positions of stock and bond prices. On average, middle age is the portion of the life cycle when saving rates are highest. (This type of saving behaviour is a feature of a theoretical life-cycle model and also seen empirically in household data.) Moreover, middle-aged workers generally are more able and willing to hold a riskier portfolio,[11] that is, one weighted more heavily towards stocks than bonds.

This is a consequence of two factors: first, while still working, a stockholder is better able to make up for any bad equity returns; second, middle-aged workers have a longer time-horizon and thus are willing to accept more risk in exchange for the expectation of higher returns. In this case, the ageing of OECD populations will tend to increase the price of stocks and bonds, decreasing their rates of return. Moreover, higher demand for stocks, relative to bonds, should increase the price of stocks relative to bonds, thus *decreasing* the equity premium. It is generally held that risk aversion increases with age (holding length of life constant). Thus, some have hypothesized that an ageing population would cause the equity premium to increase. But if the age of the population is increasing, at least in part because the life-span is increasing, and thus time horizons are lengthening, then the ageing of the population does *not* necessarily imply that average risk aversion should be increasing and risk premium on stocks should be rising. In addition, it has been argued that bringing savings under the control of pension funds and other institutional investors would also reduce the risk premium because these professional investors have a longer-term horizon and a better capacity to absorb and manage risks than individuals.

After the baby-boomers begin to retire en masse in the period 2010-2030, saving rates would tend to *fall*, stock and bond prices to *decline*, and the equity premium to *rise,* as baby-boom retirees shift their portfolios away from stocks toward bonds. However, the prediction that asset prices would fall in a dramatic way (the "asset-market-meltdown-hypothesis") when the baby-boomers start to retire, has been questioned in recent research.[12] Empirical evidence suggests a very weak link between population-age structure and asset returns. According to this view, retirees don't simply reverse the process of accumulating assets during working years, and spend everything they have. Instead, they reserve some of their savings in response to *lifetime uncertainty,* and they hold additional assets for their children and other heirs (*bequest motive*). As a result, the baby boomers' demand for equity will *not* be reversed in retirement at the same speed as it grew during the working years.

---

11 The real return on United States stocks, for example, averaged 9% over the period 1947–96 with a standard deviation of 17%. This implies that there is about a 30% probability of a decline bigger than minus 8% or a rise bigger than 26% in any given year. The average real return on long-term United States government bonds over 1953-96, however, is much lower - 3% - but also less volatile; - these returns have a standard deviation of 2%.

12 Poterba (2000).

8 Impact on International Systemic Crises

The last financial implication of global ageing and pension reform for policy is that the scope for possible international systemic crises will become larger as pension funds and other institutional investors continue to diversify into international markets. One lesson from the dramatic events in the second half of the 1990s is that the abrupt loss of access by individual countries to the global capital market may continue to occur. This is due to two factors: divergent macroeconomic conditions in capital-exporting and capital-importing countries and crises in individual capital-importing countries. In such cases, the currency of the capital-importing country will be "tested" through a sustained speculative attack, leading to a sudden drying-up of capital inflows and major capital outflows. The growing participation of institutional investors in international markets and the improved access of emerging markets to the international capital market have led to the growth of highly-leveraged hedge funds, and of proprietary traders who are prepared to tolerate significant risk in their search for weaknesses in foreign-exchange arrangements.

The globalisation of financial markets, driven in part by population ageing, pension reform and other structural factors, is reflected in the quicker international transmission of short-term price movements in financial markets, as occurred in the Mexican crisis in 1994-95 and the Asian crisis in 1997-98. Financial integration has also increased the potential intensity and duration of speculative attacks. There is evidence that pension funds and other institutional-investors (including hedge funds) have played a crucial role at times in determining asset prices in emerging and advanced financial markets, with shifts in institutional-investor sentiment occasionally contributing to increased volatility in markets. The crisis triggered by the near collapse of the US hedge funds LTCM in the second half of 1998 demonstrated that also advanced markets are not immune to serious financial turmoil. Davis (2001) distinguishes two particular types of financial instability that institutional investors give rise to.[13] (1) extreme market-price volatility after a shift in expectations of groups of pension funds and related changes in the asset mix (institutional herding); (2) a protracted collapse of market liquidity and issuance. A third type is related to the use of *similar* market-sensitive risk-management tools (such as VAR) by groups of institutional investors. This interaction of herding and the wide-spread use of these tools may manifest itself as extreme market-price volatility and/or a collapse of market liquidity. The tendency to adopt market-wide performance benchmarks and risk-management tools could potentially lead to destabilising feedback effect in periods of market stress. In particular in the case of markets prone to gapping, such as high-yield corporate debt markets, this may lead to disequilibrating behaviour of market participants.

13 Davis (2001).

Against this backdrop, the key challenge to financial policy-makers is how to deal effectively with periods of financial turmoil without creating moral hazard situations. Bailing-out investors should be avoided because it would encourage excessive risk-taking. Regulatory and supervisory regimes for banks, pension funds and other institutional investors will have to be reviewed in the light of the new or stronger linkages between and among these financial institutions and the associated use of financial instruments, including derivatives. Clearly, this review will also have to include the size and changing nature of cross-border operations. In this context, the proposed changes under a recent proposal to replace the 1988 Basel Capital Adequacy Accord with a more risk-sensitive framework (Basel II) can be considered a crucial part of this required review. In various jurisdictions one can observe a move toward the integrated or better-co-ordinated supervision of banks, investment companies, pension funds and life-insurance companies. Also, this policy response can be interpreted as part of the review for improving the monitoring of these financial institutions as well as for making the handling of the stronger or new linkages between them more effective.

Capital-importing countries should implement sound macroeconomic and structural policies (a modern financial-securities-market infrastructure, a healthy banks sector, high accounting and disclosure standards, etc.) so as to restore confidence to investors and to curb unnecessary volatility. It is fair to say that much work has been done in strengthening financial systems in both advanced and emerging markets during the last two years. Nonetheless, the financial-sector-reform agenda remains full.

References

Abel, A. B. 2000. *Will Bequests Attenuate the Predicted Meltdown in Stock Prices When Baby Boomers Retire?* mimeo, Wharton School, University of Pennsylvania.

Blommestein, H. J. 1997. *Institutional Investors, Pension Reform and Emerging Securities Markets.* Inter-American Development Bank, Working Paper 359.

Blommestein, H. J. 1998a. Pension Funds and Financial Markets. *OECD Observer*, (June/July).

Blommestein, H. J. 1998b. Impact of Institutional Investors on Financial Markets. In: H.J. Blommestein and N. Funke, eds. 1998. *Institutional Investors in the New Financial Landscape,* OECD.

Blommestein, H. J. 1998c. Ageing-Induced Capital Flows to Emerging Markets Do Not Solve OECD's Basic Pension Problem. In: H.J. Blommestein and N. Funke, *ibid.*

Bodie, Z. 1990. Pension Funds and Financial Innovation. *Financial Management*, Autumn, pp.11-21.

Bohn, H. 2001. Retirement Savings in an Ageing Society: A Case for Innovative Government Debt Management. This Volume.

Börsch-Supan, A., A. Ludwig and J. Winter. 2001. Ageing and International Capital Flows. This volume.

Davis, P. 2001. Ageing and Financial Stability. This Volume.

Merton, R. C. 1992. *Continuous-Time Finance.* Oxford, Basil Blackwell.

Merton, R. C. and Z. Bodie. A Conceptual Framework for Analysing the Financial Environment. In: D. B. Crane, K. A. Froot, S. P. Mason, A. Perold, R. C. Merton, Z. Bodie, E. R. Sirri and P. Tufano. 1995. *The Global Financial System: A Functional Perspective*, Harvard Business School Press, Boston, Massachusetts.

OECD. 1998. *Maintaining Prosperity in An Ageing Society*. Paris.

Poterba, J. M. 2000. Demographic Structure and Asset Returns. Delivered as Review of Economics and Statistics Lecture at Harvard University.

Walker, E. and F. Lefort. 2000. *Pension Reform and Capital Markets: Are There Any (Hard) Links?* Manuscript.

Olivier Davanne

In this introduction, I would like to address two rather different sets of issues. First, a broad overview of the state of pension reform in France. The French pension system is a rather complicated one, the reform process itself follows a tortuous route, and an assessment of the current state of the pension system in one of the largest euro economies might be useful. Second, I would like to enlarge upon some concerns I have regarding the way pension money is managed, not only in France but in other countries as well. Indeed, while I fully support the current move towards increased funding for French pension schemes, I must confess that I am worried by the rising risks posed by what I see as inappropriate investment guidelines.

The current state of French pension reforms

The (convoluted) French pension system has three main characteristics:

- First, it is a very fragmented system with many different regimes, more than 20, and different rules depending on the worker's status.
- Second, the normal retirement age is still 60 in most cases (except for special groups of public-sector workers, such as railway workers, who can take early retirement with full benefits).
- Third, the system is mainly on a pay-as-you-go basis. Most employees entitled to full pensions still receive, from the basic and complementary regimes, around 80% of their last salaries upon retirement. Such a high replacement ratio leaves little room for «pension funds» or other forms of retirement savings. In the private sector, contributions to the pay-as-you-go system already amount to more than 30% of net wages.

This "generous" and fragmented pay-as-you-go system needs to be reformed, to cope with the doubling of the number of individuals above 60 in the next 40 years. France is often considered to be the European laggard concerning pension reform. However, the truth is that significant reforms were decided in 1993 and 1996 for the private sector schemes, and their implementation has begun. The real French specificity lies in the complete standstill in pension schemes covering public sector employees.

As a result of those reforms, the replacement ratio at the age of 60 in the private sector is expected to fall significantly in the future, from around 80% currently to

60% or less in 2040 (even for those who started to work early and have the required 40 years of contribution). The attempt to implement similar reforms in public sector schemes failed in 1995, after causing some impressive strikes and ultimately the demise of the government.

While underestimating the reforms already decided upon in the private sector, many foreign observers criticize France for its failure to set up "pension funds". This is a rather debatable criticism, since the equivalent of defined-contribution pension funds has existed for decades in France. France has created strong tax incentives for life-insurance contracts and other long-term savings, which already attract the bulk of French household saving. In particular, there is a form of savings plan, the PEE (Plan d'Epargne d'Entreprise), which shares many similarities with the popular US 401(k) savings plans. That is a vehicle for workers to acquire mutual funds. Such plans require agreement between the workers' representatives, the company and a fund manager, and they benefit from various tax-breaks. Recently, the Jospin government has significantly increased those tax incentives for plans with a long-term horizon. Various measures have also been taken to facilitate access to such schemes for small and medium-sized enterprises.

In fact, the French specificity is definitely not the lack of "pension funds", but the lack of incentive to save, owing to the high replacement ratio expected of the pays-as-you-go system. The 1993 and 1996 reforms in the private sector have already started to change this situation.

Overall, France is pursuing a strategy of its own, but still has a lot to do in order to improve the sustainability of its pension system, especially in the public sector. These reforms will certainly have to wait until after the 2002 general elections. However, and it seems that some foreign observers have partly missed this point, as in most countries, the trend is definitely towards more pension saving in a defined-contribution setting.

Financial risks of a funded pension system

Overall, I am happy with these developments, and I am only concerned that the debate in France is sometimes too ideological, with certain participants in the public sector refusing even to acknowledge the looming deterioration on the demographic front. That being said, I am equally concerned that while many economists have been very vocal advocates of fully-funded pension systems, they have spent very little time trying to think about the financial risks and implications of funded pension schemes.

The funding of pensions creates a sizeable amount of financial risks. Asset prices may vary sharply, and someone has to bear those risks. Risks can be shared, transformed or transferred, but they never disappear, even with the help of the most creative of accountants.

It is of the utmost importance that the incompressible and large financial risks involved in the funding of pensions are borne by people who understand them, and are able to endure the vagaries of financial markets. That is not always the case, particularly in countries with less developed financial markets.

For OECD countries, I have two specific worries.

1. Firstly, I can't help being surprised by the rather unsophisticated methods used by most long-term investors, be they households or professionals, when they assess the risk-return tradeoff of their investment strategy. I am happy to see that a recent report in the UK, the Myners report, has expressed similar concerns. More often than not, asset allocations are based on rigid benchmarks. Those benchmarks generally do not result from an in-depth analysis of current valuations and their implications for future returns, but rather from a simple extrapolation of past returns, leading to excessively rigid risk-return assessments. As a result, for example, the estimated risk premium for equities becomes fairly independent of market valuation. In short, the share of equities was similar, with Nasdaq at 5000 or at 2000! It seems hard to dispute that the demand for equities should be more sensitive to key valuation indicators, like the market price-earnings ratio or the value-to-book ratio.

   For the time being, this rigidity provides a fillip to the equity market, as benchmarks are based on the assumption that future returns will look much like those in the past. But, this naive faith would be shattered by a long period of underperformance. This backward-looking methodology obviously leads to a situation where investors are overexposed at the peak of the cycle and underexposed at its bottom. This is clearly detrimental to the performance of those investors. More importantly, **financial stability will be at risk as long as asset allocations are based on the extrapolation of the past, rather than on a fundamental analysis of current valuations.**

2. Secondly, I am also worried by the lack of transparency and robustness of some risk-sharing arrangements in financial markets. Some institutions claim that they can shield workers and pensioners from most, if not all, of the financial risks involved by the funding of pensions. These institutions can be either traditional defined-benefit pension funds or, in some countries, life-insurance companies, which provide various forms of guarantees on long-term saving. It is very hard to assess whether such institutions have the financial clout to fulfil their commitments in all sorts of market environments, and I do not think the regulatory framework is protective enough for the policy-holders. In particular, it is well-known that there are loopholes in the regulation of life-insurance companies in some countries, and not only in Japan. This worry is exacerbated for defined-benefit pension funds sponsored by private companies. These are indeed very unusual financial institutions, which seem endowed with the quasi-magical ability to make financial risk disappear! Indeed, workers

assume that pension promises will be fulfiled whatever happens in the stock market. At the same time, financial analysts and shareholders looking at companies' valuations don't really care about the pension-fund situation as long as there is a large surplus in the fund. In other words, as long as the market value of assets exceeds the pension liabilities, neither workers, pensioners nor shareholders show much interest in the pension fund's financial situation. But let's repeat that financial risks seldom disappear. Indeed, in my view, shareholders and the financial analysts working on their behalf considerably underestimate the financial risk involved in the business of providing defined-benefit pensions. And this is true even of companies with well-funded pension funds. Pension fund surpluses, when they exist, are a key asset for companies as they warrant investment in riskier financial instruments. Benefiting from higher returns, they may provide pensions at a lower cost. This puts these companies in a comfortable position relative to their competitors, which have to pay high contributions to pay-as-you-go public regimes. But the opposite holds when the pension-fund surplus falls owing to adverse market conditions: this competitive advantage declines sharply as well. **Like it or not, financial risk does not disappear, and the ups-and-downs in pension-funds surpluses directly affect the true, fundamental value of the sponsoring companies.**

This analysis has various interesting implications. I will briefly mention three of them. I hope that we will have an opportunity to discuss these issues further during the course of the panel discussion.

- First, a bear stock market could send a wake-up call to financial analysts and shareholders investing in companies with large defined-benefit pension funds. As long as fund surpluses are high and the resulting costs of pensions are low, they are unlikely to improve the way they value the pension obligations of the companies. But a falling surplus would put under scrutiny the pension funds' situation, and the risks involved in providing defined-benefit pensions.
- Second, it is rather difficult to form an opinion on the changes in investment strategies that such a combination of bear market and falling surplus could trigger. On the one hand, if shareholders have a better understanding of the risk they are exposed to, they may ask for more conservative investment policies, and as a result, contribute to a downturn in the financial markets. On the other hand, there is a kind of "moral hazard" at work, and the funds' sponsors may gamble that they will be able, if their risky investment strategies were to fail, to renege on some of their pension promises and thus to share the losses with the workers.
- Third, defined-benefit pension funds deserve a specific set of regulations. There is a widespread view that pension funds are absolutely right to invest in risky assets since, in the long term, these investments provide higher returns. But I really do not think that a pension fund is the right place for a corporation

and its shareholders to locate the risks they wish to bear in order to boost expected returns. There are more efficient and more transparent ways of taking risks. In general terms, it is a better choice to locate leveraged positions in a corporation's main balance sheet, rather than partially hidden in the pension fund. Therefore, I am very sympathetic to the principle of stringent regulations severely limiting the mismatches between pension-fund assets and liabilities.

Jürgen Pfister

1 The present state of pension reform in Germany

Undoubtedly, reform of provision for old age is one of the most important political challenges in Germany today. The demographic trend and persistently high unemployment pose a huge threat to the long-term efficiency of the pay-as-you-go state pension system. It is obvious, therefore - and has become the consensus view in the meantime - that a thoroughgoing reform of the system of old-age provision in Germany is inevitable.

Politicians have also recognized this and - following the action already taken by many other European countries - they have enacted legislation to reform statutory pension insurance and promote funded provision for old age. The following major changes have emerged as a result:

- In order to reflect the demographic trend, the level of benefits provided by the statutory pension-insurance scheme is to be reduced. On average, pensions will be about 8% lower in 2030 than they would have been without the reform. At present, the state pension for "standard pensioners" in Germany amounts to 70% of their final net pay; in 2030, it will be around 65% without private provision.
- To compensate for this, a funded component is being introduced in the form of voluntary private provision, encouraged by either tax incentives or direct grants. Such provision will be exempted from income tax for up to 4% of the income threshold for assessing contributions to the statutory pension scheme, which currently stands at €53,400 per year (in western Germany). In 2002, the highest possible deductible amount will start at 1%; it will then increase every second year by a further percentage point, finally reaching 4% in 2008. To this extent, income tax will be levied when the capital that has been accumulated is consumed (taxation of outflows). Those compulsorily insured in the statutory pension-insurance scheme, i.e. not employees in the public sector and the self-employed, are entitled to tax deductions or grants. This means that practically 28 million employees will be able to benefit. If the participation rate of 85% assumed by the Federal Finance Ministry is applied, an annual €35bn or so of funds will be generated by the final phase. A good 3% annual increase in wages has been assumed. Under these circumstances, old-age provision covered by the new scheme will make up around one-fifth of private monetary wealth formation (2000: €116bn) and about 3% of households' gross monetary wealth (2000: €3,644bn).

- Eligible products for state support are above all pension and life-insurance schemes, as well as investment-fund and bank-savings plans. At the same time, pension funds are being admitted as a new type of company-pension scheme. As their scope for manoeuvre will be restricted by regulations, and pure defined-contribution schemes by employers are not envisaged, these can hardly count as pension funds along Anglo-American lines. The pension funds have to be certified by the Federal Supervisory Office for Insurance Companies (*Bundesaufsichtsamt für das Versicherungswesen*). Germany's rules may come under pressure in a few years' time if the draft EU pension-fund directive is approved (EU directive on the co-ordination of laws, regulations and administrative provisions relating to institutions for occu-pational retirement pensions, proposal by the EU Commission of October 2000).

Last but not least, as a large number of more detailed regulations have to be resolved, some important details of Germany's pension reform are uncertain.

2 Regulations overdone

Ideally, the reform and further amendments of the system of old-age-provision should fulfil the following conditions:

- It should be possible to make funded provision for old age both privately and through occupational schemes. The new legislation meets this condition.
- However, up to now only collective pension-fund schemes have been envisaged. In an individual pension-fund system of the Anglo-American type, however, employees could adjust provision to suit their individual plans and would have a direct statutory claim on the fund's assets. In this way, they would be encouraged to assume responsibility, and flexibility, e.g., when changing jobs, would be increased.
- Companies should be permitted to offer their staff pure defined-contribution schemes. As hedging costs would be removed and companies would be able to plan their future obligations better, occupational pension schemes would become more attractive to them. This would mark an important step towards overcoming stagnation in occupational pension schemes. What is more, defined-contribution plans, if they allow employees to make their own investment decisions, give participants more options to choose from, while encouraging people to assume more responsibility for their old-age provision.
- The government should not insist upon a formal guarantee for the contributions that have been paid in. Such guarantees considerably reduce the yields that can be achieved. As the typical savings plans for old-age provision have long maturities, this leads to a lower level of benefits. In the case of

have long maturities, this leads to a lower level of benefits. In the case of funds, for instance, a full guarantee reduces the return by about 2 percentage points a year. A formal guarantee is not only expensive, but also superfluous. Given a sufficiently long savings period, the use of investment funds for private old-age-provision purposes *de facto* guarantees the contributions paid in. At present, the legislation could be interpreted such that an undertaking towards the investor is sufficient. Yet this would not have the same consequences for the provider as a formal guarantee. In particular, hedging costs - in the form of liable equity or as reflected in the relevant design of the products - are not necessarily incurred for the contributions that have been made. The defined benefits option is therefore the better approach. The uncertainties with regard to the legal consequences of such a commitment have to be removed.

- Other state regulations as well - such as investment rules - should be kept to a bare minimum in order to achieve the highest degree of efficiency, for a more flexible investment approach makes higher yields and a better diversification of risk possible. But also in order to ensure that the providers of German pension funds are competitive, qualitative investment guidelines - in particular the "prudent-man" principle - should be given preference over quantitative rules. If the EU pension fund directive is adopted in the years ahead, the German pension-fund market will experience liberalization. At the moment, the proposal of the EU Commission includes not only the "European passport" for pension funds - which enables providers with a licence in one member state to launch pension funds in all the other member states - but also the prudent-man principle for investments. While it is planned that member states will still be able to impose additional quantitative rules, these may restrict the flexibility of pension funds in their investments only to a very limited extent. For example, pension funds always have to be able to invest up to 70% of their assets in equities. The pension reform in Germany commits pension funds to the prudent-man principle. At the same time, it authorizes the federal government to introduce quantitative investment rules by means of an ordinance. For the reasons mentioned, though, the government should for the most part avoid quantitative rules even now, rather than finding itself obliged to liberalize by the EU. Otherwise, fund managers from the United Kingdom and the Netherlands, which already run successful pension funds, would be given a serious competitive advantage. In addition, it is to be hoped that regulation will be considerably simplified in the future, and above all that the insured are offered more options, such as the choice of achieving a higher yield by foregoing a guarantee on the contributions made.

3 Saving habits now and in the future

The pension reform will have less impact on the level than on the structure of households' saving in Germany. Provision for old age is already an important motive for saving, which means that the reform will have little influence on the level of savings. The households most likely to increase their saving are those which currently have either a negative or a low positive savings ratio, i.e. those with net incomes of less than €1,500 per month. By contrast, households in the medium to high-income brackets merely avail themselves of the promotion measures to a growing extent, but do not save more as a result.

As regards the restructuring of private monetary-wealth formation, reflecting the greater importance of old-age provision as a motive for saving, certain tendencies have been noted empirically since 1992, and not only since the debate on pension reform began. In this connection, life-insurance policies seem to have a head start. Monetary-wealth formation with insurers has virtually doubled since 1992 relative to overall monetary-wealth formation, though it tended to be somewhat weaker in 2000 than previously. On the other hand, the significance of long-term financial investments with banks has declined over the same period, as more wealth has been accumulated in the form of equities and equity-based investment funds. This reflects an orientation geared to yields, which will probably be reinforced by old-age provision as the motive for saving.

In general terms, the restructuring produced by the pension reform is likely to strengthen the existing tendency towards longer average investment periods and higher returns. Both will encourage investment in equities and are positive from the macroeconomic standpoint. It is less evident which groups of providers of financial services will ultimately be the main beneficiaries.

4 Impact on the German financial system

In Germany, the first pillar of retirement provision, the state pay-as-you-go system, accounts for 80-85% of overall retirement income. In countries with a longer tradition of funded old-age provision, such as the United Kingdom and the Netherlands, the respective share lies between 50% and 65%. Reflecting this distribution, the share claimed by occupational pension schemes in Germany is minimal, at 5%, compared with 25-40% in the two countries just mentioned. These have Western Europe's two largest pension-fund markets. In either case, the capital accumulated is equal to roughly 85% of GDP. It is extremely unlikely, however, that Germany could achieve such a dimension, even in the medium term.

For one thing, the entire capital already committed to occupational pension schemes in Germany is put at no more than roughly €310bn, which represents about 15% in GDP terms. But these funds cannot be shifted at short notice. Moreover, provisions for pensions, which – amounting to roughly €174bn – account for 57% of the capital tied up in occupational pension schemes, remain a very important and reasonable corporate financing instrument. For another, even after the reform, state pensions will still make up a much higher share of overall retirement income in Germany than in the United Kingdom and the Netherlands, removing the need to raise occupational pension schemes to the same level as in the countries mentioned.

All the same, Germany's pension reform can be expected to have the strongest impact on the second pillar, occupational pension schemes, and primarily on pension funds. It is estimated that these could accumulate a capital stock of €300bn in Germany by the end of the decade.

The introduction of pension funds and the expected shifts between the various forms of saving - from life-insurance policies into pension funds, from life-insurance policies into fund-based savings plans, from direct investment in equities and bonds into investments in pension and investment funds - will have an impact on the financial market. Compared with life insurers and private investors, pension and investment funds channel a far larger share of their capital into equities. Apart from stronger demand for equities – above all, for foreign equities as well – Germany's system of corporate governance will also be affected. As institutional investors become more important, control over listed companies by the capital market will continue to increase. What is more, the higher proportion of foreign equities will further promote the integration of the financial markets, which should also enhance the liquidity of mid-caps, in particular in the various markets or market segments. At the same time, increasing demand for equities will make it easier for companies to take up capital. There should also be a positive influence on the raising of borrowed funds, as demand for corporate bonds issued by borrowers of sound quality will probably be strong as well.

As regards the financial system, the pension reform is also likely to lead to several changes on the supply side. To differing degrees, the strengthening of the second and third pillars will give rise to greater demand, for asset-management services in particular. Germany's capital market will be considerably upgraded as a result. By contrast, traditional bank and insurance products may decline in significance. However, as the large fund managers in Germany are subsidiaries of banks, the latter will continue to play a very important role in the financial system here. But insurers will find it difficult to secure a share in the rapidly expanding fund business, as they have not focused on asset management for third parties up to now and the difficulties facing their efforts to gain a footing in this market should not be underestimated. Probably the most that insurers can hope for is to compensate for the loss of customers to investment-fund companies by generating new business from the increased inflows of funds from private provision for old age.

5 Summing up

Overall, the pension reform will hardly trigger new developments in the German financial system; rather, it will reinforce existing trends. Provision for old age, or saving in general, will become more strongly institutionalized, Germany's equity culture will be strengthened, the international character of investments will be enhanced, new financial products will appear, and the outcome will be a strengthening of the capital market. However, the positive influences will be limited in scope. If the EU pension-fund directive is approved in a few years' time, further integration of the financial markets in the European Union, and consequently a strengthening of the single market, can be expected.

Notes on Contributors

Franklin Allen, University of Pennsylvania, Philadelphia

Alan Auerbach, University of California, Berkeley

Joseph Bisignano, Bank for International Settlements, Basel

David Blake, Birkbeck College, London

Hans Blommestein, OECD, Paris

Henning Bohn, University of California, Santa Barbara

Axel Börsch-Supan, University of Mannheim

Friedrich Breyer, University of Konstanz

Agar Brugiavini, University of Venice

Olivier Davanne, University of Paris-Dauphine

E. Philip Davis, Brunel University, Uxbridge

Hiroshi Fujiky, IMES and Bank of Japan, Tokyo

Ulrich Grosch, Deutsche Bundesbank, Frankfurt

Martin Hellwig, University of Mannheim

Heinz Herrmann, Deutsche Bundesbank, Frankfurt

Claus F. Hofmann, Ministery of Labour, Bonn

Gary Hufbauer, Institute for International Economics, Washington

Alexander Ludwig, University of Mannheim

David Miles, Imperial College Management School, London

Olivia Mitchell, University of Pennsylvania, Philadelphia

Benoit Mojon, European Central Bank, Frankfurt

Jürgen Pfister, Commerzbank, Frankfurt

Philipp Rother, European Central Bank, Frankfurt

Reinhold Schnabel, University of Essen

Platon Tinios, Office of the Prime Minister, Athens

Martin Werding, CESifo, Munich

Joachim Winter, University of Mannheim

David Wise, Kennedy School of Government, Harvard

Druck:	Strauss Offsetdruck, Mörlenbach
Verarbeitung:	Schäffer, Grünstadt